VIEWS OF NATURE

VIEWS OF NATURE

ALEXANDER VON HUMBOLDT

Translated by Mark W. Person

Edited by Stephen T. Jackson
and Laura Dassow Walls

THE UNIVERSITY OF CHICAGO PRESS CHICAGO AND LONDON

The University of Chicago Press, Chicago 60637
The University of Chicago Press, Ltd., London
© 2014 by The University of Chicago
All rights reserved. Published 2014.
Paperback edition 2016
Printed in the United States of America

25 24 23 22 21 20 19 18 17 16   3 4 5 6 7

ISBN-13: 978-0-226-92318-5 (cloth)
ISBN-13: 978-0-226-42247-3 (paper)
ISBN-13: 978-0-226-92319-2 (e-book)
DOI: 10.7208/chicago/9780226923192.001.0001

Library of Congress Cataloging-in-Publication Data
Humboldt, Alexander von, 1769–1859, author.
[Ansichten der Natur. English]
Views of nature / Alexander von Humboldt ; translated by Mark W. Person ;
edited by Stephen T. Jackson and Laura Dassow Walls.
    pages cm
Includes index.
ISBN 978-0-226-92318-5 (cloth : alk. paper)—ISBN 978-0-226-92319-2 (e-book)
1. Natural history. 2. Science. 3. Physical geography. 4. South America—
Description and travel. I. Jackson, Stephen T., 1955– II. Walls, Laura Dassow.
III. Person, Mark W. IV. Title.
    QH45.H82 2014
    508—dc23

                                                            2013050786

♾ This paper meets the requirements of ANSI/NISO Z39.48-1992 (Permanence of Paper).

Contents

Preface

This volume represents an unlikely professional intersection between a natural scientist and a humanities scholar. It is no coincidence that our professional paths crossed while pursuing the thinking of Alexander von Humboldt, who integrated these now-remote fields two centuries ago. And we were preadapted to find ourselves at this intersection: if not for a few chance events and influences, each of us could have spent a life pursuing a career in the other's field of study—respectively ecology, and American literature and history.

The book arose from our mutual recognition that Humboldt's *Ansichten der Natur* represented a uniquely important and influential component of his oeuvre, linking the sciences and humanities in his most personal and passionate published writings. We also shared a hunch that the two existing English translations, both published in 1849–50, took Victorian liberties with Humboldt's prose and didn't do justice to his vision or to his artistry. All these views were reinforced as the translation proceeded, and our appreciation of Humboldt's intellectual riches steadily expanded as we read each chapter anew as it arrived from the translator.

We had the extraordinary good fortune to enlist the interests and talents of Mark Person, a gifted scholar of early nineteenth-century German literature and culture, in taking on the translation. Mark embraced the project from the start, and his fine ear for the beauty as well as the content of Humboldt's prose is apparent throughout the translation. Comparing his translation with the two previous English versions, we couldn't help but think that he was scraping a mask of thick, sometimes garish Victorian paint off a profoundly beautiful piece of art. The translation will speak for itself, but we think that it may be as close to an embodiment of Humboldt's voice as we may see in the English language.

The University of Chicago Press continues to set the publishing standards in the history of science and has committed itself to a series of Humboldt translations and related works. We thank Christie Henry for her continuing enthusiasm and support for our Humboldt and related projects, and we look forward to seeing more of his important works become broadly available.

Progress on the translation and editing took a great leap with support from the Ucross Foundation. The editors and translator spent a week in residence at Ucross, which provided not only wonderful working spaces, accommodations, and meals but a splendid, inspirational landscape. It was thrilling to read Humboldt's accounts of the Rocky Mountains and Great Plains while situated among the "smoking hills" at the foot of the Bighorn Range.

Humboldt often expressed a feeling of being awestruck in the face of nature. We share those feelings, and experienced them daily in the landscape at Ucross and the nearby Bighorns. The spectacular skies—whether clear in day or night, colorfully clouded at dawn and dusk, or violent with thunderstorms—provided steady reminders of nature's power and grandeur. But we were also awestruck at the breadth, depth, and splendor of Humboldt's writings. We hope that the reader will share these feelings and that this translation will bring to a twenty-first-century audience a greater appreciation for the scientific aesthetic—the sensual, emotional, and spiritual elements underlying our pursuit of knowledge of the world in which we find ourselves.

In preparing this volume, we made a deliberate, pragmatic decision not to add further explanatory footnotes and appendices. We determined that doing full justice to the work, particularly the rich, fine-textured footnotes, would require a near-doubling of the volume's length and a gargantuan scholarly effort. Instead, we chose to let Humboldt's words speak for themselves, without annotation beyond our brief introductory essay, which aims to orient the reader and draw attention to the significance of the volume. *Views of Nature* is sufficiently important and manifold that it deserves a companion volume updating and explicating the footnotes, and perhaps also a critical companion volume, setting the core essays in context and exploring the countless themes they raise. We hope this translation and accompanying essay will reawaken Humboldt's vision for the unity of science, knowledge, and human experience.

Stephen T. Jackson   Laura Dassow Walls
Laramie   South Bend

Introduction:
Reclaiming Consilience

LAURA DASSOW WALLS, STEPHEN T. JACKSON
MARK W. PERSON

I. The Writing of *Ansichten der Natur*

One of the ironies of *Ansichten der Natur*, Alexander von Humboldt's most popular book and always his favorite, is that this seven-part hymn to the consilience of art and science, conceived in the tropics, was born in the bitter cold of conquest and defeat. In 1799, Humboldt had left Paris with his French traveling companion, the botanist Aimé Bonpland, on a journey of scientific discovery to the New World. In 1804 the two had returned triumphantly to Paris, laden with enough specimens, measurements, notes, and insights to launch the old inquiries of natural history into the new sciences of the modern world. When they left, France had been a young republic, and wherever they traveled, from Venezuela all the way to Peru, to Cuba and Mexico—with a final stop in Washington, DC, to compare notes with Humboldt's hero, Thomas Jefferson—they had seemed to find new republics in the making, only to discover on their return that the French Republic had collapsed into imperial monarchy. But Paris was still the scientific center of the world, and Humboldt stayed long enough to enjoy being lionized before heading to Italy to visit his brother Wilhelm (then the Prussian minister to the Vatican), then finally, late in 1805, returning home to Berlin.

There he was welcomed, of course, but he also faced harsh accusations that he had lost touch with his German roots and been so entirely absorbed in French culture that, it was rumored, he planned to write up his scientific results not in German but in French. Humboldt in turn complained bitterly of the cold and the isolation of Berlin, which, he confessed, was "a foreign land to me."[1] But he stayed for many months, working, writing, planning his return to Paris—until, suddenly, he was trapped. On October 14, 1806, the French forces attacked Jena, where they routed and destroyed the Prussian army, and

---

1. Douglas Botting, *Humboldt and the Cosmos* (New York: Harper and Row, 1973), 185.

less than two weeks later, Napoleon marched into Berlin at the head of an occupying army. The king and his court, on which Humboldt depended for his patronage, fled; the Prussian bureaucracy, in which he had himself been trained, collapsed; food prices soared, and many starved as the war reparations demanded by the French crippled the economy. The Humboldt estate at Tegel was looted, and most of the family's possessions were lost or destroyed. Berlin had fallen, and Prussia, Humboldt's homeland, had effectively ceased to exist.

Humboldt, the scion of Berlin who had adopted Paris as his home, was deeply torn. The Prussians called on him to mediate between the two warring sides, a diplomatic role that finally allowed Humboldt to return to Paris early in 1808, where he remained for the next nineteen years before returning to Berlin once again, for good, in 1827. But before he left Berlin, he completed a work dedicated "to embattled minds particularly," that is, to his fellows in defeat: a book of nature essays, drawn from his favorite moments in the tropics, brimming with hope and beauty and the aspirations of science. This book was written not in French (the language he did, indeed, use for nearly all the thirty-odd volumes of his American writings) but in his native German. Published in 1808, *Ansichten der Natur* became an instant classic, translated into a half-dozen languages; and in 1826, just before he left Paris, he reissued it in a second edition, adding two more essays. Finally, in 1849, on the heels of the worldwide success of his multivolume popular science book *Kosmos*, Humboldt revised and expanded the annotations in *Ansichten* (adding observations from his 1829 expedition to northern Asia), inserted the final essay, and reissued this collection of essays in its final form. By now he had a large and enthusiastic English-speaking audience as well, and starting in 1849, both *Kosmos* and *Ansichten* were available in English in competing translations issued by two different publishers, the latter as *Aspects of Nature* by Elizabeth Sabine (London, 1849) and as *Views of Nature* by Elise Otté (London, 1850).[2] Suddenly the Anglo-American market was flooded by a wave of inexpensive translations of Humboldt's most popular works. In the United States, the Humboldt craze (which had been simmering steadily since

2. For publication of *Ansichten der Natur*, see the translator's note below. *Kosmos* was published in Germany in five volumes from 1845 to 1862. Both major English translations bore the same main title, *Cosmos*; neither was complete (the fifth German volume has never been translated into English). Elizabeth Juliana Sabine's translation of the first two volumes was published in London, in two volumes, by Longman, Brown, Green and Longmans, starting in 1846; it went through multiple editions. Elise C. Otté's translation of Humboldt's first four volumes was published in London by Henry B. Bohn in five English volumes, from 1849 to 1858; it was reprinted in New York by Harper and Brothers, 1850–59.

Jefferson's presidency) took off, not to subside until after the 1869 centennial celebrations of his birth.[3]

Today *Cosmos* remains Humboldt's best-known work, but it is a challenging book to read: the first volume opens with a formidably long and philosophical essay staking out his claim to be developing a new kind of science, "the physical description of the world, the science of the *Cosmos*" (1:42). This launches a thrilling but densely written journey through deep interstellar spaces, the near solar system, the earth as a dynamic geological and biological system, and thence (in volume 2) through the history of the very concept of "nature" from the ancient Greeks to the present, in literature and culture, the arts, the sciences, exploration, and technology. In succeeding volumes Humboldt retraces in authoritative detail the ground he had sketched in volume 1, presenting the results of modern natural science for all to see. Not surprisingly, *Cosmos* was never finished—nor could it have been, as Humboldt himself realized. He died in 1859, midway through the fifth volume. While he had ranged from the stellar bodies of outermost space all the way earthward to the granite rocks of Russia, he left the organic and the human realms still unwritten.

By contrast, *Ansichten der Natur* was not only finished but polished and carefully composed, a gem whose seven crafted facets were designed to reflect the whole of nature. Where *Cosmos* attempted an analytical survey of the entirety of the known universe, *Ansichten* selected a handful of features, centered on places Humboldt had experienced, savored, and treasured in his memory—Wordsworthian spots of time that had helped him keep faith through the turmoil of war and conflict, offered to the broad public as "Views" or "Aspects" of nature that implied, rather than detailed, the whole: in essence a pocket *Cosmos*.

The German keyword in Humboldt's title, *Ansichten*, gives a clue to his method, which may strike modern readers as peculiar. The German word means something like "view" or "aspect," but with the connotation, largely absent in English, of active and evolving composition rather than a completed prospect open to sight. One doesn't merely look at a Humboldtian "view"; one must work actively to compose it by using the materials he presents to us, and building on the active work he performs for us through his own immersion in both nature and whole libraries of writing about nature. The "onlooking" is thus to be enacted as a way-of-seeing, not just scenery; *Ansichten*, in Humboldt's hands, isn't a version of nature-viewing suitable for consumer

---

3. For further details, see Laura Dassow Walls, *The Passage to Cosmos: Alexander von Humboldt and the Shaping of America* (Chicago: University of Chicago Press, 2009).

consumption, but rather an exercise in *Bildung* or self-education conducted under the impress of nature's forces and beauties. The task of consilience starts here, with the active work of interweaving immediate, direct, sensuous perception of nature with careful and rigorous intellectual cognition, together with a certain emotional coloring or valence, all composed into a many-faceted whole.

The seven main essays are careful compositions in just this sense, actively composed, selected, and orchestrated; but while they can stand alone as works in their own right, Humboldt, remarkably, refuses to hide the hard, long, and collective work of composition. This labor is revealed in his extraordinary annotations at the end of each essay, which are sometimes so extensive that they overwhelm the essay itself. If the essays are the front side of the tapestry, in the annotations Humboldt flips over the fabric to show us the back side, the normally hidden process of fabrication, where we can trace the threads of knowledge and the labor required to weave them into a whole. These seven "views" are to be pictures, *Gemälde* or paintings, which display both the sensibilities of the landscape artist and the portraitist's evocation of essence. Yet in effect, each of them arrives in our hands with the entirety of the artist's studio still attached, a studio that includes the artist's friends and associates with their own related paintings (some as yet unfinished), plus entire related *schools* of painting, together with a running commentary on absences, and on artworks that could exist but as yet don't. Some of these annotations grow into entire finished essays on their own account: there are disquisitions here on the cross-cultural practice of eating earth, on petroglyphs, on native peoples who build their houses in trees, on the use of camels as domestic animals, on currents, vegetation, and the phosphorescence of ocean water, even thumbnail biographies of friends. Some annotations spin off into fragments; others land us in morasses of uncertainties, questions yet unresolved, controversies still unsettled, science still too raw to package as knowledge. The result is an attempt at hypertext, envisioned long before technology made it possible. To click on a Humboldtian fact is to drop into a fractal universe where one shifts up and down scale levels, where knowledge is arrayed in self-similar layers, from the intensely local hand-held object at one end to global, wholly imaginative abstractions at the other. Objects are taken up, analyzed, made occasions for thought. Thus while there are universals in this book, they are forged, not found—forged through the interconnections Humboldt uses to yoke together the objects before him with their cognates all around the planet. His places are assemblages, his things are events.

Humboldt's experimental, multilayered writing found no imitators, and the emerging genre of the "nature essay" moved in a different and much less vertiginous direction. Perhaps only Humboldt could pull off this sort of dynamic complexity, but still, his would have been an exciting model to embrace. One person who did read Humboldt with great care and who swerved his own prose style in a Humboldtian direction was Henry David Thoreau, in whose hands it became a flexible instrument for aesthetic as well as scientific discovery. Indeed, when Thoreau was forced to put his cards on the table, he declared himself by stating, "I am an observer of nature generally, and the character of my observations, so far as they are scientific, may be inferred from the fact that I am especially attracted by such books of science as White's Selborne and Humboldt's 'Aspects of Nature.'"[4] Thoreau's linkage of Humboldt's *Ansichten* with Gilbert White's beloved classic of natural history suggests that this was indeed the most accessible and popular of Humboldt's works. Here he is ever present, at our side as guide and interpreter: since most of these essays were first developed as lectures, they retain the sense of an audience before whom the speaker, Humboldt the inimitable, gestures, points, pauses, qualifies, diverts.

And of course he was speaking in formal expository German. This fact creates for the English reader a cluster of difficulties. Humboldt's speaking presence is elevated and reticent, as befitted the Prussian lecture hall if not the demands of romantic prose style. His German sentences roll onward in massive, knotty, chewy subclauses, capturing and suspending layer upon layer of information, held aloft until at last, at the end, the verb we reach. Yet the flight of each sentence is always controlled; the line of thought, no matter how complex or dizzying, is always brought to a firm landing. Indeed, part of the pleasure of reading Humboldt is the discovery that however far he ascends or however deep he descends, he always knows exactly where he intends to go and will deliver us safely there at the end. That he is not known as a remarkable prose stylist is largely the fault of his Victorian translators, who hide this roiling precision from us. In their hands his German reticence becomes stuffy and pompous, his elegant, rolling, firmly controlled passages become strained and wordy, his innovative flights are tamed into conventional slogs. One would never guess, reading the nineteenth-century translations, that Humboldt was intentionally pushing the resources of German, of language itself, to the outermost limits. Hence the task of this new translation is to render his innovative prose style into English that is clear and readable,

4. Henry David Thoreau, *The Correspondence of Henry David Thoreau*, ed. Walter Harding and Carl Bode (New York: New York University Press, 1958), 310. Thoreau was responding to the questionnaire that accompanied an invitation from the American Association for the Advancement of Science, issued in March 1853, to become a regular member; Thoreau politely declined.

yet to retain something of the strangeness of the German original, in hopes of making available to modern readers Humboldt's capacity to push language to new places.

II. Seven Ways of Viewing Nature

If *Cosmos* is one pole of Humboldt's popular writing, the other is his *Personal Narrative*, one of the most famous travel narratives of his day, in which he takes his readers from his wartime departure from Spain to the Canary Islands, thence to the coast of modern-day Venezuela, up the Orinoco, by ship to Cuba, then, in an unexpected swerve, to the port of Cartagena—and, abruptly, the narrative breaks off virtually in midstride. Part of the charm of *Views of Nature* is that it holds to the travel narrative form without deploying the forward engine of movement: it lingers, and savors, and thinks. (It also supplies some of the material missing from the *Personal Narrative*, which never reaches the Andes.)[5] The order of these seven essays is tightly structured: where the first, "Concerning the Steppes and Deserts," turns from the ocean inland, to face and traverse the great plains, or *llanos*, of Venezuela, the last, "The Plateau of Cajamarca," concludes by turning back outward to face the ocean once again as the travelers descend from the Andes to the coast of Peru. At the very center of the book is Humboldt's keystone essay, "Ideas for a Physiognomy of Plants," which surveys the entirety of the "planetary organism"[6] as a grand global tapestry of visual forms, grouped into landscape economies indexed by plant types. This central, theoretical, broad-reaching essay is flanked by two groups of two: the first pair, "Concerning the Waterfalls of the Orinoco" and "The Nocturnal Wildlife of the Primeval Forest," narrows the lens to specific scenes and places; the second pair, "Concerning the Structure and Action of Volcanoes" and "The Life Force," both reach below the surface of natural appearances to explore deep, dynamic causes, first of planetary geological forces, then of vitality or life force. The seven essays thus describe an arc: Beginning with the shape of the planet as it structures

---

5. Some of this material is also supplied in *Vues des Cordillères, et monumens des peuples indigènes de l'Amérique* (Paris 1810); trans. Helen Maria Williams as *Researches concerning the Institutions and Monuments of the Ancient Inhabitants of America, with Descriptions and Views of Some of the Most Striking Scenes in the Cordilleras!*, 2 vols. (London, 1814). This innovative work, which integrates text and visual illustration, attempts yet a different kind of experimentation; for a discussion, see Walls, *Passage to Cosmos*, 1–20, 85–98. Also see the complete translation *Views of the Cordilleras and Monuments of the Indigenous Peoples of the Americas: A Critical Edition*, ed. Vera M. Kutzinski and Ottmar Ette, trans. J. Ryan Poynter (Chicago: University of Chicago Press, 2013).

6. See "Physiognomy of Plants" in this book, 206n13.

the history of the living beings on its surface, Humboldt turns to the flow of waters across that surface and then the flow of sounds and voices through the atmosphere. The book climaxes with a summation of visual natural pattern and form, followed by two summations of the causes of natural forms, first geological, then organic; and it closes with a turn to human history, a portrait of European conquest and indigenous survivance.[7]

One might also name this "Seven Ways of Viewing Nature," for each essay addresses the problem of seeing in a different way. "Steppes and Deserts" takes as its condition a world nearly barren of life, of water, even of elevation. The endless unadorned earth asserts itself as the necessary, if not sufficient, condition for sheer existence, leading Humboldt to a planetary view: first to ask what geological and climatic conditions lead to such lifeless expanses worldwide, then to ask how the planet itself structures the possibility for life, for the movement and migration of living beings, even the origins and fate of human civilizations. The essay presents a grim and warring world, brutal, glaring, hot, relentless, and it is climaxed with one of the most famous set-pieces of all Humboldt's writing, the gruesome and violent capture of electric eels by driving horses into the pond where they live. In the ensuing battle, the eels are drained of their deadly electric force, killing many of the maddened horses.

The opposite appears in "Concerning the Waterfalls of the Orinoco": the Orinoco River is presented as a lush and beautiful waterworld, intimate with cataracts and cloud forests. Light sparkles through the mist, a "play of the airs" that throws off the "optical magic" of rainbows, for it is light, Humboldt thinks, that leaves the strongest emotional "impression" upon us, "now in ethereal sky-blue, now in the shadow of low-hanging clouds" (p. 117). Humboldt takes us in imagination to the uppermost sources of the river, then back downriver to the fall line, where the waterfalls arrest navigation and create a transient community of travelers, some of them European, and native river guides. Humboldt ends, however, once again on an eerie note of conflict: they visit the nearby grave site of the Atures Indians, a haunting scene of a people recently driven to extinction. Humboldt's theft of bones—a frank and troubled admission of grave-robbing in the name of science—is judged and punished by a restless, procreative nature that will in the end overrun the bones of all the races of men.

7. *Survivance* is a term coined by Gerald Vizenor to designate native resistance, "an active sense of presence over absence, deracination, and oblivion . . . the continuance of stories" which declares continued native presence, rights, and sovereignty. See his essay "Aesthetics of Survivance: Literary Theory and Practice," in *Survivance: Narratives of Native Presence*, ed. Gerald Vizenor (Lincoln: University of Nebraska Press, 2008), 1–2.

Language may be the one monument that outlasts our physical remains. Humboldt in closing invokes the curious story of the Parrot of the Atures, a bird who speaks the lost language of its human teachers long after they were driven into extinction. His third essay, "Nocturnal Wildlife," takes up the recovery of language: first our own, which careless usage has worn away into cheap commonplaces; then the language of nature, which we must learn to speak if our own words are to remain vital and alive. This essay about seeing and naming the forest ends with the voices not of men but of animals: first with a cacophonous and sleepless midnight when it seems that all the animals of the forest are in full cry, then with a scorching, slumberous noontime, when it seems that all life has withdrawn into silence. Yet in the seeming silence Humboldt's ear catches the voices of insects, literally a ground-tone that tells him that life stirs everywhere, that to the ears that can hear, "everything announces a world of active, organic powers" (p. 147).

The turn in "Physiognomy of Plants" is toward a visual language that plants may be seen to speak: plants, that is, speak in form rather than in sound. In terms of science this is the most daring of the seven essays, and Humboldt's proposals, particularly his speculations about "social plants," led directly to the founding of the science of ecology. He explored the ideas advanced here at more length in the contemporaneous volume *Essay on the Geography of Plants* with its accompanying "Tableau Physique" (1807), which vividly illustrated his groupings of plant associations by climate zone.[8] This is also the most daring of the seven essays in terms of aesthetics, for by insisting that nature has a "face," literally a "physiognomy," Humboldt argues that the face of nature is not superficial and does not mask hidden depths, but in fact reveals the deep structure of landscape forms and even evolutionary change. (His discussion of evolution, midway through the immense endnote 13, reveals just how close to anticipating Darwin's thinking Humboldt had come.) In effect, if plants "speak" through form, this essay explores a virtual grammar of plant forms, a visual language that we "hear" through our eyes, the organ through which plants "impress" us as if we were living wax and they the stylus that imprints us from childhood. The painter becomes not a superficial illustrator of nature but the true carrier of insight into the heart of natural form and beauty: "To the artist," wrote Humboldt, "is left the task of separating the groups, and under his hand the great, magical image of Nature (if I may venture to use the expression) reveals itself, much like the written works of men, in a few

8. See Alexander von Humboldt, *Essay on the Geography of Plants*, ed. Stephen T. Jackson, trans. Sylvie Romanowski (Chicago: University of Chicago Press, 2009).

simple strokes" (p. 168). Humboldt's words had profound effect, inspiring painters such as Frederic Edwin Church and Charles Johnson Heade to take their brushes to the tropics and found a new, scientifically exact school of landscape art.[9] In their paintings one sees Humboldt's joyous scientific ode to biodiversity, the visual poetry of plants, manifested as art.

"Concerning the Structure and Action of Volcanoes," a scientific address Humboldt added to the 1826 edition, shifts focus yet again, from surfaces to depths. As Humboldt remarks in his opening, "Organic Nature gives to each stretch of land a physiognomic character of its own," but underneath, inorganic nature, visible wherever the hard crust thrusts through the covering of plants, shows itself the same from equator to poles. Thus the seafarer on a distant isle, surrounded by strange plants under strange stars, recognizes with joy "the clay slate of home, the comfortably familiar type of rock he knows from his fatherland" (p. 243). This planetary consistency discloses a deeper process of planetary formation, one that will require comparative observations from all across the globe to unearth the great laws "by which the layers of the Earth's crust alternately support one another, break apart into channels, or are lifted by elastic forces" (p. 244). Since the earth's depths cannot be directly seen, they must be imagined, using reason based on the subterranean forces revealed in volcanic fires—and what reason and imagination reveal is the startling truth that the solid-seeming earth, like organic life, changes and evolves, shaping itself from the core outward, having begun in the depths of time as a molten body that even still today has only partially cooled. Here, too, surfaces evince deep depths, as volcanoes become "intermediary earth-

9. Church's debt to Humboldt is well documented; see, for example, Stephen Jay Gould, "Church, Humboldt, and Darwin: The Tension and Harmony of Art and Science," in *Frederic Edwin Church*, ed. Franklin Kelly et al. (Washington, DC: Smithsonian Institution Press, 1989), 94–107. The relationship between Heade and Humboldt is less direct, but the following observation by Humboldt is richly evocative of Heade's work, with its close focus on hummingbirds and orchid blossoms seen against misty backdrops of tropical mountain vistas: "The orchid blooms sometimes resemble winged insects, sometimes the very birds which are attracted by the scent of the nectaries. The life span of a painter would be insufficient, even if concentrating upon only a narrow area, to depict the glorious orchids that adorn the deep-cut valleys of the Peruvian Andes chain" (p. 165). The oft-repeated claims that Humboldt writes only of vast cosmic panoramas, or of a serene, stable, unchanging, and harmonious "pre-Darwinian" nature, are abundantly disproved by the essays in the present volume. It is also often asserted that Humboldt was imperially unaware of indigenous peoples, a statement also abundantly contravened by the essays and annotations herein, which extensively document indigenous lives and memorials; moreover, Humboldt sponsored several artists who sought to document American indigenous peoples, including Karl Bodmer and George Catlin (see Walls, *Passage to Cosmos*, 132–34).

springs" (p. 255), the points where the earth's visible surface and its invisible depths "communicate" and so speak to us of the deep, dynamic forces that govern the very substructure of the planet.

But in "The Life Force," when Humboldt turns to the organic world and asks the same kind of question, the answer eludes him. What is the nature of life? What is its deep structure; what governing forces shape the appearances we see? We simply do not know. Science is arrested here, for the springs of vitality are not yet accessible to its methods. So, remarkably, Humboldt turns instead to poetry, inserting an allegory that he first wrote as a young man and published, in 1795, in Schiller's journal *Die Horen*.[10] Humboldt's poetic allegory hypothesizes a Life Force that, withdrawn, allows the chemical constituents of life to follow "their natural attractive forces." However, as Humboldt explains in the annotation that follows, soon after publishing this essay he changed his mind, and he no longer condones such phrases as "life force" and "intrinsic forces." Instead, he believes that life must be actuated by the interaction of matter and material forces to create a self-determining whole that, like meteorological processes, is too complex (or as we might say today, nonlinear) to be predicted. Why, then, in 1849, did he reach back in time and insert this obsolete essay? It serves as a placeholder, a visible marker that makes evident not what is known but what is not. The poetic allegory—alleged to be a reading of a pair of paintings recovered from classical Greece—is used to frame the question *as* a question, thereby opening a visible gap in the fabric of knowledge. Here is not knowledge attained, or knowledge in the process of being shaped, but knowledge still to come. Humboldt's allegory is a sign and a provocation: "Watch This Space."

The final essay, "The Plateau of Cajamarca," was the last one to be added, and it provides narrative closure by bringing Humboldt back out from his inland travels to face, once again, the ocean. With this moment, his narrative arc closes on a bittersweet note: this wasn't part of the plan; he shouldn't even be here; but as his last paragraph explains, news had earlier arrived that the Baudin Expedition that he had contracted to join was landing in Lima. This news diverted him from his original plan to cross Mexico and sail the Pacific to the Philippines; instead, he led his party down the Andes to Lima to meet Baudin, only to learn, too late, that Baudin had sailed around Africa instead. Thus was his lifelong hope and goal defeated, never, as he knew by 1849, to be realized. Humboldt keeps himself emotionally controlled to the

10. Humboldt and Friedrich Schiller were friends and correspondents from the early 1790s until Schiller's death in 1805. The poetic line "Who saves himself from life's stormy wave" and the four-line stanza ("In the mountains . . .") at the end of Humboldt's Preface to the First Edition are spoken by the chorus in Schiller's 1803 play *The Bride of Messina* (act 4, scene 7).

end, barely alluding to this bitter disappointment, gesturing instead to what he gained: an afternoon's miraculous clearing of the Andean mist, revealing the entire western slope of the cordillera all the way to the seashore. "For the first time, we were seeing the Pacific Ocean; we saw it clearly: next to the littoral, a great body of light, reflecting, ascending to a horizon now more than merely sensed" (p. 283–84). Slender compensation, perhaps, for losing the great journey he had planned, but the entire essay carries a similarly complex emotional charge. The travelers move with difficulty along a swampy, muddy road, in full view of the magnificent, imposing dry and paved road built by the Incas and destroyed by the Spanish. The Inca's ruined civilization haunts the essay, as Humboldt drops down the eastern slope of the Andes into the Amazon valley and climbs back to the western slopes by way of "old Cajamarca, where, 316 years ago now, the bloody drama of the Spanish conquista was played out" (p. 274).

Here, in the most moving human passage in all his writings, Humboldt is led through the ruins and their history by the seventeen-year-old son of the cacique Astorpilco, a direct descendant of the Inca monarchs. After their hours together, Humboldt asks the boy, who still dreams of the vast wealth he is certain lies buried just underfoot, why he does not dig it up. "The boy's answer was so simple, so much the expression of the quiet resignation that characterizes the aboriginal people of this land, that I put it down in my journal in Spanish: 'Such a desire (*tal antojo*) does not come to us; my father says that it would be a sin (*que fuese pecado*).'" For were they to dig up the treasures, "'then our white neighbors would hate us and harm us. We possess a little field and good wheat (*buen trigo*).' I believe that few of my readers will fault me for remembering here the words of the young Astorpilco and his golden dreams" (pp. 280–81). In place of a scientific description, a very personal anecdote; in place of a king's ransom, a little good wheat; in place of a voyage around the world, a moment's view of the Pacific. In this book born out of war and defeat, the victories may be small, but small gains nurtured in the face of sorrow, small beauties realized in the throes of violence and pain, form the warp of these essays, even as the hard-won triumphs of science weaves into this warp a story of the compensations of knowledge, knowledge that is built thread upon thread by the loom of history.

III. The Consilience of Knowledge and the Scientific Aesthetic

As the annotations to *Views of Nature* make clear, Humboldt's style of science drew on a huge network of observers, explorers, and scientific amateurs and professionals in every field of knowledge and from every inhabited continent on the globe. One of the most important figures on the extended Humboldt network was the British scientist and philosopher William Whewell, who is memorable today for coining two words. One of them is so standard that it's astonishing to discover it had to be invented, and the other so unusual that until recently it was practically forgotten. The standard word is *scientist*, which Whewell proposed in 1833 as a parallel for *artist*. The term took decades to catch on, which explains why no one in his day called Humboldt a "scientist": not only did the word not yet have wide currency, but "science" as a single, unified professional endeavor was still taking shape, largely through Humboldt's own work. The second word, *consilience*, was Whewell's term for the process by which two or more previously separate streams of research merge to make a new, transformed whole.[11] A classic example is Darwin's theory of evolution, by which the fossil record, ecological adaptation, developmental biology, natural variation in form, and geographic distribution were seen to "leap together," in Whewell's phrase, to make something wholly new, the emergent field of evolutionary biology. Humboldt's work can be seen as a massive project in consilience, as he strove to weave the separate strands of geology, physiology, physical geography, geophysics, geographic distribution and movements of plant and animal species, vegetation patterns, anthropology and ethnography, linguistics, weather, climate, and more into a whole that today is most closely approached by the field called "earth systems science." But *Views of Nature* insists on a second level of consilience as well, not merely of different fields of science but of science itself with the humanities, notably poetry and art. Science could not be a "sterile accumulation of facts" ("Volcanoes," p. 257) but had to be animated by imagination.

Humboldt comes to this consilience of science and aesthetics from several levels. At his most cosmic level, he assumes humans do in fact inhabit a single, coherent, and ultimately knowable universe, out of which we emerged, part and parcel. Accordingly, our minds, composed of the very stuff of the

---

11. The term *consilience* was popularized by E. O. Wilson in his book *Consilience: The Unity of Knowledge* (New York: Alfred A. Knopf, 1998). Although Wilson acknowledges Whewell as the source of his ideas, Wilsonian consilience is very different from Whewell's conception. In Wilson's explicitly reductionist epistemology, "consilience" represents a hierarchy of explanatory subsumptions. For example, attributes of human culture are subsumed as special cases of evolutionary adaptation and sexual selection. Whewellian consilience, in contrast, describes a process by which seemingly disparate phenomena are brought together under a single explanation. For example, Darwin applied Whewell's "consilience of induction" by bringing together a series of observations, including uplift of the seashore in the 1835 Concepción earthquake, the series of elevated terraces along the Pacific Coast, the increasing antiquity of fossil molluscan faunas and decreasing shell pigmentation in fossil molluscs with elevation, and biogeographical patterns. These phenomena united under a single explanation: the Andes were rising.

universe itself, are capable of discovering in the plenitude and confusion of natural phenomena the real and workable universal laws underlying superficial appearances. A second level lies at the opposite extreme: the sheer delight experienced by the individual working scientist who discovers an unanticipated connection linking together pieces of the cosmic puzzle. As every working scientist knows, even the smallest links can be exhilarating to find. A third level lies in the linkage between the two: Humboldt sensed that the first stirring of the desire to know begins with delight in nature's marvels, and that the pleasures of learning the causes of those wonders reinforce the determination to investigate natural forces and phenomena, creating a virtuous circle of pleasure and knowledge.

Revisiting Humboldt can help the scientist reclaim the long-neglected aesthetic dimension of science, from delight in witnessing some consilience of facts to awe in tracing the deep structural interconnections of the natural world. But from his cosmic level, Humboldt points us toward the larger dimension of this experience: the vital interdependence between science and art, knowledge and aesthetics. Not that his every poetic metaphor is scientifically true: when Humboldt imagined life as a *Genius* animating every particle of organic creation with the Law of unrequited desire, and death as their wild, orgasmic reunification, the allegory pleased as art, but it failed as science. Here no consilience has yet occurred, although the value of even a failed metaphor is manifest in Humboldt's own discontented rejection of it and the hesitating, tentative questions he continued to ask in the absence of any scientific answers. By contrast, his metaphor of volcanoes as "intermediary earth-springs" connecting the earth's solid surface with its roiling depths provided not only a vivid image but a consilient truth. Reliable knowledge of the deep structure of the planet was emerging in his thought, and though it was more than a hundred years before plate tectonics would explain why, as Humboldt observed, a long chain of volcanoes rimmed the Pacific Ocean, his fundamental insight that all the globe's volcanoes were interlinked by some subterranean connection helped to radically reimagine the planet as a dynamic, restless, still evolving and transforming entity. That the one metaphor turned out to be a dead end, and the other a fruitful pathway to whole fields of research, points to the fact that science, too, like the planet itself and the life forms that cover it, is constantly in a state of change.

The changing face of scientific knowledge has carried us in some respects further away from Humboldt than from, say, his protégé Darwin; some of Humboldt's statements may strike us today as ludicrous and outdated, closer to the world of Buffon than to that of Watson and Crick. Nevertheless, even as he draws on the resources of science and poetry in his cross-pollinating, experimental way, he sees each as aiding the purposes of the other, while never confusing them. His first goal—and the primary goal, he believes, for the artist as well—is to get the physical phenomena right. For Humboldt, born before the concept of "scientific objectivity" took hold,[12] getting phenomena right did not mean exempting the role of the mind that perceives and considers nature, or the shaping imagination that theorizes it, or even the emotional charge that colors it and brings us fear, or joy, or delight. Rather, "objectivity" meant embracing mind and imagination—intellect, poetry, and emotion—as multiple modes of access that allow humans to participate in the grand unfolding of the cosmos. For Humboldt, there is more than one way to look at nature: the hydraulics of the Orinoco river system are important, but so are the rainbows that play through its misty cataracts, the human communities that gather on its banks, the tangle of plants that form its impenetrable walls, the animals whose clamor awakens the travelers in the night, the travelers themselves, who witness and record, and attempt, through poetry and art and science, to locate the place of the human in the great fabric of the cosmos. Only a scientific aesthetic can point the way to weaving the braided streams of so many viewpoints into a coherent whole. *Views of Nature* is Humboldt's demonstration that not only is such a scientific aesthetic possible, but it is the only way forward in a world of force, and contingency, and sudden unanticipated clearings of the mist.

12. The contemporary notion of scientific objectivity has evolved considerably, and was largely unknown in Humboldt's time, as amply discussed by Lorraine Daston and Peter Galison in *Objectivity* (New York: Zone Books, 2007).

Translator's Note

In translating any nineteenth-century text from German to English, one may fairly expect certain difficulties, and the same can be said when facing the challenge of translating any scientific text. Having tackled both of these types of translation, I fancied myself to be well prepared for the writings of Alexander von Humboldt.[1] The expected difficulties were certainly there—the often antiquated phrases or syntax of older texts, the arcane vocabulary and field-specific usage of scientific writing, even the combination of phrases being simultaneously arcane and antiquated—but these were minor matters compared to the real challenge, one I had not anticipated. It is the exquisite architecture of Humboldt's sentences that demands the most from the translator, for I soon found that it was not merely a matter of turning a complex German sentence into a complex English sentence, though his sentences are indeed complex. What one finds in Humboldt's prose is the capacity to create a sweeping view from the breathtakingly immense to the small and intimate, acknowledging through the use of the German language's versatile extended modifiers all the things along the way that fascinate his penetrating eye, accomplishing this—and here lies the rub—with faultless clarity. It was imperative that this clarity be preserved if the text were to be rendered accessible to a twenty-first-century reader of English, and yet it could not be at the cost of Humboldt's broad yet all-inclusive vision, nor could I allow it to dull his poetic sensitivity. To reproduce this remarkable combination of beauty, complexity, and clarity in English (a language that allows little room for the extended modifier) requires serious consideration and at least a small amount of courage. It is no easy exercise to change a sentence that is a near-perfect expression in one language in the hope of approaching that same perfection in another. Should the English use a subordinate clause where the German

1. The German text from which we worked is Alexander von Humboldt, *Ansichten der Natur* (Stuttgart and Tübingen: Cotta, 1849).

has an extended modifier? Must I break this sentence into two, or even three, to make it work in English? As I worked on these sentences, I developed, in a small way, some sympathy for the man who cut the Koh-i-noor. If any beauty remains in these sentences, I ascribe it more to Humboldt's skill than to my own.

The nuts-and-bolts labor of constructing new sentences can sometimes cause the translator to lose sight of the fact that each sentence is part of a whole. It is in this area that I am most indebted to Stephen Jackson and Laura Dassow Walls. Bringing to these translations their extensive knowledge of Humboldt and their sensitivity regarding his style, not only have Laura and Steve imbued them with a much greater degree of the clarity that is peculiar to Humboldt, but they have brought the translations closer to the great scientist's poetic sensibilities as well, sensibilities evinced not so much by subjective exploration of the self (which Humboldt, in his preface to the first edition, calls "a lack of composure") as by a razor-sharp articulation of profoundly perceptive observation.

Although the most daunting part of this project was deciding how best to preserve the integrity of Humboldt's style, there were also a great number of other decisions to make, where, again, the input from Steve and Laura was invaluable. In the case of names (of rivers, lakes, mountains, native tribes, etc.), where Humboldt's spelling was idiosyncratic even for his time, we usually chose the most common modern variant used by speakers of English. If the modern names either did not exist or had not yet come into general use in Humboldt's day (Sri Lanka, for example, has since replaced Ceylon), we tried, in the English, to stay close to Humboldt. Regarding the many units of measure that Humboldt employs but are now obsolete, we felt that they have intrinsic historical value and decided to leave them as they are. (We have, however, compiled a short list of these units of measure and their conversions to more familiar units.) The same is true for some of the terms that Humboldt consistently uses but that are not common today. We do not hear of *geognosy* today, but for Humboldt's time, it is a fitting title for the science that explored elements of the two fields that we today call *geography* and *geology*. We receive a more accurate view of Humboldt's world with terms like *orology* (the physical geography of mountains) and *oryctognosy* (now called mineralogy), and even by letting him refer to a plantain or banana as a *pisang*, a term more frequent in Asia. On those occasions when the English and German common names are inconsistently equated with the scientific names—such as Humboldt's referring to a tree as a *Fichte* (spruce), for instance, where in English we would call it a fir—we usually went with the English equivalents of the terms Humboldt employs, trusting the scientific names to remove doubt.

These decisions would have been much more vexing without the wisdom of Steve and Laura.

I am also grateful to Randall Conrad, who compiled the book's index and whose painstaking precision and impressive erudition smoothed out many a wrinkle and prevented many errors from making it into the final version. I would like to thank my colleagues in the University of Wyoming Department of Modern and Classical Languages: Hannelore Mundt and Rebecca Steele in German; Angela Camino, who translated the French passages; Laura De Lozier for translating the Latin; Kevin Larsen, Dwight Hicks, and Sonia Rodriguez-Hicks for helping me with the Spanish; and Mohammed Alyousif for his help with the Arabic. And as ever, I am grateful to my wife Debbie for her sage advice and unfailing support.

In critiquing his own writing in the preface to the first edition of these essays, Humboldt writes, "The richness of Nature encourages an agglomeration of individual images that disturbs the calmness and the overall impression of the portrait. In appealing to feeling and fancy, style easily degenerates into poetic prose." As I worked on this text, I became convinced that I need not take this admonition to heart. There is indeed poetry in Humboldt's prose, but I felt that my task was not to change or remove it but, as best I could, to preserve it for a new audience.

Humboldt's *Ansichten der Natur*
Measurement Units

Temperature

Réaumur. 1° Réaumur = 1.25° C. (For ° Fahrenheit, multiply ° Réaumur by 2.25 and add 32.)

Length, Height, Depth

ell. H. is not specific as to which ell; probably the ell equal to the cubit (the length from a man's elbow to the tip of the middle finger) and not the later "cloth ell," which is roughly twice this measurement and equally variable.

fathom. Equal to the toise (6 old French feet, or 1.949 m). H. generally uses the toise for height and the fathom for depth.

foot. H. uses the old French *pied du roi* (32.48 cm), which is 1.066 English feet (30.48 cm), using the term *Parisian foot* interchangeably with it. The Roman foot is 29.79 cm; the Prussian foot is 31.38 cm.

li. Chinese unit of distance measurement, now defined as 500 meters but of variable length in H.'s day.

line (*ligne*), **Parisian.** ½₂ Parisian inch, or 2.255 mm. Similarly, the English line is ½₂ English inch, or 2.115 mm.

league (*legoa, legua*). Used in many cultures but not an official unit of measure anywhere; specific length varies greatly, though most definitions refer to the distance one may walk in one hour.

mile. H. generally uses the German/Danish geographical mile (four minutes of arc at the equator), which is exactly 4 times the modern geographical mile (1 minute of arc), 4.6 statute miles, or 7.42 km. H.'s nautical mile (3 minutes of arc) also differs from the modern nautical mile (1 minute of arc).

toise. 6 Old French feet; 1.949 m.
yard. 3 English feet, or 0.9144 m.

Area

feet, square. 1 sq. Parisian foot = 1.136 sq. English feet.
miles, square. 1 sq. German/Danish geographical mile = 21.16 sq. statute
miles.

Volume

feet, cubic. 1 cubic Parisian foot = 1.211 cubic English feet.

Weight

pound, Spanish. 460 grams (the modern avoirdupois = 453.6 grams).
centner. 100 of any basic weight unit; for H., at least in these essays,
100 Spanish lbs.

Geographic Distance

h. Hour of longitude; 1 h = 15°. This can become confusing, however,
because minutes and seconds by this formula are then also bigger by a
factor of 15 than their counterparts when locating by degrees—thus H.'s
conversion of $5^h 18' 18''$ to $79°34' 30''$, since $5^h = 75°$, $18' = 4°30'$, and
$18'' = 4'30''$. (H. far more frequently expresses longitude in degrees than
in hours.) Whether by hours or by degrees, H. reckons longitude from
the Paris meridian. To convert to Greenwich values, add $2°20' 14.025''$
to Parisian reckoning of east longitude; subtract the same from Parisian
west longitude. H. often does not state whether cited longitudes are east
or west of Paris, nor whether latitudes are north or south of the equator,
presuming that the reader will already know this.

Views of Nature

Alexander von Humboldt

Translated by Mark W. Person

Views of Nature

with
Scientific Annotations
by

Alexander von Humboldt

Corrected and Expanded Third Edition

Stuttgart and Tübingen
J. G. Cotta Publishing
1849

The author dedicates this work to
Wilhelm von Humboldt,
his very dear brother in Rome.

Berlin, May 1807

Preface to the First Edition

I humbly extend to the public a series of works that came into being within the contemplation of great objects in Nature, upon the ocean, in the forests of the Orinoco, in the steppes of Venezuela, in the wilderness of the Peruvian and Mexican mountains. Individual fragments were recorded at the place and time and only later forged together as a whole. A far-reaching overview of Nature, proof of the cooperation of forces, and a renewal of the delight that direct experience of the tropics gives to a person of feeling are the goals to which I strive. Each essay should of itself constitute a complete whole, but in each should be the equal expression of one and the same tendency. This aesthetic treatment of matters of natural history, despite the wonderful power and flexibility of the language of our native land, carries with it tremendous difficulties of composition. The richness of Nature encourages an agglomeration of individual images that disturbs the calmness and the overall impression of the portrait. In appealing to feeling and fancy, style easily degenerates into poetic prose. These ideas are here in need of no development, as the following pages offer manifold examples of such aberrations, of such a lack of composure.

May my *Views of Nature*, in spite of these flaws which I, myself, can more easily criticize than improve, provide the reader with but a portion of the enjoyment that a receptive mind finds in immediate inspection. Since this enjoyment is increased with insight into the interconnectedness of natural forces, each essay is accompanied by scientific annotations and additions.

Throughout, I have indicated the eternal influence that physical Nature exerts upon the moral disposition of Humanity and upon its fate. To embattled minds particularly, these pages are dedicated. *"Who saves himself from life's stormy wave"* will follow me gladly into the thickets of the forest, into the immeasurable steppes, and out upon the spine of the Andes range. Unto him speaks the world-directing chorus:

In the mountains is freedom! The breath of the tomb
Cannot climb up to the purest air's home,
The world is perfect anywhere,
If Humanity's anguish has not entered there.

Preface to the Second
and Third Editions

The dual direction of this text (a painstaking effort to heighten the enjoyment of Nature through living depictions, while simultaneously increasing insight into the harmonious cooperative effect of forces according to the state of scientific understanding of the time) was delineated in the preface to the first edition nearly a half-century ago. The manifold hindrances inherent in the aesthetic treatment of great scenes in Nature were already indicated then. The combination of a literary with a purely scientific goal, the desire to occupy the imagination and at the same time, through the increase of knowledge, to enrich life with ideas, renders the ordering of the individual parts and the demands of unity of composition difficult to achieve. Despite these unfavorable conditions, the public has long granted the imperfect execution of my undertaking their charitable goodwill.

I oversaw the second edition of *Views of Nature* in Paris in the year 1826. Two articles—an essay "Concerning the Structure and Action of Volcanoes in Various Regions of the Earth" and "The Life Force, or The Rhodian *Genius*"—were added at that time. During my long stay in Jena, Schiller, in memory of the medical studies of his youth, enjoyed discussing physiological matters with me. My work on the disposition of stimulated muscle and nerve fibers through contact with chemically differing materials often led our discussions in a more serious direction. The short article on life force originates from this time. Schiller's fondness for the "Rhodian *Genius*," which he accepted for publication in his journal *Die Horen*, gave me the courage to have it printed again. In a letter that has just recently been published (Wilhelm von Humboldt's *Briefe an eine Freundin*, part II, p. 39), my brother gently touches upon the same matter, but aptly adds: "The development of a physiological idea is the goal of the entire essay. At the time in which it was written, there was more love than one would find now for such half-poetic adornment of serious truths."

In my eightieth year I have had the joy of completing a third edition of my book and completely recasting it in accordance with the needs of the time. Nearly all scientific annotations have been either amended or replaced by new, more comprehensive ones. It has been my hope thereby to enliven the drive to study Nature such that, even in the smallest of spaces, the most diverse results of thoroughgoing observation might be concentrated, that the importance of exact numerical data and the thoughtful collation thereof be recognized and steered toward the dogmatic half-truths as well as the proper skepticism that have long had their place among the so-called higher circles of society.

The expedition to northern Asia (in the Urals, the Altai, and to the banks of the Caspian Sea) that I undertook in 1829 in cooperation with Ehrenberg and Gustav Rose at the command of the Emperor of Russia falls between the epochs of the second and third editions of my book. It contributed substantially to the broadening of my views in all matters concerning surface formation, the course of mountain ranges, the connection between steppes and deserts, and the proliferation of plant life in relation to measured temperature influences. The ignorance that has so long prevailed regarding the two great snow-covered mountain ranges between the Altai and the Himalayas, the Tien-Shan and the Kunlun, along with the unjustified disregarding of Chinese sources, has obscured understanding of the geography of inner Asia and, in widely read texts, has passed off fantasy as the result of observation. In just the last few months, important and rectifying elucidations have come back which were nearly unexpected in the hypsometric comparison of the culminating peaks of both continents and which will be presented for the first time in the following work (*Concerning the Steppes and Deserts*, notes 5 and 10). Now free of earlier error, the height measurements of two mountains in the eastern Andes range of Bolivia, the Sorata and the Illimani, have not yet with certainty restored to Chimborazo its standing among the highest mountains of the new continent, while in the Himalayas, new trigonometric measurement of Kinchinjinga (26,438 Parisian feet) open for this peak the next place behind Dhawalagiri, which now has also been measured with greater trigonometric accuracy.

To ensure numerical consistency with the two earlier editions of *Views of Nature*, the temperature data in this work, unless specifically stated otherwise, are expressed in the 80-degree Réaumurian thermometer. The foot measure is the old French, where the toise counts as 6 Parisian feet. The miles are geographical, 15 of which constitute one equatorial degree. Longitude is reckoned from the first meridian of the Paris Observatory.

Berlin, March 1849

Concerning the Steppes and Deserts

At the foot of the high granite spine that, in the early days of our planet, de-fied the incursion of the waters during the formation of the Antillean Gulf, there begins a broad, immeasurable plain. Upon leaving behind the valleys of Caracas and the island-rich Lake Tacarigua,[1] which reflects in its surface the trunks of the pisang trees, leaving behind fields resplendent with the delicate light green of Tahitian sugarcane or the solemn shade of cacao plants, one's gaze toward the South comes to rest upon steppes that, seeming to climb, dwindle into the distant horizon.

From the luxuriant fullness of organic life, the astonished wanderer comes to the barren edge of a sparse and treeless desert. No hill, no cliff rises as an island in this incalculable space. Only broken, stratified slabs two hundred square miles in area, lying here and there, show themselves visibly higher than the parts bordering them. The natives call these phenomena "banks,"[2] indi-cating instinctively through the half-awareness of language the state of things that once were, for these elevations were the shallows, and the steppes them-selves the bed, of a great inland sea.

Even now, the disguise of night often calls back these pictures of the past. When the guiding celestial bodies in their rapid rising and setting illuminate the edge of the plain, or when they create a quivering double image of it in the lower layer of the undulating haze, one believes he sees before him the boundless ocean.[3] Like the ocean, the steppe fills the mind with the feeling of infinity, and through this feeling, as if pulling free of sensory impression, with intellectual and spiritual inspiration of a higher order. But while the clear ocean surface in which ripples the graceful, softly foaming wave is a friendly sight, dead and stiff lies the steppe, stretched out like the naked rocky crust[4] of a desolate planet.

In all zones of the globe, Nature offers this phenomenon of immense plains; each has a character of its own, a physiognomy determined by the individual-ity of its terrain, by its climate, and by its distance above sea level.

One can view as true steppes the heathlands of Northern Europe, which, covered by one single, all-supplanting variety of flora, stretch from the point of Jutland to the outlet of the Scheldt, although smaller and hillier than the Llanos and Pampas of South America or the grassy plains near the Missouri[5] and Coppermine Rivers, on which the shaggy Bison and the small Musk-ox abound.

The plains of the African interior offer a greater and more serious vista. As with the broad expanses of the Pacific Ocean, only recently have attempts to explore them begun. These plains are part of an ocean of sand that in the east separates fertile strips of land from one another, or encompasses them, forming islands, like the deserts below the basalt mountain range of Harutsch,[6] where the ruins of the Ammon Temple in the date-rich Oasis of Siwa mark the noble spot of early human civilization. No dew, no rain moistens these desolate areas to nurture the germination of plant life in the glowing womb of the earth. For everywhere, columns of heated air climb upward, dispelling the vapors and chasing away the fleeing clouds.

Where the desert approaches the Atlantic Ocean, as between *Wadi Nun* and the White Cape, the moist sea air streams in to fill the emptiness created by these upward winds. Even the mariner who steers for the mouth of the Gambia through a sea covered like a meadow with kelp senses, when the tropical east wind suddenly leaves him,[7] the nearness of the sand, expansive and radiant with heat.

Herds of gazelle and fleet-footed ostrich run about the immeasurable space. Except for the recently discovered groups of water-rich islands in this ocean of sand, upon whose green shores swarm the nomadic Tibbos and Tuaryks,[8] the remainder of the African desert may be viewed as uninhabitable by Man. Indeed, the civilized peoples of bordering regions only periodically dare to enter it. Upon paths that trade traffic has inalterably determined for millennia moves the long caravan from Tafilet to Timbuktu or from Murzuk to Bornu: bold undertakings whose very possibility depends upon the existence of the camel, the Ship of the Desert,[9] as he is called in the old legends of the Eastern world. These African expanses fill an area that surpasses that of the nearby Mediterranean Sea threefold. They lie in part within the tropics themselves, in part near them, and this situation gives rise to the individual character of their nature. But in the eastern half of the Old Continent, this same geognostic phenomenon is more typical of the temperate zone.

Upon the ridge of Central Asia between the Golden Mountains, or Altai, and the Kunlun,[10] from the Chinese Wall to the far side of the Celestial Mountains and around the Aral Sea to the northwest, over a length of several thousand miles, are scattered the largest, if not the highest, steppes in the world. I myself have had the opportunity to see, full thirty years after my South American journey, a part of these: the Kalmykian and the Kyrgyz steppes, which lie between the Don, the Volga, the Caspian Sea, and the Chinese Lake Dsaisang, that is to say, over a stretch of almost 700 geographical miles. The occasionally hilly Asiatic Steppes, interrupted now and again by forests of spruce, have a vegetation that, grouped in different areas, is much more variegated than that of the Llanos and Pampas of Caracas and Buenos Aires. The fairer part of the plain, populated by Asiatic shepherd folk, is adorned with shrubs of abundantly blooming white Rosaceae, with Crown Imperial (Fritillaria), tulips, and Cypripedia. Just as the torrid zone is consistently distinguished by the striving of all vegetation to grow in arborescent form, so too are some steppes of the Asiatic temperate zone characterized by the wondrous heights reached by their blooming herbs: Saussureae and other Synanthereae, leguminous plants, especially a host of various types of Astragalus. When one travels in the low-slung Tartar carriages over the trackless parts of these herb-covered steppes, only by standing upright can one orient oneself and thus see the densely packed forest of plants that bow before the wheels. Some of these Asiatic steppes are grass plains; others are covered with succulent, evergreen articulated alkali plants, many of them gleaming with salts that sprout up like lichens, unevenly covering the clay-rich soil like new-fallen snow.

These Mongolian and Tartar steppes, interrupted by numerous mountain ranges, separate the long-civilized humanity of Tibet and Hindustan from the barbarous peoples of Northern Asia. The existence of these steppes has been of tremendous influence on the changing fate of the human race. They have forced human population southward; they have hindered the intercourse of nations more than the Himalayas, more than the snow-peaked ranges of Srinigar and Gorka, and they have set unchallengeable limits in Northern Asia to the dissemination of milder customs and the creative artistic spirit.

But history must not view the plain of Inner Asia only as an impeding barrier. It has also on numerous occasions brought calamity and devastation across the globe. Shepherd peoples of these steppes—the Mongols, the Getae, the Alani, and the Uysyn—have shaken the world. Over the course of the centuries, whenever early intellectual culture has traveled like revitalizing sunlight from east to west, so too have barbarism and rawness of custom subsequently threatened to creep over Europe like a fog. A brown shepherd tribe[11] (Tukiuish, i.e., Turkic), the Xiongnu, populated the high steppes of the Gobi in leather tents. Long a threat to the might of the Chinese, part of the tribe was forced southward into Inner Asia. These dislodged people spread inexorably as far as the old home of the Finns on the Ural. From there poured forth Huns, Avars, Chasars, and numerous mixed Asiatic races. Hunnish armies first appeared on the Volga, then in Pannonia, then on the Marne and on the banks of the Po: ravaging the cultivated farmlands where, since the time of Antenor, creative humanity had heaped monument upon monument. Thus blew forth from the Mongolian deserts a pestilential breath of wind that choked the tender, long-cultivated blossoms of art in lands south of the Alps.

From the salt steppes of Asia, from the European heathlands, resplendent in summer with red, honey-rich flowers, and from the greenless deserts of Africa, we return to the plains of South America, whose portrait I have already begun to sketch with crude strokes. The interest, however, that such a portrait can provide the viewer is purely an interest in Nature. Here no oasis evokes memories of earlier inhabitants; no chiseled stone,[12] no fruit tree gone wild reminds one of the efforts of bygone races. As though foreign to the fates of humanity, latching only to the present, there lies in this corner of the globe a wild showplace of free animal and plant life.

From the coastal range of Caracas, the steppe extends to the forests of Guyana, and from the snowy peaks of Mérida (on whose slopes salty Lake Urao is an object of religious superstition for the natives), it reaches down to the great delta that the Orinoco forms at its mouth. To the southwest, it stretches like a long, narrow gulf[13] beyond the banks of the Meta and the Vichada to the unseen sources of the Guaviare, and thence to the lonely massif that the Spanish warriors, giving free rein to their active imagination, dubbed the *Paramo de la Suma Paz*, the beautiful spot, as it were, of eternal peace.

This steppe covers an expanse of 16,000 square geographical miles. Due to ignorance of the geography, it has often been described as extending, uninterrupted and with a consistent breadth, as far as the Strait of Magellan, disregarding the level, forested region of the Amazon River which is bordered north and south respectively by the grassy steppes of Apure and the River Plate. Between the Chiquitos Province and the isthmus of Villabella, the Andes range of Cochabamba and the mountain groups of Brazil extend individual spurs toward one another.[14] A narrow plain joins the hylaea of the Amazon River with the Pampas of Buenos Aires. These latter surpass the area of the Llanos of Venezuela threefold. Indeed, their expanse is so extraordinarily great that they are bordered on the north by palm groves, while their southern reaches are almost covered with eternal ice. The tuyu (*Struthio rhea*), similar to the cassowary, is unique to these Pampas, as are the colonies of feral dogs[15] that live sociably together in underground burrows, yet often launch bloodthirsty attacks on the very humans for whose defense their ancestors fought.

Like the greatest portions of the Sahara Desert,[16] the Llanos, or the northern plains of South America, lie in the Earth's torrid zone. Nevertheless, each half-year they appear in a different form: first desolate, like the sand ocean of Libya; then as a grassy expanse, like so many of the steppes of Central Asia.[17]

It is a rewarding (if difficult) exercise in general regional geography to compare the natural properties of remote regions to one another and to describe the results of this comparison in concise terms. Many diverse and to some extent still unestablished causes diminish the aridity and warmth of the New Continent.[18]

A great number of conditions provide the flat parts of America with a climate that, in terms of moisture and coolness, contrasts marvelously with that of Africa: the narrowness of the extensively indented mainlands in the northern part of the tropics, where a liquid surface presents to the atmosphere a cooler, ascending current of air; a great latitudinal distance from both ice-capped poles; an open ocean, over which the cooler tropical seawinds blow; flatness of the eastern coasts; currents of cold ocean water from the Antarctic region which, though originally directed from southwest to northeast, strike the coast of Chile below the 35th parallel of south latitude and advance northward along the coast of Peru up to Cabo Parina, turning then suddenly to the west; the number of mountain ranges rich in water sources whose snow-covered peaks aspire to heights above all cloud strata, and upon whose flanks begin descending currents of air; the abundance of rivers of enormous width which, after many windings, always seek the most distant coast; sandless and thus less heat-retaining steppes; the impassable forests that fill the plains at the equator and in the land's interior, where mountains and ocean are farthest apart, protecting the ground from the sun's rays or dispersing heat by the breadth of their leaves and exhaling colossal amounts of water, some of which they have drawn in from outside and some of which originates within them—in these characteristics lies the cause of that luxuriant, succulent flora, that frondosity which is the peculiar characteristic of the New Continent.

If one side of our planet is thus said to be more humid than the other, then the observation of the present state of things is sufficient to solve this problem of inequality. The physical scientist need not wrap the explanation of such natural phenomena in the garb of geologic myths. It is not necessary to assume that the destructive battle of the elements upon the ancient earth was settled at different times in the Eastern and Western Hemispheres, or that America emerged from the chaotic covering of water later than the other parts of the world, as a swampy island, home to alligators and snakes.[19]

Certainly South America, by the shape of its outline and the direction of its coasts, has a conspicuous similarity to the southwest peninsula of the Old Continent. But the inner structure of the land and its location relative to bordering landmasses gives rise in Africa to that astonishing aridity which in immeasurable spaces stands opposed to the development of organic life. By contrast, four-fifths of South America lies beneath the equator, in a hemisphere that, due to greater proportions of water and to many other causes, is cooler and more humid than is our Northern Hemisphere.[20] To the latter hemisphere, however, belongs the greater part of Africa.

The South American steppes, the Llanos, when measured east to west, are only one-third as extensive as the African deserts. The former receive the tropical sea winds; the latter, lying beneath a latitudinal circle with Arabia and Southern Persia, are touched by layers of air that move over hot, radiating continents. Indeed, the honorable and long-unappreciated Father of History, Herodotus, in the true spirit of a broad view of Nature, depicted all the deserts of North Africa, in Yemen, Kerman and Mekran (the Gedrosia of the Greeks), even to Multan in Upper India, as a single connected ocean of sand.[21]

Associated with the effect of hot land breezes in Africa, to the extent that we are familiar with it, is the lack of large rivers, of forests that exhale water vapor and create a cooling effect, and of high mountains. One may find ice year round only on the western portion of the Atlas range,[22] whose narrow spine, when viewed from the side, appeared to ancient travelers along the coast as a single lonely and airy pillar beneath the heavens. The range runs easterly up to the area near Dakul, where, now sunk in rubble, Carthage once lay, commanding the sea. As a chain stretched along the coast, as a Gætulean outer wall, the mountains hold back the cool north winds and with them the mists that rise from the Mediterranean.

Once imagined to soar above the lower limit of the snow are the Mountains of the Moon, Djebel-al-Komr,[23] which were said to form a mountainous parallel between that "African Quito," the high plain of Habesh, and the sources of the Sénégal. Even the cordillera of Lupata, which stretches along the eastern coast of Mozambique and Monomotapa, like the Andes chain on the western coast of Peru, is capped with eternal ice in gold-rich Machinga and Mocanga. But these water-rich ranges lie far distant from the tremendous desert that stretches out from the southern flank of the Atlas and on to the Niger flowing eastward.

Perhaps all of these enumerated causes of aridity and heat would not be sufficient to convert such considerable portions of the African plains into a terrible sea of sand, had not some sort of revolution in Nature, perhaps the irrupting ocean, once robbed this flat region of its covering of vegetation and nutrient topsoil. Exactly when this occurrence took place, what force might have brought it about, is cloaked deep in the darkness of prehistory. Perhaps it was an effect of the great rotational current[24] that drives the warmer Mexican waters over the bank of Newfoundland and onward to the Old Continent, and by which coconuts of the West Indies and other tropical fruits are able to reach Ireland and Norway. At least an arm of this ocean current is still present, directed southeast from the Azores on and striking the western dune shore of Africa, bringing calamity to mariners. For all seacoasts (the Peruvian coast between Amotape and Coquimbo comes to mind) show how centuries, indeed millennia, pass before the mobile sand, in hot, rainless stretches of the Earth, where neither lecidea nor other lichens sprout,[25] is able to provide the roots of vegetation a secure purchase.

These observations are sufficient to explain why, in spite of the superficial similarity of the shape of the lands, Africa and South America feature the most divergent climatic conditions, the most differing characteristics of vegetation. Even though the South American steppe is covered with but a thin crust of fertile soil, it is also periodically saturated with pouring rain and then bedizened by luxuriantly growing grass; it was, however, unable to entice the neighboring tribes to leave the beautiful valleys of Caracas, the seacoast, and the river world of the Orinoco, to lose themselves in this treeless, springless waste. Thus the steppe, upon the arrival of European and African settlers, was found to be nearly devoid of people.

Certainly the Llanos are suitable for raising livestock; but the care of milk-giving animals[26] was nearly unknown to the original inhabitants of the New Continent. Hardly any of the American tribes knew to make use of the advantages that Nature had also afforded them in this regard. The Native American race (all one and the same from 65° north to 55° south latitude, excepting possibly the Eskimos) moved from a hunting existence directly to an agricultural, skipping the stage of a herding life. Two types of native cattle graze in the grasslands of Western Canada and in Quivira, as well as around the colossal ruins of the Aztec Castle, which (like an American Palmyra) rises abandoned in the desert near the Gila River. A long-horned mouflon, similar to the so-called father of the sheep, abounds on the dry and naked limestone cliffs of California. To the southern peninsula belong the vicuñas, guanacos, alpacas, and llamas. But of these useful South American animals, only the first two mentioned have preserved for millennia their natural freedom. The enjoyment of milk and cheese, like the possession and cultivation of grasses that yield meal or flour,[27] is a differentiating characteristic of the nations of the Old World.

If, from among these, some groups passed via Northern Asia to the west coast of America and, preferring the cold,[28] pursued the ridge of the Andes to the south, this migration must have taken place along pathways upon which neither herds nor grains could accompany the newcomers. Could it be, perhaps, that when the long-unstable empire of the Xiongnu collapsed, the rolling forth of this mighty race set in motion migrations from Northeastern China and Korea, through which civilized Asians passed over to the New Continent? If these newcomers had been inhabitants of the steppes, where agriculture was not pursued, then this bold hypothesis, which through language comparison has heretofore garnered little favor, would at least explain the conspicuous absence of actual cereal grains in America. Perhaps there landed on the coasts of New California, battered by storms, one of those Asiatic priest colonies whose mystic reveries induced them to venture on long sea voyages, and of whom the history of the populating of Japan[29] at the time of Qin Shi Huang-ti provides a memorable example.

If the pastoral life, this beneficial middle stage that binds throngs of nomadic hunters to the grass-covered lands while preparing them for agricultural life, thus remained unknown to the ancient Americans, then the reason for the sparse presence of humanity on the South American steppes lies in this ignorance. All the more freely, then, the forces of Nature manifested themselves upon the steppes in many diverse types of animal life: free and limited only by themselves, like the vegetation in the forests of the Orinoco, where the *Hymenaea* and the huge-trunked bay tree are never threatened by the hand of Man, but only by the luxuriant crush of entwining growth. Agoutis; small bright-spotted harts; plated armadillos that, like rats, frighten the hares from their subterranean burrows; herds of lethargic capybara; beautifully striped civet-cats that befoul the air; the great maneless lion; spotted jaguars (usually called tigers), which are capable of dragging a young steer they have killed to a hilltop—these and many other creatures[30] wander the treeless plain.

Inhabitable by almost none but these animals, the plains would not have been able to keep hold of any of the nomadic human throngs who, in any case (in the Indo-Asiatic way), prefer vegetable nourishment, had not the fan palms, *Mauritia*, stood scattered here and there. The virtues of this beneficial Tree of Life are widely known. At the mouth of the Orinoco, north of the Sierra de Imataca, it alone sustains the unvanquished nation of the Guarani.[31] When they were more numerous and densely populated, they not only elevated their huts on posts hewn from the palms, which bore a horizontal platform as a floor, but also (so legend tells) artfully stretched hammocks woven from the leaf-stalks of the Mauritia from trunk to trunk, so that during the rainy season when the delta is flooded, they might live in the trees in the manner of the monkeys. These suspended huts were partially covered with clay. On the damp lower level, the women stoked fire for household use. Those who passed by the river at night saw the fires blazing in rows, high in the air, detached from the earth. Even now, the Guarani owe the preservation of their physical and perhaps even their moral independence to the loose, semifluid bog soil across which they light-footedly move about, and to their residing in the trees: a high refuge to which religious enthusiasm would likely never lead an American Stylite.[32]

The Mauritia, however, provides not only safe living quarters but abundant food as well. Before the tender blossom spathe bursts forth on the male palm, and only in this phase of the plant's metamorphosis, the pulp of the trunk contains a meal similar to sago, which, like the meal of the Jatropha root, is dried into thin, breadlike wafers. The fermented juice of the tree is the sweet, intoxicating palm wine of the Guarani. Like pisang and the juice of all fruits of the tropical world, this fruit, covered in small scales and looking like reddish pine cones, provides nourishment in different ways, depending on whether one partakes of it after the full development of its sugar content or earlier, in its meal-rich state. Thus do we find, at this most basic level of human intellectual development, the existence of an entire people bound (like the insect that is restricted to certain parts of a blossom) almost solely to a single tree.

Since the discovery of the New Continent, the Llanos have become inhabitable to humans. In order to ease the traffic between the coast and Guyana (the Orinoco country), cities[33] have been built here and there on the rivers of the steppes. All across the immeasurable space, animal husbandry has begun. Single huts, woven from reeds and cords and covered in cow hides, stand a day's journey from one another. Innumerable bands of wild bulls, horses, and mules (estimated, at the peaceful time when I made my trip, at one and a half million head) swarm about on the steppe. The enormous procreation of these Old World animals is even more extraordinary given the many dangers with which they must contend in this place.

When the charred grass falls to dust under the vertical rays of the sun, the hardened earth gapes open as if shaken by powerful tremors. Should it then be touched by opposing air currents whose conflict equalizes in circular motion, the plain takes on a curious appearance. In the form of funnel clouds[34] with their tips gliding across the earth, the sand rises like steam through the airless, electrically charged center of the vortex, like the hissing waterspouts feared by experienced boatmen. A hazy, almost straw-colored half-light is thrown by the seemingly low-hanging heavens upon the desolate plain. The horizon draws suddenly nearer. It constricts the steppe and the mood of the wanderer as well. The smothering heat of the air[35] is increased by the hot, dusty earth floating in the atmosphere, which is veiled as if by fog. Instead of cooling, the east wind brings only more warmth as it blows across the long-heated ground.

The pools that the yellow-bleached fan palms protected from evaporation also gradually disappear. As animals in the North become dormant with the cold, so do the crocodile and the boa snake slumber motionlessly here, buried deep in dry clay. Everywhere the drought announces death; and yet everywhere, in the play of distorted light-rays, the illusion[36] of the waving surface of water hounds those who thirst. A narrow strip of air separates the distant grove of palms from the ground. Through the contact of layers of air of differing temperature and density, it floats, lifted by the mirage. Hidden in darker clouds of dust, spooked by hunger and burning thirst, ramble horses and cattle, the cows lowing hollowly, the horses with outstretched necks snuffling the wind, that through the moisture in the air current they might guess the nearness of a water hole not yet completely evaporated.

More deliberate and wily, the mule seeks to relieve his thirst in a different way. A spherical and many-ribbed plant, the melon cactus,[37] encloses within its prickly exterior a pulp rich in water. With his forefoot the mule strikes the needles away and only then dares to bring his lips carefully nearer and drink the cool cactus juice. But drawing drink from this living vegetable spring is not without danger; one often sees animals lamed in the hoof by cactus needles.

When the burning heat of the day is followed by the cooling of the night, which here is always of the same duration, even then the cattle and horses can enjoy no peace. Monstrous bats, vampirelike, suck the blood from them while they sleep or attach themselves firmly to their backs, where they irritate ulcerous sores in which mosquitoes, horseflies, and a host of biting insects settle. Thus do the animals live a painful life when the water disappears from the ground under the blaze of the sun.

When, after the long drought, the beneficent rainy season arrives, the scene on the steppe suddenly changes.[38] The deep blue of the previously cloudless sky grows lighter. By night, one hardly recognizes the black space in the constellation of the Southern Cross. The gentle, seemingly phosphorescent shimmer of the Magellanic Clouds goes out. Even the stars of Aquila and Ophiuchus directly overhead glow with a trembling, less planetary light. In the South, like a remote mountain, a single cloud rises vertically from the horizon. Gradually, the many vapors spread like fog across the zenith. The distant thunder announces the coming of the invigorating rain.

The surface of the Earth is hardly moistened before the fragrant steppe becomes covered with Kyllinga, with many-panicled Paspalum and a number of diverse grasses. Enticed by the light, herbaceous Mimosas unfold their leaves sunken in slumber and greet the rising sun like the morning song of the birds and the opening blossoms of the water plants. Horses and cattle graze in glad enjoyment of life. The grass shooting upward hides the beautifully spotted jaguar. Keeping watch from a safe hiding place and carefully measuring the length of his spring, he snatches passing animals, catlike as the Asiatic tiger.

Occasionally one sees (so say the natives) the now-wet clay on the edge of the swamps slowly lift itself in clods.[39] With a fierce roaring, as with the eruption of a mud volcano, the churned-up earth is flung high in the air. Those familiar with the sight flee at its occurrence, for a gigantic water snake or an armored crocodile is climbing forth from its crypt, awakened from seeming death by the first rainfall.

As the rivers that border the plain to the south (the Arauca, the Apure, and the Payara) gradually begin to swell, so then does Nature force the same animals that during the first half of the year languished in thirst on the waterless, dusty ground to live now as amphibians. The steppe now has the appearance of a vast inland sea.[40] The mares move with their foals back to the higher banks, which protrude in the form of islands above the surface of the sea. With each day, the dry spaces grow smaller; without pasture, the huddled animals swim about for hours and meagerly nourish themselves on the grass panicles that lift themselves above the brown-colored, bubbling water. Many foals drown; many are taken by crocodiles, shattered by their serrated tails, and swallowed. One often sees horses and cattle that, having escaped the jaws of these bloodthirsty, gigantic lizards, bear on their haunches the scars of those pointed teeth.

Such a sight irresistibly reminds the serious observer of the flexibility with which all-providing Nature has endowed certain animals and plants. Like the farinaceous fruits of Ceres, so have the cow and horse followed man around the entire globe: from the Ganges to the Rio Plata, from the African seacoast to the mountainous plain of Antisana, which lies higher than the cone peak of Tenerife.[41] Here the northern birch, there the date palm protect the exhausted bull from the rays of the sun. The same animal species that in Eastern Europe battle with bears and wolves are threatened under another patch of sky by the attacks of the tiger and the crocodile!

But not only crocodile and jaguar prey upon the South American horses; among the fishes they also have a dangerous enemy. The swamp waters of Bera and Rastro[42] are filled with innumerable electric eels whose slimy, yellow-flecked bodies can, from every part, discharge the jolting energy at will. These gymnotids have a length of 5 to 6 feet. They are powerful enough to kill the largest animals if they can discharge their nerve-laden organs all at once in a propitious direction. The road across the steppe at Uritucu once had to be changed because the gymnotids had massed together in a small river in such numbers that each year, several horses drowned in the ford after being stunned. All other fish in the area also flee these terrible eels. They even frighten anglers on the high banks when the wet line conducts the shock to them from afar. Thus does the electrical fire erupt from the bosom of the waters.

The capturing of the gymnotids affords a picturesque spectacle. Mules and horses, encircled by Indians, are driven into a swamp, until the bold fish are excited by the unaccustomed noise into attacking. One sees them swimming like snakes in the water and slyly crowding under the bellies of the horses. Of these, many succumb to the power of invisible strikes. With bristling mane and snorting, wild panic in their eyes, others flee from the raging storm. But the Indians, armed with poles of bamboo, drive them back to the middle of the pool.

Gradually, the rage of the mismatched fight abates. Like empty clouds, the exhausted fish disperse. They need long rest and plenty of food to regain the galvanic energy they have spent. The jolt of their strikes grows gradually weaker and weaker. Frightened by the noise of the stamping horses, they timidly draw near to the banks, where they are wounded by harpoons and pulled onto the steppe with dry, nonconductive sticks of wood.

This is the wondrous struggle of the horses and fish. That which, invisible, is the living weapon of this denizen of the water; which, awakened by the contact of moist and dissimilar parts,[43] races through all organs of animals and plants; which thunderingly inflames the broad roof of the heavens, binds iron to iron, and steers the silent, returning motion of the guiding needle—all, like the colors of the refracted beam of light, flow forth from One Source; all melt together in an eternal, all-encompassing power.

I could close here this bold attempt at a Nature-portrait of the steppe. But as one's fancy likes to occupy itself while on the ocean with pictures of distant shores, so too shall we, before the great plain vanishes, cast a fleeting glance upon the regions that border the steppe.

Africa's northern desert separates two races of humanity who originally belonged to the same part of the globe and whose evenly matched feud seems to be as old as the myth of Osiris and Typhon.[44] North of the Atlas live tribes of yellow hue with straight, long hair and Caucasian facial features. South of Senegal, in the area of the Sudan, however, live negro tribes, who may be found at many various stages of civilization. In Central Asia, Siberian barbarism is separated by the Mongolian steppe from the ancient human civilization found on the peninsula of Hindustan.

The South American plains, too, set the boundary of the region of European semi-culture.[45] To the north, between the mountain ranges of Venezuela and the Antillean seas, lie industrious cities, tidy villages, and carefully cultivated crop fields, one after the other. Aesthetic sensibility, scientific education, and the noble love of civil liberty have long since awoken there. To the south, the steppe is bordered by a formidable wilderness. Thousand-year-old forests and impenetrable undergrowth fill this damp part of the world between the Orinoco and the Amazon. Mighty, lead-colored[46] granite massifs enclose the beds of foaming rivers. Mountains and forests echo with the thunder of the crashing waters, the roar of the tigerlike jaguar, the hollow howls of the bearded monkeys, heralding the rain.[47]

Where the shallow stream allows a sandbar to remain, there lie stretched out, with open jaws and motionless as slabs of stone, the hulking bodies of the crocodiles, often covered with birds.[48] The chessboard-patterned boa snake, its tail anchored to a tree trunk and certain of its quarry, lies in wait coiled on the bank. Quickly uncoiled and thrusting forth, it seizes the young bull or the weaker deer and forces its prey, covered in saliva, laboriously down its expanding throat.[49]

Within this great and wild place of Nature live many and various races of humanity. Separated by a marvelous variety of languages, some, such as the Ottomaks and Jarures, are nomadic strangers to farming, outcasts who eat ants, rubber, and earth;[50] others, such as the Maquiritares and Macos, are settled, knowledgeable, and of gentler customs, fed by fruits of their own cultivation. Large expanses between the Casiquiare and the Atabapo are inhabited not by men but only by tapirs and colonies of monkeys. Pictures carved in stone[51] are evidence that these wastelands too were once places of higher culture. These images give witness to the changing fates of peoples, as do the variously developed and pliable languages that belong among the oldest and most everlasting historical monuments of humanity.

Though tiger and crocodile battle horses and cattle in the steppe, we see on its forested bank, in the wildernesses of Guyana, man forever armed against man. With unnatural desire, some tribes here drink the blood drained from their enemies; others, seemingly unarmed and yet equipped for murder,[52] strangle the enemy with a poisoned thumbnail. The weaker tribes, when they take to the sandy bank, carefully brush away the traces of their timid steps with their hands.

Thus does man, at the lowest level of animal brutality or in the vainglory of his elevated civilization alike, ever make for himself a wearisome life. Thus is the wanderer pursued across the wide world, over land and sea, like all historians throughout the centuries, by the monotonous, comfortless spectacle of the sundered human race.

He who seeks spiritual peace amidst the unresolved strife between peoples therefore gladly lowers his gaze to the quiet life of plants and into the inner workings of the sacred force of Nature, or, surrendering to the instinctive drive that has glowed for millennia in the breast of humanity, he looks upward with awe to the high celestial bodies, which, in undisturbed harmony, complete their ancient, eternal course.

Annotations and Additions

1. Lake Tacarigua

When one advances through the interior of South America, from the coast of Caracas or Venezuela to the Brazilian border, from the 10th degree of northern latitude to the equator, one first crosses a high mountain range (the coastal range of Caracas) which is oriented from east to west, then the great, treeless steppes or plains (*los Llanos*), which stretch from the foot of the coastal range to the left bank of the Orinoco, then finally the row of mountains that give rise to the cataracts of Atures and Maypure. Between the sources of the Rio Branco and the Rio Esquibo runs this same row of mountains, which I call Sierra Parime, eastward from the cataracts and on toward Dutch and French Guyana. This is the place of the wondrous myths of El Dorado, a rocky mass broken into many ridges in a gridlike formation. To the south it is bordered by the heavily forested plain in which the Rio Negro and the Amazon have formed their beds. One who wishes to be more thoroughly instructed as to these geographical conditions might compare the great map of La Cruz Olmedilla (1775), from which nearly all newer maps of South America have originated, with the map of Colombia that, drawn from my own astronomical observations, I published in 1825.

The coastal range of Venezuela is, geographically considered, a part of the Peruvian Andes chain itself. This chain divides within the immense knot of mountains south of Popayan that are the sources of the Magdalena (1°55′ to 2°20′ latitude) into three ranges, the easternmost of which runs into the snow-capped peaks of Merida. These snowy mountains decline toward the Paramo de las Rosas in the hilly land of Quibor and Tocuyo, which connects the coastal range of Venezuela with the cordilleras of Cundinamarca. The coastal range runs like a wall, uninterrupted from Portocabello to the Paria foothills. Its average height is barely 750 toises. Individual peaks, however, rise to 1,350 toises above sea level, such as the Silla de Caracas (known also as Cerro de Avila), adorned with Befaria (the red-blossomed American alpine rose). The shore of the terra firma shows signs of desertification. Recognizable everywhere is the impact of the great current that runs from east to west and that, after the fragmentation of the Caribbean islands, dug out the Sea of the Antilles. The spits of land of Araya and Chuparipari, especially the coast of Cumana and New Barcelona, provide for the geologist a remarkable sight. The cliff-islands Boracha, Caracas, and Chimanas rise towerlike from the sea and bear evidence of the terrible force of the crashing floods upon the shattered mountain range. Perhaps the Antillean, like the Mediterranean, was once an inland sea that suddenly came forth to join the ocean. The islands of Cuba, Haiti, and Jamaica constitute the remnants of the schist mountains that contained this lake to the north. It is conspicuous that just at the point where these three islands are closest to one another, there too rise their highest peaks. One might suppose that the primary massif of this Antillean range lay between Cape Tiburon and Morant Point. The Copper Mountains (*Montañas de Cobre*) near Santiago de Cuba are as yet not surveyed but are probably taller than the Blue Mountains of Jamaica (1,138 toises), which slightly exceed the height of the Gotthard Pass. I have already more fully developed my assumptions regarding the valleylike form of the Atlantic Ocean and the ancient connectivity of the continents in an essay written in Cumana: *Fragment d'un Tableau géologique de l'Amérique méridionale* (*Journal de Physique*, Messidor an IX). It is worth noting that Christopher Columbus himself points out in one of his official reports the correlation between the orientation of the equinoctial current and the coastal formation of the Greater Antilles (*Examen critique de l'hist. de la Géographie*, vol. III, p. 104–108).

The northern and more cultivated part of Caracas is mountainous country. The coastal range is, like the Swiss Alps, divided into several spurs or saddle-ridges that embrace long valleys. The most famous among these is the charming valley of Aragua, which produces tremendous amounts of indigo, sugar, cotton, and, most remarkably, even European wheat. The southern end of this valley is bordered by the beautiful Lake Valencia, whose ancient Indian name is Tacarigua. The contrast of its opposite shores affords it a noticeable similarity to Lake Geneva. While the desolate hills of Guigue and Guiripa indeed have a less imposing and magnificent character than the Savoy Alps, the shores of Tacarigua, on the other hand, thickly grown over with copses of pisang trees, with Mimosas and Triplaris, surpass all the wine gardens of the Vaud region in picturesque beauty. The lake has a length of about 10 nautical miles (20 of which make one degree at the equator); it is full of small islands that are growing in size, since the rate of evaporation of the water basin exceeds that of the influx. Over the last several years, even sandbanks have emerged as true islands. They are given the name significant of something newly appeared, *Las Aparecidas.* On the Isle of Cura is cultivated the curious variety of *Solanum* of which the fruit is edible, and which was described by Willdenow in *Hortus Berolinensis* (1816, tbl. XXVII). The elevation of Lake Tacarigua above sea level is nearly 1400 feet (by my measurements exactly 230 toises) lower than the average altitude of the Caracas Valley. The lake supports its own native fish species (see my *Observations de Zoologie et d'Anatomie comparée* vol. II, pp. 179–181) and ranks among the most beautiful and inviting Nature-scenes that I know of on the face of the Earth. Whilst bathing, Bonpland and I were often startled at the sight of the Bava, an undescribed, 3-to-4-foot-long crocodilian lizard (*Dragonne?*) of a frightful appearance, though harmless to man. In Lake Valencia we found a Typha (cattail) quite identical to the European *Typha angustifolia*: an extraordinary fact, important to plant geography!

Around the Lake, in the valleys of Aragua, both varieties of sugar cane are cultivated: the common *Caña criolla*, and the newly introduced South Sea *Caña de Otaheiti.* The latter possesses a far lighter and more pleasant green, so that from a great distance one can already tell a field of Tahitian sugar cane from one of the common variety. Cook and Georg Forster first described the sugar cane of Otaheiti, but as one can see in Forster's excellent treatise on the edible plants of the South Sea Islands, they little recognized the worth of this valuable commodity. Bougainville brought it to Ile de France, from whence it came to Cayenne, and from 1792 on to Martinique, Santo Domingo or Haiti, and several of the Lesser Antilles. The bold but unfortunate Captain Bligh transplanted it along with breadfruit trees to Jamaica. From Trinidad, one of the islands close to the continent, the sugar cane of the South Seas made its way to the nearby coast of Caracas. For these regions it has become more important than the breadfruit tree, which will probably never displace such a beneficial and nutritionally rich plant as the pisang. The Otaheiti sugar cane is also much juicier than the common variety, which has been attributed an East Asian origin. It yields on the same acreage one-third more sugar than the *Caña criolla*, which has a thinner stalk and narrower articulations. Moreover, as the isles of the West Indies begin to suffer a great shortage of fuels (on the island of Cuba, the sugar pans are heated with orangewood), the new sugar cane becomes still more important, since it provides a thicker, woodier stalk (*bagaso*). Had the introduction of this new product not occurred almost simultaneously to the bloody Negro uprising of Santo Domingo, sugar prices in Europe at that time would have climbed even higher than this ruinous disruption of agriculture and trade caused them to do. An important question is whether the sugar cane from Otaheiti, torn from its native soil, will gradually devolve into common sugar cane. Experience up to the present has decided against devolution. On the island of Cuba, a *caballeria*, i.e., a surface of 34,969 square toises, yields 870 centners of sugar when it is planted with Otaheiti sugar cane. Strange indeed that this important commodity of the South Sea islands is being farmed in the very part of the Spanish colonies that is farthest from the South Seas! One may make the voyage from the Peruvian coast to Otaheiti in 25 days, and yet at the time of my travels in Peru and Chile, the Otaheiti sugar cane was as yet unknown there. The inhabitants of Easter Island, who suffer a great lack of potable water, drink the juice of sugar cane and (which is physiologically more remarkable) seawater. Upon the Society, Friendly, and Sandwich Islands, the light-green, thick-stalked sugar cane is cultivated everywhere.

Besides the *Caña de Otaheiti* and the *Caña criolla*, a reddish African sugar cane is also grown in the West Indies. It is called *Caña de Guinea*. It produces little more juice than the common Asiatic. But it is believed that the juice of the African variety is especially suited to the production of rum.

Contrasting very beautifully with the light green of the Tahitian sugar cane in the Caracas Province is the dark shade of the cacao plantations. Few trees of the tropical world are as thickly foliated as *Theobroma cacao*. This exquisite plant loves hot, damp valleys. In South America as in South Asia, great fertility of the soil and insalubrity of the air are bound inseparably to one another. Indeed, one must notice that as the culture of a land increases, as the forests diminish, the soil and the climate grow accordingly more arid—and the cacao plantations thrive ever less. They are thus becoming less numerous in the Caracas Province, while in the more easterly provinces of New Barcelona and Cumana, especially in the damp, sylvan strip of land between Cariaco and the Gulf of Triste, they are quickly increasing in number.

2. The natives call these phenomena "banks"

The Llanos of Caracas are filled with a widely distributed, abundant formation of ancient conglomerate. As one descends from out of the valleys of Aragua across the southern ridge of the coastal range from Guigue and Villa de Cura toward Parapara, one comes upon, one after the other, gneiss, mica schist, a (probably Silurian) transitional stone of clay shale and black limestone, serpentine and greenstone in separate, spheroidal pieces, and finally, close to the edge of the wide plain, small hills of augite amygdaloid and porphyritic shale. These hills between Parapara and Ortiz appeared to me to have been volcanic eruptions along the ancient seacoast of the Llanos. Farther to the north stand the famous, grotesque forms of the cave-riddled cliffs known as the Morros de San Juan, which form a sort of *Teufelsmauer* (Devil's Wall) of crystalline granules like upthrust dolomite. They may thus be viewed more as parts of the shore than as islands of the ancient gulf. I call the Llanos a gulf, for when one considers their low elevation above the current level of the sea, their form, which is open in a fashion consistent to the east-west running rotational current, and the lowness of the eastern coast between the mouths of the Orinoco and the Essequibo, one can hardly doubt that the ocean once covered this entire basin between the coastal range and the Sierra de la Parime, and struck land (as in the case of its lying across the Lombardian plains and reaching the Cottian and Pennine Alps) at the mountains of Merida and Pamplona. Also, the slope, or decline, of the American Llanos is from west to east. Their altitude at Calabozo, however, 100 geographical miles from the ocean, still barely reaches 30 toises, that is, 15 less than the altitude at Pavia, and 45 less than Milan on the plains of Lombardy, between the Swiss-Lepontine Alps and the Ligurian Apennines. The formation of the Earth here brings to mind Claudian's expression *curvata tumore parvo planities* (a wavy plain upheaved its swelling sides). The horizontality (levelness) of the Llanos is so complete that in many of its spaces, no single part in over 30 square miles seems to lie a foot higher than another. Should one also consider the absence of all shrubbery, indeed, in the Mesa de Pavones of even any isolated palm trunks, then one can form a picture of the extraordinary view offered by this sealike, desolate flatland. As far as the eye can see, it will rest on practically no object reaching above several inches in height. If the horizon, due to the condition of the lower levels of the atmosphere and the play of the refracting rays of light, were not forever indistinctly defined and shimmering like waves, one could take readings with a sextant of the sun's altitude over the edge of the plain, just as over the ocean's horizon. Beside this tremendous levelness of the ancient seafloor, the banks are even more striking. They are broken strata that abruptly rise 2 to 3 feet above the rock nearby and extend uniformly for a stretch of 10 to 12 geographical miles. These banks are the point of origin for the small rivers of the steppe.

On the return trip from the Rio Negro, as we crossed the Llanos de Barcelona, we found abundant evidence of sinkholes. Instead of high banks, we saw here single layers of gypsum 3 to 4 feet deeper than the stone around them. And farther west, near the confluence where the Caura River meets the Orinoco, a great stretch of thick forest east of the Mission of San Pedro de Alcantara sank away (in an earthquake) in the year 1790. There in the midst of the plain, a lake formed, the diameter of which is over 300 toises. The tall trees (Desmanthus, Hymenaea, and Malpighia) long remained green and leafy under the water.

3. One believes he sees before him the boundless ocean.

The outlook upon the distant steppe is even more remarkable when, after being long in the thicket of the forests, one has become accustomed to a narrow field of view, and with it, the sight of a richly adorned Nature. Indelible to me will be the impression that the Llanos presented us upon our return from the Upper Orinoco as first we saw them again, at a great distance, from a mountain that stood across from the mouth of the Rio Apure, near the Hato del Capuchino. The sun had just set. The steppe seemed to rise like a hemisphere. The light of the ascending celestial bodies was refracted in the lower layer of the air. Because the plain is excessively heated by the vertical rays of the sun, so does the play of the radiant heat, the ascending air currents, and the immediate contact of atmospheric layers of differing densities last throughout the night.

4. Naked rocky crust

Tremendous expanses of land in which only naked slabs of rock are to be seen give a particular character to the deserts of Africa and Asia. In the Schamo, which separates Mongolia (the Ulangom and Malakha-Oola mountain ranges) from Northwestern China, these banks of rock are called *tsy*. One can find them as well in the forested plains of the Orinoco, surrounded by the most luxuriant vegetation (*Relation hist.* vol. II, p. 279). These granite and syenitic slabs, bare of vegetation and only sparsely sprinkled with lichens, have diameters of several thousand feet. Amidst them are to be found small islands of soil covered with low, ever-blooming herbage. They give to these spots in the woodlands or on its borders the appearance of little gardens. The monks of the Upper Orinoco believe these perfectly flat, bare stone strata, if they are especially extensive, to be strangely conducive to the development of fever and other illnesses. Some mission villages, because of this sort of widely held belief, have been abandoned and rebuilt in other places. Do the stone slabs (*laxas*) affect the atmosphere merely through their greater radiation of heat, or is it perhaps on a chemical level?

5. Llanos and Pampas of South America or the grassy plains near the Missouri

Our physical and geognostic understanding of the mountainous western regions of North America has been greatly amended by the bold journeys of Major Long, the excellent work of his companion, Edwin James, and most of all by the broadly diverse observations of Captain Frémont. All of the information collected in my works on New Spain regarding the northern mountain ranges and plains, which I could present only as hypotheses, is now clearly illuminated. In the fields of natural description and historical investigation, individual facts stand out until, through painstaking research, one is able successfully to draw the connections between them.

The eastern coast of the United States of North America inclines from southwest to northeast, as does the Brazilian coast beyond the equator, from the River Plate up to Olinda. In both lands, a short distance from the littoral, stretch two mountain ranges that are more parallel to one another than to the westward-lying Andes chain (the cordilleras of Chile and Peru) or to the Rocky Mountains of Northern Mexico. The mountain system of the southern hemisphere, the Brazilian system, constitutes an isolated group whose highest peaks (Itacolumi and Itambe) do not exceed 900 toises. Only the eastern ridges nearer the sea incline regularly from SSW to NNE; to the west, the group increases in breadth while significantly decreasing in height. The hill country of the Paresis draws near to the Itenes and Guaporé Rivers, just as the mountains of Aguapei and San Fernando (south of Villabella) draw near to the high Andes of Cochabamba and Santa Cruz de la Sierra.

There is no direct connection between the two mountain systems on the Atlantic and South Sea coasts (the Brazilian and Peruvian cordilleras); the lowland of the Chiquitos Province, a long valley running north to south which opens upon the plains of both the Amazon and Plate Rivers, separates Western Brazil from Eastern Alto Perú. Here, as in Poland and Russia, an often indiscernible crest in the ground (Slavic *Uwaly*) delineates the surface water divide between the Pilcomayo and Madeira, between the Aguapei and the Guaporé, and between the Paraguay and the Rio Topayos. This rise (*seuil*) stretches southeast from Chayanta and Pomabamba (lat. 19°–20°) and crosses the lowland of the Chiquitos Province, which has again become virtually unknown to the geographer since the expulsion of the Jesuits. Then, in a northeasterly direction where only isolated mountains rise, it forms the *divortia aquarum* at the sources of the Baure and in the area of Villabella (lat. 15°–17°).

This water divide, so vital to the traffic of the people and their burgeoning culture, corresponds with a second one in the northern hemisphere of South America (lat. 2°–3°), which separates the Orinoco River basin from those of the Rio Negro and the Amazon. One might view these rises in the level areas, these crests (*terrae tumores* to Frontin), as undeveloped mountain systems destined to connect two seemingly isolated groups, the Sierra Parime and the Brazilian highlands, to the Andes chain of Timana and Cochabamba. Such relationships, hardly examined until now, form the basis of my proposed division of South America into three geographical depressions or river basins: those of the Orinoco (in its lower course), the Amazon, and the Rio de la Plata, of which (as mentioned above) the outer ones are steppes or grassy plains, while the middle, between the Sierra Parime and the Brazilian mountain group, can be seen to be a wooded plain (*hylaea*).

Should one wish to create in so few strokes a picture of the Nature of North America, one's eye should first be fixed upon the Andes chain: at first narrow, then increasing in breadth and height, it inclines in Panama, Veragua, Guatemala, and New Spain from southeast to northwest. This range, a home to earlier human culture, creates the same hindrances for the general tropical ocean current as it does for the expedition of trade between Europe, West Africa, and East Asia. From the 17th degree of latitude on, from the famous Isthmus of Tehuantepec, it wends its way from the coast of the Pacific and becomes an inland cordillera running from south to north. In Northern Mexico, the Crane Mountains (Sierra de las Grullas) form one portion of the Rocky Mountains. To the west of these originate the Columbia River and the Rio Colorado of California; to the east, the Rio Roxo de Natchitoches, the Canadian River, the Arkansas, and the (shallow) Platte River, which has been transformed of late by uninformed geographers into a "Plata River," a source of silver. Between the sources of these rivers (lat. 37°20′ to 40°13′) rise three granite summits, rich in hornblende with little mica and reminiscent of the Swiss Schreckhorn, known respectively as Spanish Peaks, James or Pike's Peak, and Bighorn or Long's Peak. (See my *Essai politique sur la Nouvelle-Espagne*, 2ème éd. vol. I, pp. 82 and 109.) Their elevations surpass those of all of the Andes peaks of Northern Mexico, which indeed never rise above the climatic snow line from the parallels of 18° and 19°, i.e., from the groups of Orizaba (2,717 toises) and Popocatepetl (2,771 toises) all the way to the regions beyond Santa Fe and Taos in New Mexico. James Peak (lat. 38°48′) is said to rise to 1,798 toises, but of this height only 1,332 toises have been measured trigonometrically, the remaining 463 toises being based, in the absence of all barometric observations, upon uncertain estimates of the descent of rivers. Since a trigonometric measurement can almost never be made even at sea level, the determinations of insurmountable heights are always partly trigonometric and partly barometric. The estimation of the descent of rivers, their speed, and the length of their courses is so deceptive that before the important expedition of Captain Frémont, the altitude of the plain at the foot of the Rockies nearest the peaks mentioned in the text was variously estimated to be now 8,000 feet, now 3,000 feet (Long's *Expedition*, vol. II, pp. 36, 362, 382, App. p. XXXVII). It is due to a similar lack of barometric measurements that the elevation of the Himalayas remained so long uncertain; scientific culture in the East Indies has now increased to such an extent that when Captain Gerard set foot upon the Tarhigang near the Sutlej north of Shepke at a height of 18,210 Parisian feet, he could break three barometers and still have four of the same at his disposal (*Critical Researches on Philology and Geology*, 1824, p. 144).

During the expeditions that he carried out on the orders of the United States government in the years from 1842 to 1844, Frémont discovered and barometrically measured the tallest peak in the entire chain of the Rocky Mountains. To the north-northwest of Spanish, James, Long's, and Laramie Peaks, this snow-capped mountain is part of the Wind River Range. On the large map published by the chief of the Topographical Bureau in Washington, Colonel Abert, this mountain carries the name Frémont's Peak and stands below lat. 43° 10′ and at long. 112° 35′, that is, nearly 5½° farther north than Spanish Peak. Its elevation is, by a direct measurement, 12,730 Parisian feet. Frémont's Peak is thus 324 toises higher than Long's calculation for James Peak, which according to its position on the above-mentioned map, is identical to Pike's Peak. The Wind River Mountains form the water divide (*divortia aquarum*) between the two oceans. In his *Report of the Exploring Expedition to the Rocky Mountains in the Year 1842, and to Oregon and North California in the Years 1843–44* (p. 70), Captain Frémont describes looking from the summit: "On the one side we overlooked innumerable lakes and streams, the spring of the Rio Colorado which delivers its waters to the South Seas via the Gulf of California; on the other was the deep Wind River Valley, where were the heads of the Yellowstone River, which is one of the primary branches of the Missouri, which joins the Mississippi at St. Louis. To the northwest we just could discover the snowy heads of the *Trois Tetons*, where was the true source of the Missouri, not far from the source-waters of the Oregon, or Columbia, River, namely the branch known as the Snake River or Lewis Fork." To the astonishment of the bold mountaineer, the summit of Frémont's Peak was visited by bees. Perhaps they had been carried upward against their will by the rising currents of air, like the butterflies that I had observed in even higher regions of the Andes, even above the snow line. And in the Pacific I have seen large-winged lepidoptera, borne great distances across the ocean by land winds, set down upon ships far from the coasts.

Frémont's map and geographical investigations comprise the tremendous stretch of land from the mouth of the Kansas River at the Missouri up to the falls of the Columbia and to the missions of Santa Barbara and Pueblo de los Angeles in New California: a longitudinal difference of 28° (about 340 geographical miles) between the parallels of 34° to 45° northern latitude. Four hundred points have been measured hypsometrically through barometric readings, and most of these were also measured astronomically, thus making possible the depiction in one profile above sea level of a tract of land that, factoring in its unevenness, covers some 900 geographical miles, from the mouth of the Kansas River to Fort Vancouver and to the coasts of the Pacific (nearly 180 more miles than the distance from Madrid to Tobolsk). As I believe myself to have been the first to undertake the task of depicting entire countries (the Iberian Peninsula, the highlands of Mexico, the cordilleras of South America) in geognostic profile (the half-perspective projections of a Siberian traveler, Abbé Chappe, were based mostly upon simple and for the most part absurd estimates of waterfalls), it is a great pleasure to me to see employed in the most excellent fashion the graphic method of the vertical depiction of the Earth's topography, of the elevation of the solid over the liquid. Between the middle latitudes of 37° to 43°, along with the snowy summits that compare to the heights of the peak of Tenerife, the Rocky Mountains feature high plains of such immensity as can hardly be found elsewhere on Earth, and which extend from east to west nearly twice as far as the high plains of Mexico. From the mountain group that begins somewhat to the west of Fort Laramie to beyond the Wasatch Mountains, the swell of the ground maintains without interruption an altitude of five to seven thousand feet above sea level; indeed, from 34° to 45° latitude, this swell fills the entire space between the actual Rocky Mountains and the snowy range of the California coast. This space, a sort of broad, long valley like that of Lake Titicaca, has been called "the Great Basin" by Joseph Walker and Captain Frémont, those travelers who have supplied so much information about the western regions; it is a terra incognita of at least 8,000 square geographical miles—dry, nearly devoid of people, and full of salt lakes, the greatest of which lies 3,940 Parisian feet above sea level and is connected to the narrow Utah Lake (Frémont, *Report of the Exploring Expedition*, p. 154 and 273–276). Into the latter flows the water-rich Rock River (*Timpan Ogo* in the Ute language). Father Escalante discovered Frémont's Great Salt Lake in 1776 during his migration from Santa Fé, New Mexico, to Monterey in New California and, confusing river and lake, gave it the name Laguna de Timpanogo. I included these as such in my map of Mexico, which gave rise to a good deal of uncritical debate concerning the presumed nonexistence of a large inland salt lake, a subject already settled by the knowledgeable American geographer Tanner. (Humboldt, *Atlas Mexicain*, pl. 2; *Essai politique sur la Nouv. Esp.*, vol. 1, p. 231, vol. II, pp. 243, 313 and 420; Frémont, *Upper California*, p. 9; cf. Duflot de Mofras, *Exploration de l'Orégon*, 1844, vol. II, p. 40.) In the *Archaeologia Americana*, vol. II, p. 140, within the treatise concerning indigenous tribes, Gallatin expressly states: "General Ashley and Mrs. J. S. Smith have found the Lake Timpanogo in the same latitude and longitude nearly as had been assigned to it in Humboldt's Atlas of Mexico."

I deliberately dwell upon these contemplations of the extraordinary swell of the ground in the region of the Rocky Mountains, for it is beyond a doubt that due to its extent and elevation, it must exert a great, though previously unacknowledged, influence upon the climate of all of North America toward the south and east. In the great and uninterrupted high plains Frémont saw how ice would cover the surface of the water every night in the month of August. And no less significant is the importance of this topography to the social conditions and the progress of culture in the large North American states. Despite the fact that the water divide reaches an altitude that approaches that of the passes of Simplon (6,170 ft), Gotthard (6,440 ft), and Great St. Bernhard (7,476 ft), the upward slope is so elongated and gradual that between the Missouri and Oregon territories, between the Atlantic states and the settlements on the Oregon or Columbia River, between the coasts that lie across from Europe and China respectively, there is nothing to impede traffic in coaches and wagons of all sorts. The distance from Boston to old Astoria on the Pacific at the mouth of the Oregon, reckoned in a straight line and considering longitudinal differences, is 550 geographical miles, approximately one-sixth less than the distance from Lisbon to the Ural near Katharinenburg. As there is such a gentle slope to the high plain that extends from the Missouri to California and into the Oregon Territory, determining the high divide-point of the *divortia aquarum* was not done without difficulty. (From Fort Laramie and the river of the same name near the northern branch of the Platte River all the way to Fort Hall on the Lewis Fork of the Columbia River, all the campsites that were measured lay at five to seven thousand Parisian feet—indeed, the campsite in Old Park lay at 9,760!) The water divide is located south of the Wind River Mountains, quite nearly midway between the Mississippi and the Pacific littoral, at an altitude of 7,027 ft, that is, only 450 ft lower than the Great St. Bernhard Pass. Immigrants refer to this divide as South Pass (Frémont's *Report*, pp. 3, 60, 70, 100, and 129). It lies in a charming area where many Artemisia, especially *A. tridentata* (Nuttall), varieties of Aster, and cactus cover the ground of mica slate and gneiss. Astronomical observations determine a latitude of 42°24′ and a longitude of 111°46′. Adolf Erman has pointed out that the line of the great East-Asiatic Aldan mountain range that separates the Lena region from the influx of the Great Ocean (the Pacific), if extended as the largest circle on the Earth's sphere, will pass through several peaks in the Rocky Mountains between 40° and 55° latitude. "An American mountain range and an Asiatic one seem in this way to be but parts of the same fissure, which followed the shortest path in breaking open." (See Erman, *Reise um die Erde*, Part I, vol. 3, p. 8, Part II, vol. 1, p. 386, with his *Archiv für wissenschaftliche Kunde von Russland*, vol. 6, p. 671.)

Very different from the Rocky Mountains, which decrease in height in the region of the often ice-covered Mackenzie River, and from the high plains, upon which individual snow-capped summits rise up, are the higher and more westerly littoral mountains, the range of Californian coastal alpine peaks called the Sierra Nevada de California. As inconceivably chosen as the unfortunately universally adopted name "Rocky Mountains" may be for the northern extension of the Mexican central chain, it still seems to me unadvisable to name them the "Oregon Chain," as is often tried. It is certainly true that from within these mountains spring the source-waters of the three main branches (Lewis', Clark's, and North Fork) that form the mighty Oregon or Columbia River, but this same river also runs through the Californian range of perpetually snow-covered coastal mountains. The name Oregon District is also politically and officially applied to the smaller areas of land west of the littoral range, where Fort Vancouver and the Walahmutti (Willamette) settlements lie, and it is prudent to give the name Oregon to neither the central nor the littoral ranges. This name, by the way, misled a famous geographer, Mr. Malte-Brun, to a misunderstanding of a most curious sort. On an old Spanish map, he read ". . . and it is still unknown (*y aun se ignora*) where the source of this river (the one now called the Columbia) is . . ." and thought that in the word *ignora* he recognized the name "Oregon." (See my *Essai polit. sur la Nouv. Espagne*, vol. II, p. 314.)

The cliffs, which at the point where the range is penetrated form the cataracts of the Columbia, are a distinguishing feature of the continuation of the Sierra Nevada de California from the 44th to the 47th parallel (Frémont, *Geographical Memoir upon Upper-California*, 1848, p. 6). In this northern continuation lie the three colossi Mount Jefferson, Mount Hood, and Mount St. Helens, which rise up to 14,540 Parisian feet above sea level. The height of this littoral "Coast Range" far surpasses that of the Rocky Mountains. "In our eight months' circuit along the coastal mountains we were never out of sight of snow; while we were able to surmount the Rocky Mountains at South Pass at an elevation of 7,027 feet, we found in the coastal mountains, which are divided into several parallel ranges, that the passes were fully 2,000 feet higher," that is, just 1,170 feet lower than the peak of Aetna. It is also especially noteworthy, and reminiscent of the situation as regards the eastern and western cordilleras of Chile, that only the mountain range closest to the sea, that is, the Californian, still features active volcanoes. The conical mountains Rainier and St. Helen's can be seen to smoke nearly without interruption; on the 23rd of November, 1843, the latter volcano spewed forth ash that covered the banks of the Columbia like snow for 10 miles. Also belonging to the volcanic Californian range, in the distant north of Russian America, are Mt. St. Elias (1,980 toises high according to La Pérouse, as much as 2,792 according to Malaspina) and Mt. Fairweather (2,304 toises according to Cerro de Buen Tiempo). Both conical peaks are believed to be still-active volcanoes. In the Rocky Mountains, Frémont's expedition, being both botanical and geognostic, also collected volcanic products (vitreous basalt, trachyte, even true obsidian); an old, burned-out crater was found somewhat to the east of Fort Hall (lat. 42°2′, long. 114°50′), but of still-active volcanoes ejecting lava and ash there was no trace. One must not mistake for active volcanoes the as yet barely explained phenomenon of "smoking hills," as the English settlers called them, or *côtes brûlées*, *terrains ardens* in the language of the French-speaking natives. "Rows of conical hills," says a careful observer, Mr. Nicollet, "are often, at nearly periodic intervals, covered for two to three years with thick black smoke. Flames are not visible upon them. The phenomonon is to be found primarily in the regions of the upper Missouri, and still nearer to the eastern slope of the Rocky Mountains, where there is a river that the natives call *Mankizitah-watpa*, i.e., 'River of the Smoking Earth.' Vitreous pseudo-volcanic products, a sort of porcelain-jasper, can be found in the area of the smoking hills." Since the Lewis and Clark expedition, the belief has widely spread that the Missouri washes up true volcanic pumice upon its banks. Fine whitish alveolar masses were mistaken here for pumice. Professor Ducatel wished to attribute this occurrence, which is to be observed primarily in chalk formations, to "a displacement of water by iron pyrite, and a reaction upon beds of brown coal." (Compare Frémont's *Report*, pp. 164, 184, 187, 193, and 209 with Nicollet's *Illustration of the Hydrographical Basin of the Upper Mississippi River*, 1843, pp. 39–41.)

On concluding these general observations upon the configuration of North America, if we again cast our gaze upon the regions of the continent that separate the two diverging coastal ranges from the central chain, we will find a stark contrast: to the west, between the central chain and the Pacific alpine mountains of California, is a dry and uninhabited high plain of five-to-six thousand feet elevation above sea level, while to the east, between the Alleghenies—whose highest peaks, Mount Washington and Mount Marcy, ascend to 6,240 and 5,066 feet—and the Rocky Mountains, is the water-rich, fertile, densely populated Mississippi basin, the greatest part of which lies at more than twice the elevation of the plain of Lombardy, or four to six hundred feet. The hypsometric constitution of this eastern lowland, that is, its relation to sea level, has only recently been elucidated by the outstanding work of the talented French astronomer Nicollet, whose early death was a loss to Science. His map of the upper Mississippi, completed from 1836 to 1840, is based upon 240 astronomical measurements of latitude and 170 barometric measurements of elevation. The surface that makes up the basin of the Mississippi is identical to the more northerly Canadian; one and the same depression stretches from the Gulf of Mexico to the Arctic Ocean. (See my *Relation historique*, vol. III, p. 234, and Nicollet, *Report to the Senate of the United States*, 1843, pp. 7 and 57.) Where the lowland surface is wavelike and hills (*Côteaux des Prairies, Côteaux des Bois* in the still non-English nomenclature of the natives) appear in contiguous rows between the 47th and 48th parallels of latitude, these rows and gentle swells of the ground separate the waters of the Hudson Bay and the Gulf of Mexico. Such a watershed is created by the Missabay Heights north of the Upper Lake (Lake Superior, or *Kichi Gummi*) and farther west by the so-called *Hauteurs des Terres*, in which lie (first discovered in 1832) the true sources of the Mississippi, one of the world's greatest rivers. The highest of these ranges of hills barely reach 1,400 to 1,500 ft. From the mouth (Old French La Balize) to St. Louis, somewhat south of the confluence of the Missouri and the Mississippi, the latter experiences a decline of only 357 ft, despite covering an itinerant distance of over 320 geographical miles. The surface of Lake Superior lies 580 ft high, and as its depth in the area around Madeline Island is exactly 742 ft, the lake floor is 162 ft below the level of the ocean (Nicollet, pp. 99, 125, and 128).

Beltrami, who in 1825 had broken off from the Major Long expedition, boasted of having found the source of the Mississippi in Cass Lake. But the river, in its uppermost path, flows through four lakes, of which Cass Lake is the second. The outermost is Lake Itasca (47°13′ lat., 97°22′ long.), and was first recognized in 1832 during the expedition of Schoolcraft and Lieutenant Allen as the true source of the Mississippi. This stream, which later becomes so mighty, is only 16 feet wide and 14 inches deep at its outflow from the curiously horseshoe-shaped Lake Itasca. The local situation was first exhaustively illuminated through astronomical positioning by the scientific expedition of Mr. Nicollet in 1836. The elevation of the sources, i.e., of the last influxes that Lake Itasca receives from the ridge of the water divide known as *Hauteur de terre* is 1,575 feet above sea level. Quite nearby, and yet on the southern slope of the same dividing ridge, lies Elbow Lake, the source of the Little Red River of the North, which flows with many meanderings to Hudson Bay. The Carpathian Mountains show similarly related sources of rivers that pour their waters into the Baltic and Black Seas. To twenty small lakes clustered in tight groups to the south and west of Itasca Mr. Nicollet gave the names of famous astronomers and of intimate friends and enemies that he had left behind in Europe. The map becomes a geographical album reminiscent of Ruiz and Pavon's *Flora peruviana*, in which the names of new plant species were created to reflect the court registry and the ever-changing *Oficiales de la Secretaria*.

To the east of the Mississippi are predominantly forests, some still quite dense; to the west are only expanses of grass in which the buffalo (*Bos americanus*) and the musk-ox (*Bos moschatus*) graze in herds. Both animals, the largest of the New World, serve to feed the nomadic Indians, the Apaches Llaneros and the Apaches Lipanes. The Assiniboines hunt in so-called bison parks, which are artificial enclosures into which the wild herds are driven, occasionally killing in a few days as many as seven to eight hundred bison (Maximilian, Prinz zu Wied, *Reise in das innere Nord Amerika*, vol. I, 1839, p. 443). The American bison, which the Mexicans call *Cibolo*, is killed mostly for the tongue, a popular delicacy. He is by no means simply a variation on the Old World aurochs, although other animals such as the European elk (*Cervus elces*) and the reindeer (*Cervus tarandus*), and even the short-bodied peoples of the polar regions, are common to the northern reaches of all the continents and may thus all be seen as evidence of their former long-enduring connectedness. In the Aztec dialect, the Mexicans refer to

European cattle as *quaquahue*, "horned animal," from *quaquahuitl*, "horn." Immense cattle horns found in old Mexican buildings not far from Cuernavaca, southwest of the capital, Mexico City, seem to me to have come from bull bison. It is possible to train the Canadian bison for fieldwork. It can be bred with European cattle; it was long uncertain whether the hybrid offspring would be fertile and capable of reproducing. Albert Gallatin, who through his own observations amassed an extraordinary knowledge of the uncultivated regions of North America before distinguishing himself in Europe as an excellent diplomat, asserts that the fertile crossbreed of the American buffalo with European cattle is undeniable: "The mixed breed was quite common fifty years ago in some of the northwestern counties of Virginia; and the cows, the issue of that mixture, propagated like all others. . . . I don't remember," Gallatin adds, "grown bison being tamed, but occasionally dogs would catch young bison calves, which people would raise and herd with European cows. For some time at Monongahela, all cattle were of this mixed breed. One heard the complaint that they gave little milk." The favorite food of the bison is *Tripsacum dactyloides* (called "buffalo grass" in North Carolina) and an as yet unclassified sort of clover related to *Trifolium repens*, which Barton has designated *Trifolium bisonicum*.

I have pointed out elsewhere (*Kosmos*, vol. II, p. 488) that as late as the sixteenth century, according to the report of the very credible Gómara (*Historia general de las Indias*, chap. 214), an Indian tribe lived in the northwestern part of Mexico, below the 40th parallel of latitude, and that their greatest wealth consisted of herds of tamed bison (*bueyes con una giba*). And in spite of this capacity of the bison to be tamed, in spite of the great deal of milk that it produces, in spite of the herds of llamas in the Peruvian cordilleras, no shepherd folk, no pastoral existence, was found upon the discovery of America. No evidence from history bespeaks this intermediate stage of human development ever having existed here. It is also remarkable that the North American buffalo or bison influenced geographical discoveries in pathless mountainous regions. In winter, the bison, seeking a milder climate, travel in herds of several thousand into the lands south of the Arkansas River. During these migrations, their size and awkward shape make it difficult for them to cross high mountains. Wherever one finds a well-trodden buffalo path then, one must follow it, for it surely offers the most comfortable passage over the mountains. Thus did buffalo paths lay out the best passes through the Cumberland Mountains in the southwest portions of Virginia and Kentucky, in the Rocky Mountains between the sources of the Yellowstone and Platte Rivers, and between the southern branch of the Columbia River and the Rio Colorado of California. From the eastern regions of the United States (the migrating beasts once trod the banks of the Mississippi and the Ohio far beyond Pittsburgh), European settlement has gradually driven back the bison. (*Archaeologia Americana*, vol. II, 1836, p. 139.)

From the granite cliffs of the Islas Diego Ramirez, from the many-fissured Tierra del Fuego which, containing Silurian shale in its eastern regions, contains to the west this same shale metamorphosed by subterranean fire into granite (Darwin, *Journal of Researches into the Geology and Natural History of the Countries Visited 1832–1836 by the Ships "Adventure" and "Beagle,"* p. 266)—from these lands all the way to the northern polar seas, the cordilleras extend to a length of more than 2,000 geographical miles. They are not the tallest, but they are the longest mountain range on our planet, pushed upward along a fault that, like a meridian, runs along our planet from pole to pole, exceeding in length the number of miles one would cross upon the Old Continent in traveling from the Pillars of Hercules to the ice cap of Chukchi in Northeast Asia. Where the Andes are broken into several parallel ranges, those closest to the sea offer, on the whole, the most active of volcanoes; it has often been noted, however, that when the appearances of the subterranean fire cease in one range, the fire then breaks out in a range running parallel. As a rule, the volcanic cones follow the directional axis of the range, but in the Mexican highlands, the active volcanoes stand upon a perpendicular fault that is oriented in an east-west direction from sea to sea (Humboldt, *Essai politique*, vol. II, p. 173). Where access to the molten center is opened by the lifting of the mountainous masses at this ancient wrinkle of the Earth's crust, the center proceeds by way of the network of cracks to work upon the lifted, wall-like masses. That which we call a mountain chain is not lifted and shaped into its present appearance all at once. Various mountains of greatly differing ages have overlaid one another, forcing themselves through paths opened earlier. Variation in the types of mountains comes into being through the outpouring and lifting of rock by eruption, as well as through the slow and complex transformation processes introduced by steam-filled, heat-bearing fissures.

For some time, from 1830 to 1848, the highest culmination points in all of the cordilleras of the New Continent were thought to be these:

The Nevado de Sorata, also known as Ancohuma or Tusubaya (15°52′ south latitude), somewhat south of the village of Sorata or Esquibel, in the eastern range of Bolivia, altitude 3,949 toises or 23,692 Parisian feet;

The Nevado de Illimani, west of the Mission of Irupana (16°38′ south latitude), 3,753 toises or 22,518 Parisian feet, likewise in Bolivia's eastern range;

Chimborazo (1°27′ south latitude) in the Province of Quito, 3,350 toises or 20,100 Parisian feet.

Sorata and Illimani were first measured by Pentland, an excellent geographer, in 1827 and 1838. However, since the appearance of his large map of the basin of the Laguna de Titicaca in June of 1848, we know that the above-listed values of the heights of the Sorata and Illimani are overstated by around 3,718 and 2,675 feet, respectively. The map attributes to Sorata 21,286, to Illimani 21,149 English feet, i.e., only 19,974 and 19,843 Parisian feet. A more exact calculation of the trigonometric operations of 1838 yielded Mr. Pentland these new results. He also measured four peaks in the western cordillera at heights between 20,360 and 20,971 Parisian feet. The summit of Sahama would thus be 871 feet higher than Chimborazo but 796 feet lower than Aconcagua.

6. The basalt mountain range of Harutsch

Near to the Egyptian natron lakes, which in Strabo's time had not yet been divided into six basins, there rises a range of hills. They ascend proudly in the north and extend from east to west and beyond Fezzan, where finally they appear to connect to the Atlas range. In the northeast of Africa (like the Atlas range in the northwest) they separate Herodotus's populated Libya, which lies near the sea, from the Berber lands, or Biledulgerid, home to many animals. At the borders of Middle Egypt, the entire tract of land south of the 30th parallel of latitude is an ocean of sand in which are scattered oases, islands where water and vegetation are abundant. The number of these oases, of which the ancients counted but three and which Strabo compares to the spots on the panther's pelt, has increased considerably due to the discoveries of more recent travelers. The third oasis of the ancients, now known as Siwa, was the *nomos* of Ammon, a priestly state and a rest stop for the caravans as well as the home of the temple of horned Ammon and the Spring of the Sun, the waters of which were purported to grow cool at certain intervals. The ruins of Ummibida (Omm-Beydah) indisputably belong to the fortified caravansary of the Ammon Temple and are therefore among the oldest memorials to have come down to us from the age of the dawn of human civilization (Caillaud, *Voyage à Syouah*, p. 14; Ideler, in *Fundgruben des Orients*, vol. IV, pp. 399–411).

The word *oasis* is Egyptian and has the same meaning as *Auasis* and *Hyasis* (Strabo, lib. II, p. 130, lib. XVII, p. 183 Cas.; Herod. Lib. III, cap. 26, p. 207 Wessel). Abu'l-Fida calls the oasis *el-Wah*. In the later times of the Caesars, criminals were sent to the oases. They were banished to islands in the sand-ocean, just as the Spanish and English send their criminals to the Falklands or to Australia. It is almost easier to escape across the ocean than through the desert that surrounds the oases; even the oases diminish in fruitfulness as their sandiness increases.

It is said that the mountainous area of Harutsch consists of basalt hills of grotesque shape (*Ritter's Africa*, 1822, pp. 885, 988 and 1,003). This is Pliny's *Mons ater* (black mountain); its western reaches, where it is called Soudah Mountain, were explored by my unfortunate friend the bold traveler Ritchie. These eruptions of basalt in tertiary limestone, these rows of hills pushed up like a wall from crevasses, seem analogous to the basalt eruptions of the Vicentinian Alps. Nature repeats this phenomenon in the most remote regions of the Earth. In the White Harutsch (*Harudje el-Abiad*), among the region's limestone formations (which likely belong to the old chalk), Hornemann found a tremendous number of petrified fish heads. Ritchie and Lyon also noted that the basalt of Soudah Mountain, like that of Monte Berico, was in many places thoroughly mingled with calcium carbonate, a phenomenon that is probably an effect of eruption through layers of limestone. Lyon's map indicates that there is even dolomite in the area. In Egypt, recent geologists have indeed discovered syenite and greenstone, but not basalt. Could it be, then, that the ancient vessels one occasionally sees that are made from true basalt can thank this western mountain for their material? Could there also be *obsidius lapis* there? Or could one seek basalt and obsidian on the Red Sea? The band of volcanic eruptions of Harutsch on the edge of the African desert, incidentally, remind the geologist of the porous, maclurite-containing amygdaloids, phonolites, and greenstone porphyry that may be found only on the northern and western borders of the Venezuelan steppes and on the Arkansas plains, in both cases on the ancient river ridges. (Humboldt, *Relation historique*, vol. II, p. 142; Long's *Expedition to the Rocky Mountains*, vol. II, pp. 91 and 405.)

7. Through a sea covered with kelp, when the tropical east-wind suddenly leaves him

There is an occurrence well known to sailors, yet still remarkable, that near the African coast (between the Canary Islands and the isles of Cape Verde, especially between the Cape of Bojador and the mouth of the Senegal River) there is often, instead of the subtropics' generally prevalent east-wind, or trade wind, a wind that blows from the west. The cause of this wind is the vast Sahara Desert. Over the expanse of hot sand, the air grows thin and rises vertically to the heavens. The ocean air streams in to fill this region of thin air and thus is a west wind occasionally generated on Africa's western coast, a wind that works against ships bound for America. These ships, without seeing the continent, feel the effect of the heat-radiant sand. Clearly, the exchange of land- and sea-winds that blow alternatingly on every coast at certain hours of the day and night are due to the same cause.

The accumulation of kelp in the region of Africa's western shores is frequently mentioned in ancient writings. The particular position of this accumulation is a problem that is profoundly connected to suppositions about the extent of Phoenician shipping. The *periplus* that is attributed to Scylax of Caryanda, and that, according to the investigations of Niebuhr and Latronne, was quite probably compiled during the time of Philip of Macedonia, describes a sort of kelp sea beyond Cerne, a *Mar de Sargasso*, a great abundance of fucus; the designated location, however, seems to me to be quite different from the one given in the work *De mirabilibus auscultationibus*, which has long been incorrectly ascribed to the great Aristotle (cf. *Scyl. Caryand. Peripl.* in Hudson, vol. II, p. 53 with Aristot. *De mirab. Auscult.* in *Opp. omnia* ex rec. Bekkeri, p. 844 § 136). "Driven by the east wind," says the pseudo-Aristotle, "Phoenician ships came after a four-day voyage out of Gades to a region where they found the sea covered with reed and seaweed [θρύον καὶ φῦκος]. The seaweed is uncovered by the ebb and immersed at high tide." Does this not speak of a shoal between the 34th and 36th parallels? Has a shallow area there disappeared as the result of volcanic revolution? Vobonne speaks of cliffs north of Madera (cf. Edrisi, *Geogr. Nub.*, 1619, p. 157). In Scylax it is stated so: "The sea beyond Cerne, due to extreme shallowness and to mud and seaweed, is no longer navigable. The seaweed lies a span thick and is pointed at its upper end, so that it stabs." The kelp that one may find between "the Green Cape" and Cerne (the Phoenician cargo station of Gaulea—according to Gosselin, the small island of Fedallah on the northwest coast of Mauritania), today in no way forms a great meadow, a cohesive group, or a *mare herbidum*, as it does beyond the Azores. And in the poetic description of the coast given by Festus Avienus (*Ora maritima*, v. 109, 122, 388, and 408), which, as Avienus himself very clearly states (v. 412), was composed using Phoenician ships' logs, the obstacle of the seaweed is described in great detail, but Avienus places the obstacle much further north, near to Ierne, the Holy Island:

> Sic nulla late flabra propellunt ratem,
> Sic segnis humor aequoris pigri stupet.
> Adjicit et illud, plurimum inter gurgites
> Exstare fucum, et saepe virgulti vice
> Retinere puppim . . .
> Haec inter undas multa caespitem jacet,
> Eamque late gens Hibernorum colit.

If the seaweed (*Fucus*), the mud (πηλός), the shallowness of the sea, and the everlasting stillness of the wind are continually declared by the ancients to be characteristic of the western ocean beyond the Pillar of Hercules, then one might well be inclined (especially as regards the lack of wind) to make the presumption of Punic artifice, the tendency of a great merchant people to discourage competitors from sailing westward through the use of scare tactics. But in his actual books as well (Aristotle, *Meteorol.* II. 1, 14), the Stagirite stands fast upon this notion of the absence of the wind and seeks the clarification of a falsely observed fact, or, to express myself more correctly, of a mythical sea story, within a hypothesis on ocean depth. The stormy sea between Gades and the Isles of the Blessed (between Cadiz and the Canaries) surely may not be compared with that sea which, touched only by gentle trade winds, lies between the tropics and which, truly characteristic of the Spanish, has been given by them the name *el Golfo de las Damas* (Acosta, *Historia natural y moral de las Indias*, lib. III, cap. 4).

According to my careful observations and through comparison with several English and French ship's journals, the old and so very indefinite expression *Mar de Sargasso* refers to two fucus banks, of which the greater, more extended, and more easterly lies between the parallels of 19° and 34° on a meridian 7 degrees west of the island of Corvo in the Azores, while the smaller, rounder, and more westerly bank is situated between the Bermudan and Bahamian Islands (25°–31° lat., 68°–76° long). The central axis of the small bank, which is bisected by the ships sailing from Baixo de Plata (Caye d'Argent) north of Santo Domingo to the Bahamas, seems to me to be oriented to N60°E. A transverse strip of *Fucus natans* extending east and west between lat. 25° and 30° connects the large and small banks. I have had the pleasure of seeing that these observations have been accepted and corroborated through further investigation by my late friend Major Rennell in his outstanding work on ocean currents (cf. Humboldt, *Relation historique*, vol. I, p. 202, and *Examen critique*, vol. III, p. 68–69, with Rennell, *Investigation of the Currents of the Atlantic Ocean*, 1832, p. 184). Both groups of seaweed, taken together with the transversal strip under the old name of Sargasso Sea, cover a surface area that is between six and seven times the area of Germany.

Thus does the ocean's vegetation offer the curious example of social plants of one particular variety. On dry land, the savannas or prairies of America, the heathlands (*Ericeta*) and forests in the northern reaches of Europe and Asia, where conifers, birches (*Betula*), and willows (*Salicineae*) grow side by side, present considerably less uniformity than these thalassophytes. In the North, along with the predominant *Calluna vulgaris*, our heathlands offer *Erica tetralix*, *Erica ciliaris*, and *Erica cinerea*; to the south, *Erica arborea*, *Erica scoparia*, and *Erica mediterranea*. No other association of socially occurring species compares with the vision of uniformity afforded by the *Fucus natans*. Oviedo calls the *Fucus* banks meadows, *praderias de yerva*. When one considers that Pedro Belasco, born in the Spanish port of Palos, discovered as early as 1452 the island of Flores by heading out from Fayal following the direction of the flight of certain birds, it seems impossible, given the proximity of the great *Fucus* bank of Corvo and Flores, that at least a part of the oceanic meadow should not have been discovered before Columbus by Portuguese ships, known to have been driven westward by storms. But the astonishment of the Admiral's shipmates as they sailed surrounded by seaweed from the 16th of September to the 8th of October, 1492, indicates that the immensity of the phenomenon was at that time still unknown to mariners. Columbus, to be sure, mentions neither the worries aroused by the great accumulation of kelp nor the grumbling of his companions in the ship's log excerpted by Las Casas. It speaks only of the complaining and grumbling regarding the danger of the consistently weak east wind. Only his son, Fernando Colon, would make the effort to depict, somewhat dramatically, the worries of the mariners in the biography of his father.

By my calculations, Columbus crossed the great Fucus bank at lat. 28½ ° in 1492, at 37° in 1493, in both cases at long. 40°–43°. These figures can be arrived at with considerable accuracy due to the estimates of speed and "distance sailed daily" written down by

Columbus—measurements made, admittedly, not by heaving the log but by the running off of sand in a 30-minute glass (*ampolletas*). I do not find a certain and distinct account of the log, the *Catena della poppa*, until Pigafetta's 1521 travel journal of the Magellan global circumnavigation (*Kosmos*, vol. II, pp. 296, 469-472). The determination of the ship's position during the days when Columbus crossed the kelp meadow is all the more important to us in that it shows that for three and a half centuries the great accumulation of socially growing thalassophytes (be they the result of the composition of the local seabed or the result of the direction of the returning Gulf Stream) have remained in the same place. Such demonstrations of the persistency of great natural phenomena capture the attention of the physical scientist in two ways, whenever we find them in the consistently moving oceanic elements. Although the edges of the Fucus banks will noticeably oscillate after winds that, for a long enough period, maintain a consistent strength and direction, one may for now (that is for the middle of the 19th century) consider the meridian of 41° longitude west of Paris as the main axis of the great bank. In Columbus's lively imagination, the idea of the position of this bank became attached to the notion of the great physical dividing line that, according to him, supposedly "separates the globe into two parts and is intimately connected to the configuration of the Earth, to variances in magnetic declination, and to climatic conditions." Columbus, when he is uncertain of his longitude, orients himself (February 1493) by the first appearance of floating strips of kelp (*de la primera yerva*) at the eastern edge of the Great Corvo Bank. Due to the powerful influence of the Admiral, the physical dividing line transformed into a political one on May 4th, 1493, in the form of the famous Line of Demarcation between Spanish and Portuguese rights of occupation (comp. my *Examen critique*, vol. III, pp. 64-99, and *Kosmos*, vol. II, pp. 316-318).

8. The nomadic Tibbos and Tuaryks

These two nations inhabit the desert between Bornu, Fezzan, and Lower Egypt. They first became better known to us through Hornemann's and Lyon's travels. The Tibbos, or Tibbous, abound in the eastern, and the Tuaryks (Tueregs) in the western, regions of this ocean of sand. The former, due to their nomadic ways, are called "birds" by other tribes. The Tuaryks may be divided into two groups, those from Aghadez and those from Tagazi. They are often caravan leaders and merchants. Their language is that of the Berbers, and they indisputably belong to the primitive peoples of Libya. A strange physiological peculiarity may be found among the Tuaryks. Particular strains among them may be white, yellow, even nearly black, depending on the conditions of the climate, but always without woolly hair or negroid facial features (*Exploration scientifique de l'Algérie*, vol. II, p. 343).

9. Ship of the Desert

In oriental poems, the camel is called the "Landship" or "Ship of the Desert" (*seyfnet el-badyet*—Chardin, *Voyages, nouv. éd. par Langlès 1811*, vol. III, p. 376).

But the camel is not only the desert's beast of burden and a means of conveyance that interconnects countries; it is also, as Carl Ritter has laid out in his excellent treatise on the species' sphere of distribution, "the main prerequisite of nomadic existence at the patriarchal stage of cultural development in those hot regions of our planet that receive little or no rain. No other creature has been so closely and naturally connected to a particular primitive stage of human development, with a connection that may be observed throughout the history of so many millennia, as in the case of the camel's place in the lives of the Bedouins" (*Asien*, vol. VIII, sec. 1, 1847, pp. 610, 758). "Through all the centuries of their most flourishing existence and even to the collapse of their merchant state, the culturally advanced Carthaginians knew nothing of the camel; it first appears in a military capacity among the Marusians of Western Libya in the days of the Caesars, perhaps as a consequence of its commercial use by the Ptolemies in the Nile Valley. The Guanches, inhabitants of the Canary Islands probably related to the Berbers, were not familiar with camels before the 15th century, when they were introduced by Norman conquerors and settlers. In the undoubtedly very limited traffic between the Guanches and the African coast, the transport of large animals must have been hindered by the small size of their boats. Even up to the present day, the actual Berber tribe, which is dispersed throughout the interior of North Africa and to which belong the aforementioned Tibbos and Tuaryks, may owe not only its mobility back and forth but its very salvation from ruin and its continued existence as a people to the use of the camel throughout the Libyan Desert and its oases. In contrast, the use of the camel has remained foreign to the negro tribes; it has been only with the conquering hosts of the Bedouins through all of North Africa and with the missions of their religious proselytizers that the useful beast of the Nadj, the Nabateans, and the whole Aramaic zone has been able to advance here, as it has everywhere, to the west. As early as the fourth century, the Goths brought camels to the lower Istros (Danube), as the Ghaznavids transplanted them in even greater numbers into India, up to the Ganges." In this dispersal across the African continent, one must differentiate two epochs: that of the Lagids, which exerted its influence across all of Northwest Africa through Cyrene, and the Mohammedan epoch of the conquering Arabs.

Whether the domesticated animals that have accompanied humans the longest—cattle, sheep, dogs, camels—are still to be found in their original wild state has long remained problematic. The Xiongnu of East Asia are among the first people to tame wild camels to domesticity. The compiling editor of the great Chinese work *Si-yu-wen-kien-lo* (*Historia regionum occidentalium quae Si-yu vocantur, visu et auditu cognitarum*) asserts that in the middle of the 18th century in East Turkestan, besides wild horses and asses, there was also an abundance of wild camels. Hadschi Chalfa, too, writes of very frequent hunts for wild camels in the high plains of Rashgar, Tursan, and Khotan in the Turkish geography he completed in the 18th century. Schott translates from a Chinese author, Ma Xi, that wild camels may be found in the lands north of China and west of the Huang He (Yellow) River basin in Ho Si or Tangut. Only Cuvier (*Règne animal*, vol. 1, p. 257) doubts the current presence of wild camels in Inner Asia. He believes they have reverted to the wild state, since Kalmyks and other relatives of the Buddhist religion "in order to perform a service to this world" set camels and other animals free. According to Greek accounts from the time of Artemidorus Ephesius and Agatharchides of Cnidus, the homeland of wild Arabian camels was the Ailanitic Gulf of the Nabateans (Ritter, pp. 670, 672, 746). Especially remarkable is the discovery in 1834, by Captain Cautley and Doctor Falconer, of prehistoric fossilized camel bones in the Sivalik Hills (the foothills of the Himalayas). They may be found alongside the prehistoric bones of mastodons, true elephants, giraffes, and a gigantic tortoise 12 feet long and six feet high, *Colossochelys* (Humboldt, *Kosmos*, vol. I, p. 292). The prehistoric camel has been named *Camelus sivalensis*, without having displayed, however, any substantial differences from the still extant Egyptian and Bactrian, one- and two-humped camels. Quite recently, 40 camels were introduced to Java out of Tenerife (*Singapore Journal of the Indian Archipelago*, 1847, p. 206). The first attempt was made in Samarang. In the same way, reindeer were not introduced to Iceland out of Norway until just the previous century. They were not found there upon the first migration, despite the proximity of the eastern coast of Greenland and the presence of the swimming ice floes (Sartorius von Waltershaufen, *Physisch-geographische Skizze von Island*, 1847, p. 41).

10. Between the Altai and the Kunlun

The great highland, or as it is usually called, the mountain plateau of Asia, which comprises Little Bucharia, Congaria, Tibet, Tangut, and the Mogol lands of the Chalcas and the Olotes, lies between the 36th and 48th degrees of latitude and between the longitudinal meridians of 79° and 116°. It is an erroneous view that perceives this portion of Inner Asia as a single, undivided mountainous elevation, as one hump-backed swelling, continuous as the high plateau of Quito and Mexico, lifted to between seven and nine thousand feet above sea level. I have already expanded upon the view that there is, in this sense, no undivided mountain plateau in Inner Asia in my investigations of the mountains of Northern India (Humboldt, *Premier Mémoire sur les Montagnes de l'Inde* in *Annales de Chimie et de Physique*, vol. III, 1816, p. 303; *Second Mémoire*, vol. XIV, 1820, p. 5–55).

Early on, my views on the geographical distribution of vegetation and on the median temperature that is essential to certain types of cultivation had brought me to find the idea of the continuity of a great Plateau of Tartary between the Himalayan and Altai chains to be quite dubious. Yet this plateau was continually characterized in the way it was represented by Hippocrates (*De aëre et aquis*, § XCVI, p. 74): "as the high and bare plains of Scythia, which, without a crown of mountains, extend and lift themselves up to the constellation of the Bear." To Klaproth goes the undeniable credit that he was the first to show us, in a region of Asia more central than Kashmir, Baltistan, or the holy Tibetan lakes of Manasa and Ravana-hrada, the true position and extent of two great and wholly distinct mountain ranges, the Kunlun and the Tien Shan. Pallas, of course, had already hinted at the importance of the Celestial Mountains (Tien Shan) without acknowledging their volcanic nature. Captured as he was, however, in the contemporarily prevalent hypotheses of a dogmatic geology rich in fantasy, and in the firm belief in "ranges extending in a radiating pattern," this talented natural explorer saw in Bogdo Oola (*Mons augustus*, the point of culmination of the Tien Shan) "such a central hub, from which all the other mountain ranges of Asia radiate, and which dominates the rest of the continent"!

The fallacious notion of a single, immeasurable high plain that fills all of Central Asia (*Plateau de Tartarie*) originated in France in the second half of the 18th century. It was the result of historical combinations and an insufficiently painstaking study of the famous Venetian traveler, and of the naive tales of those diplomatic monks who in the 13th and 14th centuries (thanks to the unity and extent of the Mongol Empire at that time) were able to travel across nearly the entire interior of the continent, from the harbors of Syria and the Caspian Sea to the eastern shores of China, washed by the great ocean. If we had possessed our more accurate understanding of the language and of old Indian literature for more than a half-century, the hypothesis of this central plateau in the broad area between the Himalayas and Southern Siberia would doubtlessly have been supported by some ancient and honorable authority. The poem *Mahabharata*, in its geographical fragment *Bhischmakanda*, seems to refer to the Meru not so much as a mountain but as an enormous swelling of the ground which simultaneously supplies water to the sources of the Ganges, the Bhadra-soma (the Irtysh), and the forked Oxus. In Europe, ideas from other regions, mythical dreams of the origin of the human race, mixed with these physiographical views. The high regions, from which it is supposed the waters first receded (most geologists were averse to upheaval theories) must also have nurtured the first seeds of civilization. Systems of a Hebraized geology of the Great Flood, founded upon local tradition, promoted these suppositions. The intimate connection between time and space, between the beginning of the social order and the ductile composition of the Earth's surface, endowed the supposedly uninterrupted highland, the Plateau of Tartary, with a singular importance, an almost moral interest. Positive information, which was the fruit of scientific journeys and direct measurements, as well as a fundamental study of Asiatic languages and literatures, especially of the Chinese, have gradually revealed the inaccuracies and exaggerations of these wild hypotheses. The high mountain plateaus (ὀροπέδια) of Central Asia are no longer viewed as the cradle of human civilization and the point of origin of all arts and sciences. The ancient people of Bailly's Atlantis, about whom d'Alembert happily avers "that they taught us everything but their name and their existence," have disappeared. The Oceanic Atlanteans were indeed treated with no less scorn even in the time of Posidonius (Strabo lib. II, p. 102 and lib. XIII, p. 598 Casaub).

A plateau that is considerably high, but uneven in its elevations, extends with little interruption from SSW to NNE out of Eastern Tibet up to the mountain group of Kentei, south of Lake Baikal. It is known by the various names of Gobi, Sha-mo ("sand desert"), Sha-ho ("river of sand"), and Hanhai. This swelling of the ground, probably older than the mountain ranges that it cuts through, lies, as mentioned above, between 79° and 116° of longitude east of Paris. Measured at right angles to its long axis, it is 180 geographical miles wide in the South, between Ladak, Gertop, and Hlaffa, the seat of the High Lama. Between Hami in the Celestial Mountains and the great bend of the Huang He at the In Shan range, it is barely 120 geographical miles, but in the North, between the Khangai, where once lay the great city of Karakorum, and the longitudinal range Khin-gan-Petsha (in that part of the Gobi that one must cross in order to travel from Kiachta via Urga to Peking) it is close to 190 geographical miles wide. Due to its topography, this entire swelling, which one must carefully differentiate from the much higher mountain ranges to the east, may be attributed a complete area of nearly three times that of France. The map of the volcanoes and mountain ranges of Central Asia that I drew up in 1839 (but which was first published in 1843) most clearly illustrates the hypsometric conditions relative to the mountain ranges and the Gobi Plateau. It is based upon the critical application of all available astronomical observations and the inestimably rich orographical descriptions presented in Chinese literature which Klaproth and Stanislas Julien investigated at my urging. My map presents the interior of the Asiatic continent in broad outline from 30° to 60° latitude, and between the meridians of Peking and Cherson, while indicating the mean direction and altitude of the mountain ranges. It deviates substantially from all previous maps.

The Chinese have enjoyed a threefold advantage in collecting in their earliest literature such a remarkable amount of orographic information concerning Asian high country, especially as concerns the regions (until now so little known to the Occident) between the In Shan, the alpine lake Khuku-noor, and the banks of the Ili and the Tarim Rivers, north and south of the Celestial Mountains. These are the three advantages:

> along with the peaceful conquest of the Buddhist pilgrims, the war campaigns against the West (as early as the Han and Tang Dynasties, 122 years before our current era of chronological reckoning, and again in the 9th century, when conquerors succeeded in reaching Fergana and the shores of the Caspian Sea);
> religious interests, which were tied to particular mountain summits by prescribed and periodically repeated sacrifices;
> finally, the early and generally well-known use of the compass for orientation regarding the directions of mountain ranges and rivers. This use and the understanding of the magnetic needle's indication of south, twelve centuries before the Christian era, made the Chinese orographic and hydrographic descriptions of the country far superior to those few offered by Greek and Roman authors. Even Strabo, sharp-witted Strabo, knew as little about the direction of the Pyrenees as he did about the direction of the Alps and the Apennines (cf. Strabo, lib. II, pp. 71, 128, lib. III, p. 137, lib. IV, pp. 199, 202, lib. V, p. 211, Casaub).

The lowlands comprise nearly all of Northern Asia, northwest of the volcanic Celestial Mountains (Tien Shan), the steppes in the North of the Altai and the Sayan range, and the lands that extend from Tengiz or Balkhash Lake through the Kyrgyz steppes up to the Arals and the southern end of the Urals, and from the Caspian Sea both to the north-and-south-running Bolor or Bulyt tagh Mountains ("Cloud Mountains" in the Uiguran dialect) and to the upper Oxus, whose sources were found by the Buddhist pilgrims Hiuen Thang (518) and Song Yun (629), by Marco Polo (1277), and by Lieutenant Wood (1838) to be in the Pamershian Lake Sirikol (Lake Victoria). Alongside mountain elevations of 6,000 to 10,000 feet, the term *lowland* may well be employed to describe areas of land that lie at a mere 200 to 1,200 feet above sea level. The first of these figures describes the elevation of the city of Mannheim; the second describes that of Geneva and Tübingen. If the word *plateau*, which has suffered so much abuse in recent geographies, is to be expanded to include swellings of the ground that feature a barely noticeable change in climate or the character of vegetation, then physical geography, thanks to the imprecision of the at best relative designations *highlands* and *lowlands*, abandons the idea of a correlation between altitude and climate, between surface relief and decrease in temperature. When I visited Chinese Dzungaria, between the Siberian border and Lake Zaysan (Dsai-sang), at an equal distance from the Arctic Ocean and the mouth of the Ganges, it was easy to believe that I was in Central Asia. But the barometer quickly showed me that the plains through which the upper Irtysh flows between Ust-Kamenogorsk and the Chinese Dzungarian post of Chonimailachu ("the bleating of sheep") lie only 800 to 1,100 feet above sea level. Pansner's older barometric measurements of altitude, which were first published, however, after my expedition, are supported by my own. Both refute Chappe's hypotheses regarding the elevation of the banks of the Irtysh in Southern Siberia, which he based upon the so-called estimates of the vertical descent of the river. Even farther to the east, Lake Baikal lies only 222 toises (1,332 feet) above the ocean.

In order to attach the concept of the relative comparison of lowlands and highlands and the gradual nature of the increase in ground elevation to actual examples validated by exact measurements, I offer the following table of European, African, and American altitudes, arranged in increasing order. What has now become known of the average elevation of the Asiatic plains (the actual lowlands) can then be compared to these figures:

Plateau	of Auvergne	.	.	.	.	.	170 toises
"	of Bavaria	.	.	.	.	.	260 "
"	of Castile	.	.	.	.	.	350 "
"	of Mysore	.	.	.	.	.	460 "
"	of Caracas	.	.	.	.	.	480 "
"	of Popayan	.	.	.	.	.	900 "
"	at Lake Tzana (Abyssinia)	.	.	.	950 "
"	of the Orange River (South Africa)	.	.	1,000 "
"	of Axum (Abyssinia)	.	.	.	.	1,100 "
"	of Mexico	.	.	.	.	.	1,170 "
"	of Quito	.	.	.	.	.	1,490 "
"	of Provincia de los Pastos	.	.	.	1,600 "
"	of the area around Lake Titicaca	.	.	2,010 "

No part of the so-called Gobi Desert (which in some places contains beautiful grazing land) has been so thoroughly explored in terms of differing altitudes than the broad zone (almost 150 geographical miles) between the sources of the Selenga and the Great Wall. Under the auspices of the Petersburger Academy, a very accurate barometric leveling survey was completed by two excellent scholars, the astronomer Georg Fuβ and the botanist Bunge. In 1832, they accompanied a mission of Greek monks to Peking, in order to set up one of the many magnetic stations that I had recommended. The average altitude of this part of the Gobi does not attain the 7,500 to 8,000 feet that some had rashly extrapolated from the measurements taken of nearby mountain peaks by the Jesuits Gerbillon and Verbiest, but instead barely reaches 4,000 feet (667 toises). The surface of the Gobi between Erghi, Durma, and Sharaburguna has an elevation of no more than 2,400 feet (400 toises) above sea level. It is barely 300 feet higher than the Plateau of Madrid. In the middle of this stretch lies Erghi, at 45°31′ latitude and 109°4′ east longitude. It is situated in a depression or hollow more than 60 miles wide which inclines SW to NE. Old Mongolian lore designates it as having been the floor of a great inland sea. Halophytes and species of reed can be found there, mostly of the same varieties as grow on the low-lying shores of the Caspian Sea. Here in the midst of the desert lie small saline lakes, the salt from which is exported to China. According to a peculiar belief that is widely held by the Mongolians, the ocean will one day return and assert once more its dominion over the Gobi. Such geological dreams are reminiscent of the traditional Chinese opinions on the alkali lakes of inner Siberia, which I have mentioned elsewhere (Humboldt, *Asie centrale*, vol. II, p. 141; Klaproth, *Asia polyglotta*, p. 232).

In much the same way, the Kashmir basin, so enthusiastically acclaimed by Bernier, while perhaps too moderately praised by Victor Jacquemont, gave rise to terrific hypsometric exaggerations. Through careful barometric measurement, Jacquemont found the elevation of Wulur Lake, in the Kashmir Valley not far from the capital city of Srinagar, to be 836 toises (5,016 feet). Debatable results derived by the boiling point of water gave Baron Carl von Hügel 910 toises, while yielding for Lieutenant Cunningham only 790 toises (compare my *Asie centrale*, vol. III, p. 310, to *Journal of the Asiatic Society of Bengal*, vol. 10, 1841, p. 114). The mountainous country of Kashmir, which has garnered such great interest within Germany while the acceptability of its climate is somewhat impaired by the four months of winter snow in the streets of Srinagar (Carl von Hügel, *Kaschmir*, vol. II, p. 196), does not lie upon the highest spine of the Himalayas, as is often said, but as a true *Kesselthal*, a tub-shaped valley, on their southern slope. Where it is separated from the Indian Punjab by the wall-like Pir Panjal, we are told by Vigne that the snow-capped peaks are crowned by basalt and amygdaloid formations. Quite characteristically, the natives call these formations *shishak deyu*, the devil's pox (Vigne, *Travels in Kashmir*, 1842, vol. I, pp. 237–239). The charms of its vegetation have always been described in a very inconsistent fashion, depending on whether the travelers came from the luxuriant and variegated world of plant life in India or from the North—from Turkestan, Samarkand, and Ferghana.

Only recently has a clearer insight been gained regarding the altitude of Tibet, after a long period wherein the elevation of the plain had been injudiciously confused with that of the mountain peaks that rise from it. Tibet occupies the area between the two mighty mountain chains, the Himalayas and the Kunlun; it forms the uplifted valley floor between the two chains. The land is divided, both by the natives and by Chinese geographers, into three parts from east to west. These are delineated as Upper Tibet, with the capital city of Lhasa (at an elevation probably around 1,500 toises); Middle Tibet, with the city of Leh or Ladak (1,563 toises); and Little Tibet, or Baltistan, called the Tibet of Apricots (Sari Butan), where lie Iskardo (985 toises), Gilgit, and to the south of Iskardo on the left bank of the Indus, the Deotsu Plateau, which was measured by Vigne (1,873 toises). If one seriously examines all of the reports regarding the three Tibets that have been available to us up to the present, and that will be greatly enriched this year by the outstanding work of the border-plotting expedition sponsored by Governor General Lord Dalhousie, one quickly reaches the conclusion that the region between the Himalayas and the Kunlun is not an unbroken high plain at all. It is, rather, cut through by groups of mountains belonging to distinctly different elevation systems. Very few true plains are to be found. The most notable are the plains between Gertop, Daba, Shangtung (Shepherds' Plain, the fatherland of the Cashmere goat), and Shipke (1,634 toises); those in the area around Ladak (not to be confused with the depression in which the city lies) that reach 2,100 toises; finally, the plateau of the Holy Lakes, Manasa, and Ravanahrada (probably 2,345 toises), which was visited in 1625 by Father Antonio de Andrada. Other regions are quite filled with concentrated mountainous masses, "rising" (in the words of a recent traveler) "like the waves of a vast Ocean." Along the rivers—the Indus, the Sutlej, and the Yarlung Tsangpo, which was formerly thought to be identical to the Buramputer (Brahmaputra)—points have been measured that reach an elevation of only 1,050 to 1,400 toises above sea level; the same as the Tibetan villages of Pangi, Kunawur, Kelu, and Murang (Humboldt, *Asie centrale*, vol. III, pp. 281–325). From many carefully collected measurements of elevation, I believe I may come to the conclusion that the Plateau of Tibet between 71° and 80° east longitude attains a median altitude of less than 1,800 toises (10,800 feet); this barely reaches the altitude of the fertile plain of Cajamarca in Peru, but falls short of the elevations of the Plateau of Titicaca by 211 toises, and of the paving stones of the high city of Potosi (2,137 t.) by 337 toises.

The fact that Asia, outside of the Gobi Desert (the limits of which have been discussed) and the Tibetan highland, features depressions of considerable size—indeed, actual lowlands—between the 37th and 48th parallels (that is, exactly where there arose yarns of an immeasurable continuous plateau), is demonstrated by the cultivation of plants there that demand a specific range of temperatures in order to thrive. An attentive study of the travel writings of Marco Polo, in which are mentioned viniculture and the production of cotton in the northern latitudes, brought this subject to the attention of the perceptive Klaproth quite some time ago. A Chinese work with the title *Reports on the Recently Subdued Barbarians (Sin-kiang-wai-tan-ki-lio)* states, "The Land of Aksu, somewhat south of the Celestial Mountains and near to the rivers that form the great Tarim-gol, produces grapes, pomegranates, and innumerable other fruits of excellent quality; also cotton (*Gossypium religiosum*), which bedecks the fields like yellow clouds. In summer, the heat is exceptionally great, while in the winter here, as in Turpan, there is neither great cold nor heavy snow." The region around Khotan, Kashgar, and Yarkand still pays its tribute in home-grown cotton, as it did in Marco Polo's time (*il Milione di Marco Polo pubbl. dal Conte Baldelli*, vol. I, pp. 32, 87). And at the oasis of Hami (Khamil), more than 50 geographical miles east of Aksu, orange trees, pomegranates, and delectable grapes also thrive.

The horticultural conditions described support the conclusion that extensive regions exist here at a low ground elevation. In a location that is such a great distance from the coasts and so far east as to increase the intensity of winter cold, a plateau rising to the elevation of Madrid or Munich could probably have very hot summers but could hardly have exceedingly mild winters with virtually no snow at 43° and 44° latitude. At the Caspian Sea (at Astrakhan, 46°21' lat.), 78 feet lower than the level of the Black Sea, I saw a great summer heat suitable to the cultivation of grapevines, but the winter cold there falls to −20° or −25° C. Also, the vine is sunk to a great depth in the earth after November. It is understood that cultivated plants that thrive, as it were, only in summer, such as grapes, cotton, rice, and melons, can still be successfully grown between the latitudes of 40° to 44° at an altitude exceeding 500 toises, aided by the effects of the radiant warmth. But how would the pomegranates of Aksu, the oranges of Hami, proclaimed to be excellent by Father Grosier, survive a long winter, the inevitable result of so great an elevation of the ground? (*Asie centrale*, vol. II, pp. 48–52, 429.) Carl Zimmermann (in the learned analysis of his map of Inner Asia, 1841, p. 99) has indicated that it is quite probable that the Tarim depression, that is, the desert between the Tien Shan and Kunlun mountain ranges, where the Tarim-gol, the river of the steppes, emptied into the then-alpine Lop Lake, sits at an elevation barely 1,200 feet above sea level, that is, at only twice the altitude of Prague. Sir Alexander Burnes also credits Bukhara an elevation of only 186 toises (1,116 feet). It is most greatly to be desired that all doubts regarding the plateau elevation of Central Asia south of the 45th line of latitude be finally dispelled by direct barometric measurements, or by the determination of the boiling point (which admittedly would demand greater care than is usually taken). All considerations connected to the limiting factor of year-round snow and the maximum elevation for viniculture in various climates now rest on elements that are too complex and uncertain.

In order to report with the necessary brevity what was said in the last edition of this work about the mountain systems that cut across Inner Asia, I present the following general overview. We begin with the four parallel chains that seem to be rather consistently oriented from east to west, and yet seldom connected in a latticelike formation. As in the Western European Alps, deviations of direction are indicative of the various epochs of upheaval. After the four parallel chains (the Altai, Tien Shan, Kunlun, and Himalaya), we may consider as meridional the Ural, Bolor, and Khingan ranges, as well as the Chinese ranges that cut north and south near the great bend of the Tibetan and Assam-Burmese Tsangpo chu. The Ural range separates lower Europe from lower Asia. In the writings of Herodotus (Schweighäufer, ed., vol. V, p. 204) and even in those of Pherecydes of Syros, this lower Asian region is a Scythian (Siberian) Europe that comprises all lands north of the Caspian Sea and those lying on the westward-flowing Jaxartes: thus a continuation of our Europe that may be seen as "extending in its length beyond that of Asia."

1) The Great mountain system of the Altai (called the "Mountains of Gold" by the seventh-century historian Menander of Byzantium, *Altaï-alin* in Mongolian, *Kin Shan* in Chinese) extends from 50° to 52½° north latitude and forms the southern frontier of the great Siberian depression, from the rich silver diggings of Snake Mountain and the confluence of the Uba and the Irtysh Rivers and onward to the meridian of Lake Baikal. The divisions and names Greater and Lesser Altai, taken from an obscure passage in the writings of Abulghast, should certainly be avoided (*Asie centrale*, vol. I, p. 247). The Altai mountain system contains the following:

a) the true, or Kolyvan Altai, which is completely subject to the scepter of Russia—west of the intersecting fissures of Lake Teletskoye, and in prehistoric times probably the eastern shore of the great arm of the sea by which the Aralo-Caspian basin, following the direction of the still-extant lake systems of Aksakal Barbi and Sary Kupa (*Asie central*, vol. II, p. 138), connected with the Arctic Ocean;

b) east of the Teletskoyan ranges, which run along the meridian, the Sayan, Tangnu, and Ulaangom or Malakha ranges, all of which extend from east to west and for the most part parallel to one another. The Tangnu, which dwindles into the Selenga basin, has since ancient times formed the racial division between the Turkish tribe in the South and the Kyrgyz (Hakas, identical to Σάκαι) in the North (Jacob Grimm, *Geschichte der deutschen Sprache*, 1848, part I, p. 227). It is the place of origin of the Samoyedics or Soyots, who migrated as far as the Arctic Ocean and who were long perceived by Europeans as being a people exclusive to the polar coasts. The highest snow-capped peaks of the Kolyvan Altai are the Bielucha and Katunia Pillars. The latter reaches only 1,720 toises, the height of Aetna. The Daurian highlands, to which the Kemtei mountain group belongs and along whose eastern border lies the Yablonoi Chrebet, separate the Baikal and Amur depressions.

2) The mountain system of the Tien Shan, the Celestial Mountain range, the Tengri-tagh of the Turkic peoples (Tukui) and of their tribal relatives the Xiongnu, stretches from east to west to eight times the length of the Pyrenees. On the far side, that is, west of its intersection with the longitudinal range of Bolor and Kosyurt, the Tien Shan goes by the names Asferah and Aktagh, is rich in metals, and is cut through by open fissures that emit hot vapors that glow by night and that are used to obtain sal ammoniac (*Asie centrale*, vol. II, pp. 18–20). East of the bisecting Bolor and Kosyurt range, the Tien Shan features, in succession: the Kashgar Pass

(*Kashgar-dawan*); the Glacier Pass of Djeparle, which leads to Kashi and Aksu in the Tarim basin; the Pechan volcano, which spewed forth fire and streams of lava at least as late as the middle of the seventh century by our reckoning; the great and snow-covered elevated mass of Bogdo Ula; the fumarole of Urumchi, which yields sulfur and sal ammoniac (*nao sha*) in an area rich in black coal; the Turfan volcano (the volcano of Ho-chow or Beshbalik), which is situated close to the midpoint between the meridians of Turfan (*Kune-Turpan*) and Pidjan, and is still active to-day. According to Chinese historians, the volcanic eruptions in the Tien Shan go back at least as far as the year AD 89, when the Chinese drove the Xiongnu away from the sources of the Irtysh to Kashi and Karashar (Klaproth, *Tableau hist. de l'Asie*, p. 108). The Chinese military leader Dou Xian crossed the Tien Shan and saw "the Mountains of Fire, whose boulders melt and flow for many *li*."

The great distance of the volcanoes of Inner Asia from the seacoasts is a curious and isolated phenomenon. Abel Rémusat, in a letter to Cordier (*Annales des Mines*, vol. V, 1820, p. 137) first directed the attention of geologists to this distance. For the Pechan volcano, for example, it is 382 geographical miles northward to the Arctic Ocean at the effluence of the Obi, and 378 southward to the mouths of the Indus and the Ganges—this is how centrally located these fiery eruptions in the Asiatic continent are. Heading westward from Pechan, it is 340 geographical miles to Kara Bogaz Bay in the Caspian Sea, and 255 to the eastern shores of the Aral Sea. Before this, the active volcanoes of the New World offered the most conspicuous examples of great distances from the seacoasts. In the case of Popocatepetl in Mexico, however, this distance is only 33 geographical miles, and for the South American volcanoes Sangay, Tolima, and de la Fragua, it is but 23, 26, and 29 geographical miles, respectively. All extinct volcanoes, all trachytic mountains with no permanent connection to the Earth's interior, are excluded from these reports (*Asie centrale*, vol. II, pp. 16–55, 69–77, 341–356). To the east of the Turfan volcano and the fertile, fruited oasis of Hami, the Tien Shan chain diminishes into the high plain of the Gobi, which runs from southwest to northeast. This interruption extends across 9½ degrees of longitude, but beyond the diagonally situated Gobi, the somewhat more southerly range of the In Shan (Silver Mountains) forms a continuation of the Tien Shan stretching from west to east almost to the coast of the Pacific Ocean by Peking, north of Zhili. Just as the In Shan may be viewed as an eastward continuation of the fault along which the Tien Shan rises, so might one be inclined to perceive in the Caucasus a westward continuation beyond the Aralo-Caspian depression or the lowlands of Turan. The middle parallel, or the axis of elevation of the Tien Shan, oscillates between 40⅔° and 43° latitude; that of the Caucasus, according to the map of the Russian General Staff, between 41° and 44°, lying ESE to WNW (Baron von Meyendorff in the *Bulletin de la Société géologique de France*, vol. IX, 1837–1838, p. 230). Among the four ranges that cut across the entirety of Asia, the Tien Shan is the only one in which no peak has been measured.

3) Along with the American cordillera of the Andes, the mountainous system of the Kunlun (*Kurkun* or *Kulkun*), taken together with the Hindu-Kho and its western extension in the Persian Elburz and Demavend, forms the longest line of elevation on our planet. Where the longitudinal Bolor range crosses the Kunlun chain at

right angles, the latter assumes the name "Onion Mountains" (Tsung-Ling); indeed, part of the Bolor range itself, within the eastern angle of intersection, is also given this name. Bordering Tibet to the north, the Kunlun extends consistently east and west along the line of 36° latitude; an interruption occurs at the meridian of Lhasa, due to the mighty knot of mountains that surround those bodies of water so famous in the mythical geography of the Chinese, the Sea of Stars (Sing-so-hai) and the alpine lake Khuku-noor. The Nan Shan and Qilian Shan ranges that rise somewhat to the north may practically be viewed as eastern extensions of the Kunlun. They extend to the Chinese wall at Liangzhu. To the west of the intersection of the Bolor and the Kunlun (Thsung-ling), as I believe I have established in an earlier work (*Asie centrale*, vol. I, pp. XXIII and 118–159, vol. II, pp. 431–434 and 465), the consistent axes of elevation (east-west in the Kunlun and the Hindu-Kho, as opposed to southeast-northwest in the Himalayas) prove that the Hindu-Kho is a prolongation of the Kunlun and not of the Himalayas. From the Taurus range in Lycia to Kafiristan, stretching across 45 degrees of longitude, the chain follows the Parallel of Rhodes, the Diaphragm of Dicaearchus. The tremendous geological view of Eratosthenes (Strabo, lib. II, p. 68, lib. XI, pp. 490 and 511, lib. XV, p. 689), which was expanded upon by Marinus of Tyre and Ptolemy, and according to which "the continuation of the Taurus range in Lycia extends all the way across Asia and on to India in one and the same direction," seems to have been based in part upon notions that passed from the Punjab to the Persians and Indians. "The Brahmans maintain," writes Cosmas Indicopleustes in his *Christian Topography* (Montfaucon, *Collectio nova Patrum*, vol. II, p. 137), "that a cord drawn from Tzinitza [Thinae, i.e., China] straight across Persia and Romania would lie directly along the midline of the inhabited Earth." It is remarkable, as even Eratosthenes already indicates, that this greatest axis of elevation of the Old World passes through the basin (depression) of the Mediterranean Sea between the parallels of 35½° and 36° to the Pillars of Hercules (comp. *Asie centrale*, vol. I, pp. XXIII and 122–138, vol. II, pp. 430–434, with *Kosmos*, vol. II, pp. 222 and 438). The eastern region of the Hindu-Kho is the Paropamisus of the ancients, the Indian Caucasus of the companions of the great Macedonian. As we have long since learned from the travels of the Arab Ibn-Battuta (*Travels*, p. 97), the name "Hindu-Kush," used so often lately by geographers, applies only to a single mountain pass, upon which the cold weather often killed many Indian slaves. And even at its great distance of several hundred miles from the seacoast, the Kunlun also features fiery eruptions. From the crater of the Shin-tien Mountain erupt flames that may be seen far and wide (*Asie centrale*, vol. II, pp. 427 and 483, from a passage of the Yuen-tong-ki as translated by my friend Stanislaw Julien). The highest measured peak in the Hindu-Kho northwest of Jellalabad attains an altitude of 3,164 toises above sea level; westward toward Herat, the chain descends to 400 toises, until, to the north of Tehran at the Demavend volcano, it again rises to 2,295 toises.

4) The mountain system of the Himalayas. Its mean direction is east-west, as one follows it from 79° to 95° eastward, from the colossal Mount Dhaulagiri (4,390 toises) for 15 degrees of longitude to the gap of the long-problematic Dzangbo Chu (called the Irrawaddy by Dalrymple and Klaproth) to the meridian chains, which cover the entirety of Western China and, especially in the provinces of

Szechuan, Hukuang, and Kuangsi, form the great group of mountains that provide the sources of the Yangtze. Along with Dhaulagiri, the point of culmination of this east-west extension of the Himalayan chain is not, despite earlier belief, the more easterly peak of Chomo Lhari, but Mount Kanchenjunga. Kanchenjunga, on the meridian of Sikkim between Bhutan and Nepal, between Chomo Lhari (3,750 toises?) and Dhaulagiri, attains 4,406 toises or 26,438 Parisian feet. This peak was trigonometrically measured only this year, and since the same notification to this effect which was sent to me from East India states clearly that "likewise, a new measurement of Dhaulagiri puts it in first place among all the snow peaks of the Himalayas," then Dhaulagiri must necessarily have a height greater than the 4,390 toises or 26,340 Parisian feet that has been attributed to it until now (letter from the knowledgeable botanist of the last expedition to the South Pole, Dr. Joseph Hooker, sent from Darjeeling, July 25, 1848). The turning point is not far from Dhaulagiri at 79° east longitude from Paris. West of this line, the Himalayas no longer run from east to west but from SE to NW as a mighty, thronging procession that connects to part of the Hindu-Kho between Muzaffarabad and Gilgit, south of Kafiristan. Such a turning and change in the direction of the Himalayas' axis of elevation (from E-W to SE-NW) clearly indicates, as in the case of our Western European Alps, a different epoch of upheaval. From the holy lakes Manasa and Ravanna-hrada (2,345 toises), in whose vicinity the great river springs forth, down to Iskardo and to the Plateau of Deosai that Vigne measured (2,032 toises), the course of the upper Indus adheres in the Tibetan highland to the same north-westerly direction of the Himalayas. Here the long-since well-measured Jawahir (Nanda Devi) climbs to 4,027 toises, while the completely wind-still valley of Kashmir at the Wulur Lake, which lies frozen all winter and is never rippled by a wave, sits at only 836 toises.

Besides the four great mountain systems of Asia, which in their basic geognostic character form parallel chains, one should also mention the long list of alternating meridional elevations that extent from Cape Comorin, across from the isle of Ceylon, all the way to the Arctic Ocean, alternatingly positioned between 64° and 75° longitude, extending from SSE to NNW. The Ghats, the Sulaiman range, the Paralasa, the Bolor, and the Urals all belong to this system of meridional ranges, which in their alternation give the impression of staggered veins. The interruption of the relief (of the meridional elevations) is configured in such a fashion that each new range arises at a latitude that the previous range has not reached, and that the ranges alternatingly stand opposite one another. The importance that the Greeks placed upon these meridian ranges, probably not earlier than the second century of our reckoning, persuaded Agathodaimon and Ptolemy (tbl. VII and VIII) to consider the Bolor, which they called Imaus, to be an axis of elevation that reached to 62° latitude, that is, to the depression of the lower Irtysh and the Obi (*Asie centrale*, vol. I, pp. 138, 154, and 198, vol. II, p. 367).

Since the vertical height of a mountain peak over the level of the sea ever remains, as with all things that are difficult to attain, an object of popular curiosity (regardless of how unimportant the phenomenon of the greater or lesser creasing of the crust of a planetary orb is in the eyes of the geognost), subsequent historical notice will find a respectable place for the gradual progress of hypsometric knowledge in this area. When

I returned after a four-year absence to Europe in 1804, no high snow peak of Asia (not in the Himalayas, nor the Hindu-Kho, nor the Caucasus) had yet been measured with any degree of accuracy. I was unable to compare my determinations of the heights of the everlasting snows in the cordilleras of Quito and the mountains of Mexico with any from India. The important journey of Turner, Davis, and Saunders to the Tibetan highland had indeed taken place in 1783, but the profoundly learned Colebrooke correctly observes that the elevation that Turner gives for Chomo Lhari (28°5' lat., 87°8' long., somewhat to the north of Tassisudan) rests upon foundations as weak as those of the so-called measurements of the elevations seen from Patna and Kafiristan by Colonel Crawford and Lieutenant Macartney (compare Turner, in *Asiatic Researches*, vol. XII, p. 234, with Elphinstone, *Account of the Kingdom of Caubul*, 1815, p. 95, and Francis Hamilton, *Account of Nepal*, 1819, p. 92). Only with the excellent works of Webb, Hodgson, Herbert, and the Gerard brothers has a reliable light been shed upon the heights of the colossal peaks of the Himalayas; in 1808, however, the hypsometric knowledge of the East Indian mountain chains remained so uncertain that Webb could write to Colebrooke: "The heights of the Himalayas remain problematical. I certainly believe that the peaks one may view from the high plain of Rohilkand stand 21,000 English feet (3,284 toises) above this plain, but we do not know the absolute elevation above sea level."

Not until the beginning of the year 1820 did the news spread through Europe not only that the Himalayas had much higher peaks than the cordilleras, but also that fertile meadows and lovely fields of grain had been found, by Webb in the Niti Pass and by Moorcroft on the Tibetan Plateau of Daba and the Holy Lakes, at elevations well above that of Mont Blanc. This news was received with great incredulity in England and rejected due to doubts regarding the influence of light refraction. I have shown in two treatises printed in the *Annales de Chimie et de Physique* "sur les Montagnes de l'Inde" that these doubts are groundless. The Tyrolean Jesuit P. Tiefenthaler, who in 1766 journeyed as far as the provinces of Kumaun and Nepal, had already guessed at the significance of Dhaulagiri. One may read on his map: "Montes Albi, qui Indis Dolaghir, nive obsiti." Captain Webb also consistently uses this name. Until the measurements of Djawahir (30°22' lat., 77°36' long., elevation 4,027 toises) and Dhaulagiri (lat. 28°40', long. 80°59', elevation 4,390 toises?) became known in Europe, Chimborazo (3,350 toises according to my trigonometric calculations; *Recueil d'Observations astronomiques*, vol. I, p. LXXXIII) was considered everywhere to be the highest peak on Earth. The Himalayas then, depending upon whether one was making the comparison with Djawahir or with Dhaulagiri, seemed to be either 676 toises (4,056 Parisian feet) or 1,040 toises (6,240 Parisian feet) higher than the cordilleras. Pentland's South American journeys in 1827 and 1838 brought attention (*Annuaire du Bureau des Longitudes pour 1830*, pp. 320 and 323) to two snow peaks in Upper Peru, east of Lake Titicaca, that were thought to surpass the height of Chimborazo by 598 and 403 toises (3,588 and 2,418 Par. ft). As mentioned above, the most recent calculations in the measurements of Sorata and Illimani have shown the error in this hypsometric assertion. Dhaulagiri, upon whose slopes in the Gandaki River Valley are collected the Salagrama ammonites (famed as the shell-incarnation of Vishnu in the Brahman cult), attests to a height difference in the two continents of more than 6,200 Par. ft.

The question has been raised whether there might lie still greater heights beyond the more or less accurately measured southern chain. Colonel George Lloyd, who in 1840 published the important observations of Captain Alexander Gerard and his brother, is of the opinion that there are, in that part of the Himalayas to which he attaches the somewhat uncertain name of the "Tartaric Chain" (probably in Northern Tibet approaching the Kunlun chain, perhaps in Kailasa of the Holy Lakes or beyond Leh), peaks approaching 29,000 to 30,000 English feet (4,534 to 4,690 toises), thus one to two thousand English feet higher than Dhaulagiri (Lloyd and Gerard, *Tour in the Himalaya*, 1840, vol. I, pp. 143 and 312; *Asie centrale*, vol. III, p. 324). As long as actual measurements are unavailable, no decision may be made upon such matters, for the indicator by which the natives of Quito recognized the peak of Chimborazo as the point of culmination (long before the arrival of Bouguer and La Condamine), namely, its higher penetration above the snow line, becomes very deceptive in the moderate zone of Tibet, where the plain's radiation of warmth is so effectual and where the lower limit of the region of everlasting snow is not a regular line at a consistent level, as it is in the tropics. The greatest height above sea level that men have reached on the slopes of the Himalayas is 3,035 toises, or 18,210 Parisian feet. Captain Gerard attained this height carrying seven barometers, as we have mentioned above, on Mount Tarhigang, somewhat northwest of Shipke (Colebrooke, in *Transactions of the Geological Society*, vol. VI, p. 411). It is, incidentally, almost the same height I myself (June 23, 1802) and, thirty years later, my friend Boussingault reached on the slopes of Chimborazo. The unreached peak of Tarhigang is, by the way, 197 toises higher than that of Chimborazo.

The passes that cross the Himalayas from Hindostan into Chinese Tartary, or even more, into western Tibet, especially between the Baspa and Shipke (or Langsing Khampe) Rivers, have an elevation of 2,400 to 2,900 toises. In the Andes chain, I found that the Assuay Pass between Quito and Cuenca, at the Ladera de Cadlud, also sits at 2,428 toises. A great number of the mountain plateaus of Inner Asia would lie covered year round with everlasting snow and ice, if the lower limit of this eternal snow had not been marvelously lifted to perhaps 2,600 toises above sea level by the power of the radiant warmth of the Tibetan tableland, the endless serenity of the sky, the unusual qualities of the formation of snow in the dry air, and the strong heat of the sun peculiar to the Eastern continental climate that one finds on the northern slopes of the Himalayas. Barley fields (of *Hordeum hexastichon*) may be found in Kinnaur up to 2,300 toises; another variety of barley called *Ooa*, related to the *Hordeum celeste*, may be found even higher. Wheat thrives admirably on the Tibetan highland up to 1,880 toises. On the northern slopes of the Himalayas, Captain Gerard found the upper limit of tall birch stands to reach 2,200 toises; indeed, at the northern latitude of 30.75° to 31°, smaller shrubs that serve to heat the cottages of the inhabitants reach even as high as 2,650 toises, that is, 200 toises higher than the snow line below the equator. One may conclude from the experiences thus far collected that the average figure for the snow line on the northern slopes is at least 2,600 toises, while on the southern slopes of the Himalayas the snow line drops to 2,030 toises. Without this peculiar distribution of heat in the upper atmospheric layers, the plateaus of Western Tibet would not be inhabitable by millions of people (cf. my investigation of the snow line on both slopes of the Himalayas in *Asie centrale*, vol. II, pp. 435–437, vol. III, pp. 281–326, and in *Kosmos*, vol. I, p. 483).

A letter from India that I recently received from Mr. Joseph Hooker, who is simultaneously involved in meteorological and geognostic investigations, as well as investigations involving the geography of plants, reports the following: "Mr. Hodgson, whom we here regard as the geographer who is most fundamentally familiar with the hypsometric conditions of the high mountain ranges, fully recognizes the validity of your assertion in the third part of *Asie centrale* regarding the cause of the differing heights of the lines of permanent snow on the northern and southern slopes of the Himalayan chain. In the Transsutledge region at 36° latitude we often would first observe the snow line at elevations of 20,000 English feet (18,764 Parisian ft), while in the passes south of Brahmaputra, between Assam and Birman at 27° latitude, where lie the southernmost snowcaps of Asia, the snow line drops to 15,000 English feet (14,073 Par. ft)." I believe one must differentiate between the extremes and the mean elevations; in both, however, the once-debated difference between the Tibetan and Indian slopes becomes apparent.

My figures for the average elevation of the snow line, from Asie centrale, vol. III, p. 326:	Extremes, according to Joseph Hooker's letter:
Northern Slope: 15,600 Parisian ft	18,764 Parisian ft
Southern Slope: 12,180 " "	14,073 " "
Difference: 3,420 " "	4,690 " "

Local differences vary still more, as one may observe from the list of extremes that I gave in *Asie centrale*, vol. III, p. 295. Alexander Gerard saw the snow line lying as high as 19,200 Parisian ft. on the Tibetan slope of the Himalayas, while on the southern Indian slope, north of Cursali near Jumnoutri, Jacquesmont actually found it at an elevation of 10,800 Parisian ft.

11. A brown shepherd-tribe, the Xiongnu

The Xiongnu (Hioung-nou), whom Deguignes and many historians with him held to be the Hunnish people, inhabited the tremendous Tartar territory that borders on Uo-leang-ho (now the realm of the Manchus) to the east, on the Chinese wall to the south, on the Uysyn lands to the west, and on the land of the Eleutli to the north. But the Xiongnu belong among the Turkic tribes, while the Huns are of Chinese or Ural extraction. The northern Huns, a rough shepherd-folk who knew nothing of farming, were blackish-brown (burned by the sun?); the southern, or Hajatelahs (who were called Euthalites or Nephthalites by the Byzantinians, and who lived along the eastern coast of the Caspian Sea), were of whiter facial coloration. This second group both farmed and lived in cities. They were often referred to as "White Huns," and d'Herbelot even declared them to be Indo-Scythians. For information regarding the Punu, the military leader or *Tanju* of the Huns, and the drought and famine of the year AD 46 that gave birth to the northward migration of part of this people, see Deguignes, *Histoire gén. des Huns, des Turcs, etc.*, 1756, vol. I, part 1, p. 217, part 2, pp. 111, 125, 223, 447. All statements regarding the Xiongnu taken from these famous works have been subjected by Klaproth to a strict and learned analysis. According to the results of his investigations, the Xiongnu belong among the widely dispersed Turkic tribes of the Altai and Tangnu mountains. The name Xiongnu itself was, in the third century before Christian reckoning, a general term for the Ti, Tukiu, or Turkic peoples of the North and Northwest of China. The southern Xiongnu submitted to the Chinese and, together with them, destroyed the kingdom of the northern Xiongnu. The northern Xiongnu were then forced to flee to the West, and this flight set in motion the migration of the Middle Asian peoples. According to Klaproth, though they are mistaken by many to be the Xiongnu (much as the Uyghur are mistaken for the Ugur and the *Ungarn* [Hungarians]), the Huns belong to the Finnish tribe of the separating Ural Mountains, a tribe that intermixed freely with Germanic, Turkic, and Samoyedic peoples (Klaproth, *Asia polyglotta*, pp. 183 and 211; *Tableaux historiques de l'Asie*, pp. 102 and 109). The Huns (Οὖννοι) are first mentioned by Dionysius Pariegetes, who was able to procure more accurate information on Inner Asia when he, a native of Charax on the Arabian Gulf and a learned man, was sent back to the Orient by Augustus in the company of the emperor's adoptive son Caius Agrippa. A hundred years later, Ptolemy writes Χοῦνοι, strongly aspirated, which may yet be found, as St. Martin reminds us, in the geographical name Chunigard.

12. No chiseled stone

On the banks of the Orinoco near Caicara, where the forested region borders on the plain, we did indeed find depictions of the sun and the figures of animals carved into the cliffs. But in the Llanos themselves, no trace of these memorials to earlier inhabitants was ever discovered. It is regrettable that there has been no more exact report concerning a monument that was sent to Count Maurepas in France; according to Kalm's narrative, the monument was found by Mr. de Verandrier in the grasslands of Canada, 900 French miles west of Montreal, during an expedition to the Pacific coast (*Kalms Reise*, vol. III, p. 416). In the middle of the plains, this traveler came upon massive stones, set upright by human hands, upon one of which was found what looked like a Tartaric inscription (*Archaeologia: or Miscellaneous Tracts Published by the Society of Antiquaries of London*, Vol. VIII, 1787, p. 304). How could such an important monument have remained uninvestigated? Could it really have contained writing with letters? Or was it more likely a historical picture, like the "Phoenician Inscription" (known by this name since it was thus described by Court de Gébelin) found on the stone on the banks of the Taunton River? I certainly find it very probable that cultivated people once crossed these plains. Pyramidal grave mounds and bulwarks of extraordinary length found between the Rocky Mountains and the Alleghenies, upon which Squier and Davis have shed new light in *Ancient Monuments of the Mississippi Valley*, seem to give evidence of these treks (*Relation hist.*, vol. III, p. 155). Verandrier was sent out around the year 1746 by the Chevalier de Beauharnois, the governor general of Canada. Several Jesuits in Montreal assured Mr. Kalm that they had held the so-called inscription in their hands. It was engraved into a small tablet that had been found embedded in a carved pillar. I have implored several of my friends in France to inquire after this monument, in the case that it really was to be found in the collection of Count Maurepas, but to no avail. I have found older, equally dubious examples of alphabetic writings of the primitive peoples of the Americas in Pedro de Cieza de Leon, *Chronica del Peru*, part I, cap. 87 (*losa con letras en los edificios de Vinaque*); in Garcia, *Origen de los Indios*, 1607, lib. III, cap. 5, p. 258; and in Columbus's daybook of the first voyage in Navarrete, *Viages de los Españoles*, vol. I, p. 67. Mr. de Verandrier also maintains (as other travelers before him are also to have observed) that marks of the plowshare could be seen for entire days' marches in the grasslands of Western Canada. But the complete ignorance among the early peoples of North America regarding this farming implement, the absence of domestic cattle, and the enormity of the distances that these furrows cover in the savanna lead me to assume that it was through some sort of water runoff that the surface of the ground received this extraordinary appearance of a plowed field!

13. Like a long narrow gulf

The great steppe that extends east to west from the mouth of the Orinoco to the snow-capped mountains of Merida turns to the south below the 8th parallel of latitude and fills the space between the eastern slope of the high mountains of New Grenada and the Orinoco, which here flows northward. This part of the Llanos, which is watered by the Meta, the Vichada, the Zama, and the Guaviare, connects the valley of the Amazon to the valley of the Lower Orinoco. The word *páramo*, which I use frequently in these pages, describes in the Spanish colonies all mountain regions lying 1,800 to 2,200 toises above sea level and in which a raw, unfriendly, and foggy climate predominates. Hail and snow flurries fall for hours daily in the high *páramos* and beneficially saturate the mountain plants—not that there is, at these high altitudes, a great deal of water vapor in the air; it is, rather, due to the frequency of the precipitation brought on by the rapidly shifting air currents and changes in the electrical tension of the atmosphere. The trees there are low and spread like umbrellas but are adorned with fresh and evergreen foliage on their knotted limbs. They are mostly large-blossomed alpine shrubs with leaves reminiscent of laurel or myrtle. *Escallonia tubar, Escallonia myrtilloides, Chuquiraga insignis,* aralia, weinmannia, frezieria, gaultheria, and *Andromeda reticulata* may be viewed as representative of this plant physiognomy. South of the city of Santa Fé de Bogota lies the acclaimed Páramo de la Sumapaz, a lonely mountain group in which, according to the legends of the Indians, great treasures lie hidden. From this *páramo* springs the brook that comes foaming forth beneath a remarkable natural bridge in the rock-cleft of Icononzo. In my Latin text—*De distributione geographica Plantarum secundum coeli temperiem et altitudinem montium* (*Concerning the geographic distribution of plants according to the temperature of the climate and the height of the mountains,* 1817)—I have attempted to characterize these mountain regions in the following way: "Altitudine 1700–1900 hexapod. asperrimae solitudines, quae a colonis hispanis uno nomine Paramos appellantur, tempestatum vicissitudinibus mire obnoxiae, ad quas solutae et emollitae defluunt nives; ventorum flatibus ac nimborum grandinisque jactu tumultuosa regio, quae aeque per diem et per noctes riget, solis nubila et tristi luce fere nunquam calefacta. Habitantur in hac ipsa altitudine sat magnae civitates, ut Micuipampa Peruvianorum, ubi thermometrum centes. meridie inter 5° et 8°, noctu −0°,4 consistere vidi; Huancavelica, propter cinnabaris venas celebrata, ubi altitudine 1835 hexap. fere totum per annum temperies mensis Martii Parisiis" (At an altitude of 1700–1900 *toises* are the harshest wastelands, which the Spanish colonists call by the name *páramos.* They are exceedingly subject to changes in the weather, and the snow once it has become loose and softened slips down to them. The region is in a constant state of uproar due to the blasts of the winds and the sudden downpour of rain and hail and is uniformly freezing cold throughout the day and throughout the night, as it is almost never warmed due to the gloomy light of the dim sun. At this very height are located quite large towns, like Micuipampa of the Peruvians—where I saw the thermometer come to a halt at midday between 5° and 8° centesimal degrees, at night −0.4°—and Huancavelica, which is famous on account of its veins of cinnabar, where at an altitude of 1835 *toises*, its temperature throughout nearly the whole year is like that enjoyed by the Parisians during the month of March. (Humboldt, *de distrib. geogr. Plant.*, p. 104.)

14. Extend individual spurs toward one another

The immense space that separates the eastern coast of South America from the eastern slope of the Andes chain is narrowed by two mountain masses, which in part separate the three valleys or plains of the Lower Orinoco, the Amazon, and the Plata Rivers. The northern mass, called the Parima group, lies opposite the Andes of Cundinamarca, which extend far to the east; between the 68th and 70th degrees of longitude, the Parimas take on the appearance of high mountains. By way of the narrow ridge of Pacaraima, this group connects with the granite hills of French Guyana. This connection is clearly depicted on the map of Colombia that I drew up according to my own astronomical observations. The Caribs, who have spread from the Caroni Missions to the plains of the Rio Branco and onward to the Brazilian border, cross over the spines of Pacaraima and Quimiropaca when making this journey. The second mountain mass that separates the valley of the Amazon from that of the Plata is the Brazilian group. In the province of Chiquitos (west of the rows of hills of Parecis), this mountain group approaches the foothills of the Santa Cruz de la Sierra. Since neither the Parima group, in which fall the great cataracts of the Orinoco, nor the Brazilian group definitively connects to the Andes chain, the plains of Venezuela then are immediately connected to those of Patagonia (see my geognostic sketch of South America in *Relat. hist.*, vol. I, pp.188–244).

15. Feral dogs

In the grasslands (*pampas*) of Buenos Aires, there are European dogs that have gone wild. They live sociably in dens, in which they keep hidden their young. If the social group should grow too large, single families will move out and form a new colony. The feral European dog barks as loudly as the shaggy American native breed. Garcilaso relates that the Peruvians had the *perros gozques* before the arrival of the Spanish. He calls the native dog *Allco*. In the present-day Quichua language, in order to differentiate this dog from the European variety, one employs the term *Runa-allco*, that is, "Indian dog" (dog of the natives). The shaggy *Runa-allco* seems to be merely a variety of shepherd dog. He is small with long hair, mostly of an ochre-yellow color with white and brown spots, with pointed, upright ears. He barks a great deal, but for all that, seldom bites the natives, however vicious he might be toward white people. When the Inca Pachacutec in the course of his religious wars conquered the Indians of Xauxa and Huanca (the present-day valley of Huanaca and Jauja) and forced their conversion to sun worship, he found among them a sacred veneration of these dogs. The priests blew on skeletonized dog skulls; indeed, the substantial form of the dog-deity was also consumed by the faithful (Garcilaso de la Vega, *Comentários Reales*, part I, p. 184). The worshiping of dogs in the Valle de Huancaya is probably the reason that dog skulls and even mummies of whole dogs are occasionally found in the *Huacas*, the Peruvian tombs of the earliest epoch. The author of an excellent *Fauna peruana*, Mr. von Tschudi, examined these dog skulls and believes that they come from a separate species of dog, which he calls *Canis ingae*, and which is quite distinct from the European dog. Even now, the Huancas are still mockingly referred to as "dog-eaters" by the inhabitants of other provinces. Among the native North Americans of the Rocky Mountains too, boiled dog meat is served to the stranger as a special honor. Not far from Fort Laramie (an outpost of the Hudson's Bay Company for pursuing the fur trade with the Sioux Indians), Captain Frémont was obliged to attend such a "dog-feast" (Frémont's *Exploring Expedition*, 1845, p. 42).

On the occasion of lunar eclipses, Peruvian dogs played a unique role. They were beaten until the eclipse passed. The only silent, truly wholly silent, dog was the Mexican *Techichi*, a variation of a common dog known in Anahuac as the *Chichi*. Literally, *Techichi* means "stone dog," from the Aztec *tetl*, stone. This silent dog (as in the old Chinese custom) was eaten. Even among the Spaniards, this food was so indispensable that before the introduction of cattle, the entire breed was gradually eaten to near extinction (Clavigero, *Storia antica del Messico*, 1780, vol. I, p. 73). Buffon confuses the *Techichi* with the *Koupara de Guyana* (vol. XV, p. 155). But this second animal is identical to the Procyon or *Ursus cancrivorus*, the *Raton crabier* or the mussel-eating *Aguara-Guaza* of the Patagonian coast (Azara, *Sur les Quadrupèdes du Paraguay*, vol. I, p. 315). Linné, on the other hand, confuses the silent dog with the Mexican *Itzcuintepotzotli*, an as yet not fully described type of dog distinguished by a short tail, a very small head, and a large hump on his back. The name means "hump-backed dog," from the Aztec *itzcuintli* (another word for dog) and *tepotzotli*, "humped" or "humped one." Something else that I find remarkable in America, especially in Quito and Peru, is the great number of black hairless dogs that Busson has called *chiens turcs* (the *Canis aegyptius* of Linnaeus). This variety is common even amongst the Indians, though on the whole despised and mistreated. All European dogs propagate well in South America; if one finds them less beautiful than the dogs of Europe, it is partly due to poor care and partly due to the fact that the most beautiful breeds (fine greyhounds, Danish tiger hounds) have never been introduced there.

Mr. von Tschudi has made the unusual comment that in the cordilleras, at elevations above 12,000 feet, the delicate dog breeds as well as the European house cat are susceptible to a peculiar sort of fatal sickness. "Countless attempts have been made to keep cats as house pets in the city of Cerro de Pasco (at 13,228 feet above sea level), but such attempts have ended unhappily, as after a few days the cats, and dogs as well, have died in the midst of violent convulsions. The cats are overcome with spasms, try to climb the walls, then fall back, motionless and exhausted. In Yauli, I was able to observe the chorealike illness several times. It appears to be a result of decreased atmospheric pressure." In the Spanish colonies, the hairless dog is thought to be Chinese. He is called *perro chinesco* or *Chino*, and the breed is thought to have come from Canton or Manila. According to Klaproth, the breed is indeed very common in the Chinese Empire, and has been, in fact, since the earliest days of the culture. Also native to Mexico was *Xoloitzcuintli* (Mex. *xolo* or *xolotl*, servant, slave), a completely hairless, doglike, but extremely large wolf! For information on American dogs, see Smith Barton's *Fragments of the Natural History of Pennsylvania*, part I, p. 34.

The result of Tschudi's investigations of the indigenous dog-breeds of America is that there are two specific varieties:

1. The *Canis caraibicus* of Lesson: completely hairless except for a small shock of hair on the brow and on the tip of the tail; slate-gray with no voice; found by Columbus in the Antilles, by Cortes in Mexico, by Pizarro in Peru; suffers in the

cold of the cordilleras; still commonly referred to as *perros chinos* in the warmer regions of Peru.

2. The *Canis ingae*: pointed nose and ears; barks; now protects the cattle herds and appears in a wide variety of color variations resultant from crossing with European dogs. The *Canis ingae* followed people to the cordilleras. In ancient Peruvian grave sites, his skeleton is occasionally found resting at the feet of the human mummy, the virtual symbol of loyalty, as was commonly depicted by sculptors of the Middle Ages (J. J. von Tschudi, *Untersuchungen über die "Fauna Peruana,"* pp. 247–251).

Since the beginning of Spanish conquest, feral European dogs have also been present on the islands of Santo Domingo and Cuba (Garcilaso, P. I, 1723, p. 326). In the grasslands between Meta, Arauca, and Apure, silent dogs (*perros mudos*) were still being eaten in the sixteenth century. The natives called them *Majos* or *Auries*, says Alonzo de Herrera, who undertook an expedition to the Orinoco in 1535. A very knowledgeable traveler, Mr. Gisecke, found this same voiceless dog in Greenland. The dogs of the Eskimos spend their entire lives outdoors; at night they dig dens in the snow, and they howl in the manner of wolves, sitting in a circle and responding to the first of their number to howl with howls of their own. In Mexico, the dogs were castrated so as to make them fatter and more flavorful. At the borders of the Durango Province and farther north at Slave Lake, the natives would, at least in earlier times, load their tents of buffalo leather onto the backs of large dogs when, with the change of the seasons, they moved to new lodgings. All of these things are also typical of the lives of East Asiatic peoples (Humboldt, *Essai polit.*, vol. II, p. 448; *Relation hist.*, vol. II, p. 625).

16. Like the greatest portions of the Sahara Desert

Meaningful names, especially those that refer to the shape (the relief) of the Earth's surface and that were conceived during a time of very uncertain knowledge of the land and its hypsometric characteristics, have led to numerous and long-lasting geographical misapprehensions. The harmful influence we are describing here was exerted by the old Ptolemaic naming of the Greater and Lesser Atlas Mountains (*Geogr. lib.* III, cap. 1). There is no doubt that the perpetually snow-covered peaks of the Western Moroccan Atlas range may be considered the Greater Atlas, but where is the border of the Lesser Atlas? May one continue to maintain that this division into two Atlas ranges, which thanks to the conservative tendencies of geographers has held on for 1,700 years, lies in the region of Algiers, between Tunis and Tlemcen? May one perceive a "Greater" and a "Lesser" Atlas range in the parallel littoral and interior ranges? All travelers with any familiarity with geognostic ideas who have visited Algeria (the region around Algiers) since the French occupation now dispute the sensibility of this widespread nomenclature. Among the peaks of the parallel ranges, those of Jurjura are considered to be the highest of the ones that have been measured, but the knowledgeable Fournel (who long held the position *Ingénieur en Chef des mines de l'Algérie*) maintains that the Aurés group near Batnah, which he has seen still covered with snow at the end of March, achieve a greater height. Fournel is no more convinced that there are a Greater and a Lesser Atlas than I am that there exist a Greater and a Lesser Altai (*Asie centrale*, vol. I, pp. 247–252). There is but one Atlas range, once called the Dyris by the Mauritanians, and "with this name one should include the ridges [*rides, suites de crêtes*] that form the water-divide between the sources that flow to the Mediterranean or to the depression of the Sahara." The high Moroccan Atlas range, unlike the more easterly Mauritanian range, does not extend from east to west, but from northeast to southwest. It ascends into peaks that, according to Renou (*Exploration scientifique de l'Algérie de 1840 à 1842, publiée par ordre de Gouvernement, Science hist. et géogr.* vol. VIII, 1846, pp. 364 and 373), rise to 10,700 feet, thus exceeding the height of Aetna. A curiously formed highland, nearly in the shape of a square (*Sahab el-Marga*) lies below 33° lat., bounded by mountains to the south. From this point, the Atlas range flattens downward to the sea in the West, one degree south of Mogador. This southwestern part of the Atlas has the name *Idrar N'Deren.*

The great lowland of the Sahara, in the Mauritanian North as well as in the South toward the fertile Sudan, still has but few explored frontiers. If one were to take the parallels from 16½° to 32½° latitude as the outer limits, then the desert, including the oases, would cover an area of more than 118,500 square geographical miles: 9–10 times that of Germany, and almost three times that of the Mediterranean not including the Black Sea. The newest and most exhaustive reports, for which we are indebted to the French explorers of the Sahara, Col. Daumas as well as Messrs. Fournel, Renou, and Carette, inform us that the surface of the desert consists of many individual depressions set close together, and that the inhabitation and the number of fertile oases is much greater than one was made to believe by the frightful character of the desert between In Salah and Timbuktu, or on the way from Murzuq in Fezzan to Bilma, Tirtuma, and Lake Chad. It is now generally accepted that the sand covers only the smaller portion of the lowland. This opinion was expressed previously by the sharply observant Ehrenberg, my traveling companion in Siberia, based upon his own experiences (*Exploration scientif. de l'Algérie, Hist. et Géogr.*, vol. II, p. 332). The only large animals to be found here are gazelles, wild asses, and ostriches. "The Lion of the Desert," says Carette (*Explor. de l'Alg.*, vol. II, pp. 126–129), "is a myth popularized by artists and poets. He exists only in their imagination. This animal does not leave its mountain, where it finds what it needs for shelter, food, and water. When someone speaks to the desert people about these ferocious beasts that Europeans suppose are their companions, they respond with an imperturbable composure: 'So where you come from there are lions that drink air and graze on leaves? Where we come from, lions need running water and living flesh. Therefore, lions appear in the Sahara only where there are wooded hills and water. We fear only the viper [*lefa*] and the innumerable swarms of mosquitoes, the latter being where there is some moisture.'"

While Dr. Oudney ascribes an elevation of 1,536 feet to the southern Sahara on the long road from Tripoli to Lake Chad, German geographers have ventured to increase this figure by a thousand feet; meanwhile, the engineer Fournel has shown, by way of careful barometric measurements based on corresponding observations, that part of the northern desert probably lies below sea level. The part of the desert known as *le Sahara d'Algérie* advances as far as the hills of Metlili and el-Gaous, where lies the northernmost of all oases, the date-rich oasis of el-Kantara. This deep basin touching on the 34th parallel of latitude receives the radiant heat of a chalkbed that lies at an incline of 65° and is full of Inoceramus fossils (Fournel, *Sur les Gisements de Muriate de Soudeen Algérie*, p. 6, in the *Annales des Mines* 4 ᵐᵉ Série, vol. IX, 1846, p. 546). "Once we arrived in Biscara [Biskra]," says Fournel, "there was, unfolding before us, an indefinite horizon, like that of the sea." Between Biskra and Sidi Okba, the ground is elevated only 228 feet above sea level. The downslope increases considerably to the south. In another work (*Asie centrale*, vol. II, p. 320), where I compiled all of the characteristics related to the depression of certain continental regions below the level of ocean, I have already noted that according to Le Père, the bitter lakes (*lacs amers*) of the Isthmus of Suez, at times when their water level is low, lie lower than the surface of the Mediterranean, and that the same is true of the Natron Lakes of Fayum, according to General Andréossy.

I also possess, among other handwritten notes from Mr. Fournel, a geognostic elevation profile, which depicts, with all curvatures and every inclination of the strata included, the entire land surface from the littoral region near Philippeville to the Sahara Desert not far from the oasis of Biskra. The direction of the line of barometric measurements is South 20° West, but the ascertained high points of elevation, as was the case with my Mexican elevation profile, are projected upon another surface (with a N-S direction). Climbing steadily from Constantine (332 toises), the point of culmination was found at only 560 toises, between Batna and Tizur. In the part of the desert lying between Biskra and Touggourt, Fournel succeeded in digging a series of artesian wells (*Comptes rendus de l'Acad. des Sciences*, vol. XX, 1845, pp. 170, 882, and 1,305). We know from Shaw's old reports that the inhabitants of the country are aware of the subterranean water supply and tell tales of a "sea under the ground" (*bahr tôht el-erd*). Fresh waters, which flow under pressure between the clay and marl layers of the old chalkstone and other sediment formations, will, upon breaking through, create springs (Shaw, *Voyages dans plusieurs parties de la Barbérie*, vol. I, p. 169; Rennell, *Africa* Append., p. LXXXV). Geognosts with mining experience will not find it surprising that the fresh water here is often found in the vicinity of rock salt deposits, for Europe offers many analogous examples of this phenomenon.

The wealth of salt rock in the desert, as well as the mining of salt, has been known since the time of Herodotus. The salt zone of the Sahara (*zone de salifère du desert*) is the southernmost of three zones that cross North Africa from southwest to northeast, which are believed to be connected to the rock salt deposits of Sicily and Palestine that have been described by Friedrich Hoffman and Robinson (Fournel, *Sur les Gisements de Muriate de Soudeen Algérie*, pp. 28–41; Karsten, *Über das Vorkommen des Kochsalzes auf der Oberfläche der Erde*, 1846, pp. 497, 648, and 741). The salt trade with the Sudan and the possibility of the cultivation of date palms in the many oasis-forming hollows (brought about by depressions in the tertiary, chalk, or Keuper gypsum layers) contribute equally to the increase of human traffic at many points in the desert. With the high temperatures of the air that circulates over the Sahara and makes daytime travel so arduous, the cold of night, which Denham and Sir Alexander Burnes so often lamented in the African and Asiatic deserts, is even more apparent. Melloni (*Memoria sull' abbassamento di temperatura durante le notti placide e serene*, 1847, p. 55) ascribes this cold, which is doubtless a product of the ground's radiation of the heat), not to the great clarity of the sky ("irraggiamento calorifico per la grande serenità di cielo nell' immense e deserta pianura dell' Africa centrale") but to the extreme stillness of the wind—that is, to the lack of all air movement at night. (For more on African meteorology, see also Aimé, *Exploration de l'Algérie, Physique générale*, vol. II, 1846, p. 147.)

The southern slope of the Moroccan Atlas range sends down to the Sahara at 32° latitude a river that is nearly dry for most of the year, the Oued-Drâa (Wadi-Dra), which

Renou (*Explor. D l'Alg., Hist. et Géogr.*, vol. VIII, pp. 65–78) asserts is ⅙ longer than the Rhine. It first flows from north to south down to 29° lat., then bends at 7½° longitude in almost a right angle toward the west to flow through the freshwater Lake Debaid and on to the ocean at Cape Nun (28°46′ lat., 13½° long.). This region, once made famous when discovered by the Portuguese in the 15th century, and later to slip into geographical obscurity, is now known in the littoral region as the Land of Sheik Beirouk (who is independent of the Moroccan emperor). At the behest of the king of France, the region was explored in the months of July and August of 1840 by ship's captain Count Bouet-Villaumez. The handwritten reports and drawings that I have received reveal that the mouth of the Oued-Drâa is presently quite choked with sand and has an opening of only 180 feet. At this same estuary, but slightly farther east, pours forth the still little-known Saguia el-Hamra, which flows from the south and is supposed to be 150 geographical miles long. The length of such deep and yet mostly dry riverbeds is astounding; they are ancient channels, such as I have seen in the Peruvian desert at the foot of the cordilleras, between the mountains and the Pacific coast. In Bouet's manuscript *Relation de l'Expédition de la Malouine*, the mountains north of Cape Nun are ascribed the great height of 2,800 meters (8,616 feet).

It is generally believed that at the command of the famous Infante Henry, Duke of Viseu, founder of the Academy of Sagres, over which presided the pilot and cosmographer Master Jacomè of Majorca, the promontory of Nun (Chaunar) was discovered by the knight Gilianez in 1433, but the *Portulano Mediceo*, the work of a Genoese mariner of the year 1351, already contains the term *Cavo di Non*. As was later the case with Cape Horn, the circumnavigation of this cape was feared at that time, although at 23′ north of the Parallel of Tenerife, it could be reached in but a few days travel from Cadiz. The Portuguese adage *quem passo o Cabo de Num, ou tornarà ou não* (whoever passes Cape Nun will never return) could not frighten the Infante, whose French heraldic motto, *talent de bien faire*, expresses his noble exploits and his powerful character. The name of the promontory, in which people have for a long time playfully sought a negation, strikes me as not being of Portuguese origin. On the northwest coast of Africa, Ptolemy places a River Nuius: *Nunii Ostia* in the Latin rendering. Edrisi mentions a city, somewhat to the south and three days' travel to the interior, called Nul or *Wadi Nun*, called Belad de Non by Leo Africanus. There were, incidentally, other European mariners long before the Portuguese squadron of Gilianez who sailed much farther southward than Cape Nun: the Catalan Don Jayme Ferrer, so the *Atlas Catalon* published by Buchon in Paris tells us, sailed in 1346 as far as the Gold River (Rio do Ouro) at 23°56′ lat.; Normans at the end of the 14th century reached Sierra Leone, 8°30′ lat. But the credit for having first crossed the equator on the western coast, as for so many great deeds, goes to the Portuguese.

17. *Then as a grassy expanse, like so many of the steppes of Central Asia*

The flatlands (Llanos) of Caracas, around the Rio Apure and Rio Meta, upon which graze abundant livestock, are in the strictest sense grassy plains. Found here in great numbers, from both the Cyperaceae and Gramineae families, are manifold forms of Paspalums (*P. leptostacyrum, P. lenticulare*), Kyllingas *(K. monocephala* [Rottb.], *K. odorata*), Panicums (*P. granuliferum, P. micranthum*), Antephora, Aristida, Vilfa, and Anthisthiria (*A. reflexa, A. foliosa*). Only here and there, mixed in with the Gramineae, occurs a herbaceous dicotyledon, the low-growing "sensitive" plants (*Mimosa interme-dia* and *M. dormiens*), so palatable to the cattle and wild horses. The natives character-istically call this group of plants *dormideras*, "sleeping herbs," for at every touch, the plants close their tender, feathered leaves. Where individual trees spring up (though for many square miles there may not be a single one), they will be: in damp places, the Mauritia palm; in dry regions, a Proteaceae described by Bonpland and myself, the *Rhopala complicata* (*chaparro bobo*), which Willdenow believes to be an *Embothrium*; also, the very useful *palma de covija* or *de sombrero*—our *Coryphe inermis*, a fan palm related to the Chamaerops species used in the thatching of huts. How very much more variegated and diverse is the view of the Asiatic plains! A great portion of the Kyrgyz and Kalmykian steppes that I traversed, crossing 40 degrees of longitude from the Don, the Caspian Sea, and the Ural River (the Jayiq) at Orenburg to the Obi and the upper Irtysh near Lake Dsaisang, never presents at its utmost visible limit that oceanlike horizon car-rying the vault of Heaven, as is so often seen upon the American Llanos, Pampas, and prairies. Here, the view presented is at most but one segment of the heavens, for these steppes are often crisscrossed with ranges of hills or covered with coniferous forest. The Asiatic vegetation, even in the most fertile meadows, is by no means restricted to the family of Cyperaceae; prevalent here is a terrific variety of herbs and shrubs. In spring-time, tiny, snow-white and pink blooms of Rosaceae and Amygdala (*spiraea, crataegus, Prunus spinosa, Amygdalus nana*) afford a friendly sight. The many luxuriantly thriving Synanthereae (*Saussurea ama, S. salsa*, Artemisia, and Centaurea), and the leguminous plants (forms of *Astragalus, Cytisus*, and *Caragana*) I have mentioned in an earlier work. The bright colors of Kaiser's Crown (*Fritillaria ruthenica* and *F. meleagroides*), of Cy-pripedia and tulips, are a joy to behold.

With this more graceful vegetation, the Asiatic flatlands are a contrast to the desolate salt steppes, especially the portion of the Barabinski steppe at the feet of the Altai range between Barnaul and Snake Mountain, and to the land east of the Caspian Sea. Growing side by side, Chenopods, varieties of Salsola and Atriplex, Salicornia and *Halimocnemis crassifolia* (Göbel, *Reise in die Steppe des südlichen Russlands*, 1838, part II, pp. 244 and 301) cover the clayey ground in patches. Among the 500 phanerogamic species col-lected by Claus and Göbel in the steppes, the Synanthereae, the Chenopods, and the Cruciferae were more numerous than the grasses. These last were only $\frac{1}{11}$ of the overall number, while the two former made up $\frac{1}{7}$ and $\frac{1}{9}$. In Germany, when both mountainous regions and plains are considered, the Glumaceae (i.e., the Gramineae, Cyperaceae, and Juncaceae together) constitute $\frac{1}{7}$, the Synanthereae (Composeae) $\frac{1}{8}$, and the Cruciferae $\frac{1}{18}$ of all German phanerogamic species. In the northern reaches of the Siberian flatlands, the ultimate tree and shrub limit (for conifers and Amentaceae) is found, according to Admiral Wrangell's excellent map, near the Bering Strait below $67\frac{1}{2}°$ latitude, farther west, however, toward the banks of the Lena, below 71°, that is, under the parallel of the North Cape of Lapland. The plains that border the Arctic Ocean are the region of cryp-togamic plant life. These flat lands are called tundra (*tuntur* in Finnish); they are vast stretches of boggy land that are covered partly with a thick felt of *Sphagnum palustre* and other mosses, and partly with a dry, snow-white blanket of *Cenomyce rangiferina* (reindeer moss), *Stereocaulon paschale*, and other lichens. "This Tundra," says Admiral Wrangell of his perilous expedition to the island of New Siberia, where fossilized tree trunks are found in abundance, "led me to the outermost arctic littoral region. Their ground is a realm of the Earth frozen for millennia. In the sad uniformity of the land-scape, the traveler, surrounded by reindeer moss, gratefully allows his eye to rest on any patch of green grass that might appear on some damp spot."

18. Diminish the aridity and warmth of the New Continent

I have attempted to put together a single picture of the diverse causes of America's humidity and decreased heat. It goes without saying that this is only a general discussion of the hygroscopic activity of the atmosphere and temperature of the entire New Continent. Individual regions—the island of Margarita, the coasts of Cumana and Coro—are as hot and dry as any part of Africa. Also, the maximum warmth at certain hours of a given summer day, observed over the course of many years, has been found in all regions of the planet, on the Neva, the Senegal, the Ganges, or the Orinoco, to be almost the same: between 27 and 32 degrees Réaumur; on the whole no higher, as long as one takes the readings in the shade, far from heat-radiant solid bodies, in air that is not filled with hot dust (grains of sand), and so long as one does not take them with thermometers filled with light-absorbent spirits of wine. The terrible heat of 40° to 44.8° Réaumur, measured in the shade at the Murzuk oasis, to which my unfortunate friend Ritchie (who died there) and Captain Lyon were exposed for many weeks, may be ascribed to grains of sand suspended in air (forming centers of radiant heat). The most curious example of very high temperature, seemingly taken in air free of dust, has been presented by an observer who was capable of adjusting all of his instruments with the greatest accuracy. Under conditions of an overcast sky, strong southwest winds, and an approaching storm, Rüppell measured 37.6° Réaum. at Ambukol in Abyssinia. The median annual temperature of the tropics, or of the actual climate of palms, on dry land, is 20½° Réaum., and one finds no considerable difference in observations made at the Senegal, in Pondicherry, and in Suriname (Humboldt, *Mémoire sur les lignes isothermes 1817*, p. 54, and in *Asie centrale*, vol. III, Mahlmamm's table IV.)

The great coolness (one might say coldness) that prevails for a large portion of the year on the Peruvian coast below the Tropic of Cancer, where the temperature may sink to 12° Réaum., is, as I believe I have proved in another work, in no way an effect of nearby high mountains but much more due to the sun being obscured by fog (*garua*), and to a current of cold seawater that originates in the south polar regions and flows from the Southwest until it meets the coast of Chile at Valdivia and Concepcion and, with great force, pushes northward to Cabo Pariña. On the coast at Lima, the temperature of the Pacific is 12.5 Réaum., while it is, at the same latitude but outside of the current, 21°. How strange, that such a remarkable fact remained unnoticed until my visit (October 1802) to the South Sea coast!

The temperature differences of various geographical areas are primarily the result of the characteristics of the base upon which the atmosphere rests, that is, of the characteristics of the firm or fluid (continental or oceanic) surface that comes into contact with the atmosphere. Oceans, diversely channeled by streams of warm and cold water (pelagic currents), function differently from broken or unbroken continental masses, or from islands, which may be viewed as depressions in the atmosphere and which, despite their small size, often exert a remarkable and far-reaching influence upon the ocean's climate. On the continental masses, one must differentiate between deserts devoid of vegetation, savannas (grassy plains), and forested areas. In Upper Egypt and in South

America, Nouet and I found the ground temperature of the granite sand to be 54.2° and 48.4° Réaum. According to Arago, many careful observations in Paris resulted in 40° and 42° (*Asie centrale*, vol. III, p. 176). The savannas, which are called prairies between the Missouri and Mississippi, but in the South appear as Llanos in Venezuela and Pampas in Buenos Aires, are covered with monocotyledons (of the family Cyperaceae) and with grasses whose slender stalks and lanciform leaves radiate heat to the cloudless skies and possess an extraordinary capacity for emission. Even in our latitudes, Wells and Daniell (*Meteor. Essays*, 1827, pp. 230 and 278) saw in the grass, upon decreased clarity of the atmosphere, a drop of 6.5° to 8° on the Réamur thermometer as a result of heat radiation. Recently, Melloni very cleverly educed (*Sull' abbassamento di temperature durante le notti placide e serene*, 1847, pp. 47 and 53) how, along with a state of windless circulation of air (a necessary condition for strong radiation of heat and the formation of dew), the cooling of the grass layer is also promoted by the fact that those parts of the air that have already cooled, being heavier, sink to the ground. Near the equator, under the heavily clouded skies of the Upper Orinoco, Rio Negro, and Amazon, the flatlands are covered by thick primeval forest; north and south of this wooded region, however, from the zone of palms and tall dicotyledonous trees, the Llanos of the Lower Orinoco, Meta, and Guaviare stretch out in the Northern Hemisphere, and in the Southern, the Pampas of the Rio de la Plata and Patagonia. The area covered by all of these grassy plains (savannas) of South America is at least nine times the area of France.

The forested areas affect temperature in three ways: through cool shade, through evaporation, and by cooling through the radiation of heat. The forests, which in our temperate zones consist uniformly of socially growing plant species from the families of the conifers and the Amentaceae (oaks, beeches, and birches), while in the tropics consisting of single, asocial species, protect the ground from immediate insolation, evaporate the fluids that they create within themselves, and cool the nearby layers of air through the radiation of heat by their leafy appendicular organs. The leaves, which are by no means all parallel with one another, have different inclinations to the horizon; according to the law discerned by Leslie and Fourier, however, the influence of this inclination upon the quantity of heat emitted by way of radiation (*rayonnement*) is such that the radiation potential (*pouvoir rayonnant*) of a measured surface *a*, at a specified inclined orientation, is equal to the radiation potential of a leaf of such size as would have the projection of *a* upon a horizontal surface. In the initial phase of radiation, then, of all the leaves that form the top of the tree and partially cover one another, the first to cool are those that freely face the cloudless sky. This process of cooling (or exhaustion of heat through emission) grows more appreciable the thinner the leaf is. A second layer of leaves has its upper surface facing the lower side of the first layer and will, by the process of radiation, give off more to the first layer than it can absorb from it. For the second layer of leaves, the result of this unequal trade-off is again a reduction in temperature. This effect is propagated from layer to layer until all of the leaves of the tree, with the strength or weakness of their radiation of heat modified by their position, enter a state of stable equilibrium, from which the law can be established through mathematical analysis. In this manner, through the process of radiation over the course of the long and mellow nights of the equinoctial zone, the forest air contained in the spaces between the leaf layers is cooled, and thanks to the great multitude of appendicular organs (leaves), a tree whose top might have a horizontal cross-section measuring barely 2,000 square feet effects a reduction in the air temperature over an area of naked or grass covered ground many thousand times greater than 2,000 square feet (*Asie centrale*, vol. III, pp. 195–205). In these pages, I have extensively analyzed the complex relationships involved in the effect of large forest regions on the atmosphere, because they are so often touched upon in the important question regarding the climate of ancient Germania and Gaul.

Since European civilization has its primary settlement on a western shore, it must have been noticed early on that at the same latitudes, the mean annual temperature in the opposite eastern littoral of the United States of North America was several degrees cooler than in Europe, which is to some degree a peninsula of Asia, a relationship reminiscent of Brittany's to the rest of France. It was forgotten, however, that these differences at the higher latitudes quickly decrease at the lower, until disappearing almost completely below 30° latitude. For the western coast of the New Continent there are virtually no exact thermal measurements, but the mildness of the winter in New California shows that, in terms of annual mean temperatures, the western coasts of America and Europe at similar latitudes differ very little. The following small table shows what mean annual temperatures illustrate about the same geographic latitudes on the eastern coasts of the New Continent and the western coasts of Europe.

Similar degree of latitude	America's eastern coast	Europe's western coast	Mean temperature for the year, the winter, and the summer	Difference in annual mean temp. between E. America and W. Europe
57°10′	Nain		−2.8 $\frac{-14.4}{6.1}$	
57°41′		Gothenburg	6.4 $\frac{-0.2}{13.5}$	9.2
47°34′	St. John's		2.7 $\frac{-4.0}{9.8}$	
47°30′		W. Budapest	8.2 $\frac{-0.4}{16.8}$	5.8
48°50′		Paris	8.7 $\frac{2.6}{14.5}$	
44°39′	Halifax		5.1 $\frac{-3.5}{13.8}$	
44°50′		Bordeaux	11.2 $\frac{4.8}{17.4}$	6.2
40°43′	New York		9.1 $\frac{0.1}{18.2}$	
39°57′	Philadelphia		9.0 $\frac{0.1}{18.1}$	3.4

38°53′	Washington		10.2	$\frac{1.8}{17.4}$	
40°51′		Naples	12.9	$\frac{7.8}{19.1}$	3.4
38°52′		Lisbon	13.1	$\frac{9.0}{17.4}$	
29°48′	St. Augustine		17.9	$\frac{12.2}{22.0}$	
30°2′		Cairo	17.7	$\frac{11.8}{23.4}$	0.2

In the table above, the number before the fraction expresses the mean annual temperature, the numerator of the fraction the mean winter temperature, and the denominator of the fraction the mean summer temperature. Besides the greater difference in mean annual temperature, its division into the different seasons on the opposite coasts shows a conspicuous contrast, and it is just this division that has the most profound effect on our feelings and on the process of the growth of vegetation. Dove makes the general observation that the summer temperatures in America at a given latitude are cooler than those of Europe (*Temperatur Tafeln nebst Bemerkungen über die Verbreitung der Wärme auf der Oberfläche der Erde*, 1848, p. 95). The climate of Petersburg (54°43′ lat.), or to put it more accurately, the mean annual temperature of this city, will be found on the east coast of America at 47½°, thus 12½° latitude farther south. In the same way, we find the climate of Königsberg (54°43′ lat.) in Halifax, which lies down at the latitude of 44°39′. Toulouse (at 43°36′ lat.) is comparable to Washington in terms of thermal conditions.

It would be quite bold to proclaim general results regarding the distribution of heat in the United States of North America, as there are three regions that must be differentiated:

1) the region of the Atlantic states east of the Alleghenies;
2) the western states between the Alleghenies and the Rocky Mountains that lie in the broad basin through which flow the Mississippi, the Ohio, the Arkansas, and the Missouri;
3) the high plains between the Rocky Mountains and the coastal range of New California, through which passes the Oregon, or Columbia, River.

Since as part of John Calhoun's well-known undertaking, uninterrupted observations of temperature according to a uniform schedule have been made at 35 military posts and reduced to daily, monthly, and yearly averages, we have arrived at more accurate views on climate than were generally promulgated in the days of Jefferson, Barton, and Volney. These meteorological outposts stretch from the tip of Florida and Thompson's Island (Key West) at 24°33′ lat., all the way to Council Bluffs on the Missouri, and if Fort Vancouver (45°37′ lat.) is also figured in, the longitudinal differences constitute 40°.

One may not presume from the mean annual temperature that the second region is, in general, warmer than the first, the Atlantic, region. The more northerly occurrence of certain plants west of the Alleghenies is due partly to the nature of these plants and partly to the differing distribution of the heat over the course of the four seasons. The broad Mississippi Valley is situated with its northern and southern ends under the warming influences of the Canadian lakes and the Mexican Gulf Stream. The five lakes (Superior, Michigan, Huron, Erie, and Ontario) cover an area of 92,000 square English miles (4,232 square geographical miles). The climate is so much milder and more temperate in the proximity of the lakes that, for example, the winter in Niagara (43°15′ lat.) reaches a mean temperature of only one-half of a degree below freezing, while far from the lakes at Fort Snelling (44°53′ lat.), by the confluence of the St. Peter's River with the Mississippi, there is a mean winter temperature of −7.2° Réaumur (see Samuel Forry's excellent work *The Climate of the United States*, 1842, pp. 37, 39, and 102). At this distance from the Canadian lakes, the surfaces of which are elevated five to six hundred feet above sea level, while the lakebeds of Michigan and Huron lie almost five hundred feet below sea level, the climate of the land, according to recent observations, has an actual continental character, i.e., hotter summers and colder winters. "It is proved," says Forry, "by our thermometrical data, that the climate west of the Alleghany [*sic*] Chain is more excessive that that on the Atlantic side." At Fort Gibson on the Arkansas River, which flows into the Mississippi (35°47′ lat., with a mean annual temperature that barely reaches that of Gibraltar), in August of 1834, the thermometer was seen to rise, in the shade and without reflection from the ground, to 37.7° Réaum.

The oft-repeated legend, based upon absolutely no data, that the climate has grown more moderate—milder in winter, cooler in summer—since the first European settlements of New England, Pennsylvania, and Virginia, due to the eradication of several forests on both sides of the Alleghenies, is now generally doubted. Series of reliable thermometric observations go back barely 78 years in the United States. In the observations made in Philadelphia from 1771 to 1824 a rise in annual mean temperature of only a bare 1.2° Réaum. can be discerned, which is attributed to the expansion of the city, its large population, and its numerous steam-powered machines. Perhaps the observed yearly increase is merely coincidental, for in the same period I find a decrease in mean winter temperature of 0.9°. Excepting the winter, all of the other seasons had grown somewhat warmer. Thirty-three years of observations in Salem, Massachusetts show no change at all; in the mean figure for all of the years they oscillate barely a single degree Fahrenheit, and the winters in Salem, rather than having grown milder due to the above-mentioned eradication of forests, have grown 1.8° Réaum. colder over the course of these 33 years (Forry, pp. 97, 101, and 107).

As the east coast of the United States at given latitudes is similar to the Chinese and Siberian eastern coasts of the Old Continent in terms of mean annual temperature, so too have the western coasts of Europe and America been justifiably compared to one another. I wish to present just a few examples from the western region of the Pacific which come to us courtesy of two of Admiral Lütke's voyages around the world: Sitka (New Arkhangelsk) in Russian America and Fort George, which share latitudes with Gothenburg and Geneva, respectively. Iluluk and Danzig lie on roughly the same parallel, and although the mean temperature of Iluluk, due to the island climate and the cold ocean current, is less than that of Danzig, the American winter is actually milder than winter on the Baltic.

Sitka	57°3′ lat.	137°38′ long.	5.6°	*0.6°*
				10.2°
Gothenburg	57°41′ lat.	9°37′ long.	6.4°	*−0.2°*
				13.5°
Fort George	46°18′ lat.	125°20′ long.	8.1°	*2.6°*
				12.4°
Geneva	46°12′ lat.	Elev. 203 toises	7.9°	*0.7°*
				14.0°
Kherson	46°38′ lat.	30°17′ long.	9.4°	*−3.1°*
				17.3°

On the Oregon or Columbia River, snow is almost never seen. The river is covered with ice on only a few days. The lowest temperature that Mr. Ball once observed there in 1833 was 6.5° Réaum. below freezing (*Message from the President of the United States to the Congress*, 1844, p. 160; Florry, *Clim. of the U. St.*, pp. 49, 67, and 73). A quick glance at the summer and winter temperatures recorded above shows that upon or near the west coast, a true island climate dominates. While the winter temperatures are not as cold as in the western parts of the Old Continent, the summers are far cooler. The contrast becomes most noticeable when one compares the estuary of the Oregon to Fort Snelling, Fort Howard, and Council Bluffs, which lie in the interior of the Mississippi/Missouri basin (44°–46°), where, as Buffon asserts, there is found an excessive, a truly continental, climate: winter cold on some days reaching −28.4° and −30.6° Réaum. (−32° and −37° F.), followed by summer heat that rises to between 16.8° and 17.5° mean temperature.

19. *That America emerged from the chaotic covering of water later than the other parts of the world*

An astute naturalist, Benjamin Smith Barton, said long ago, and quite correctly (*Fragments of the Natural History of Pennsylvania*, part I, p. 4): "I cannot but deem it a puerile supposition, unsupported by the evidence of nature, that a great part of America has probably later emerged from the bosom of the ocean than the other continents." I touched upon the same topic in an essay on the natives of America (*Neue Berlinische Monatschrift*, vol. XV, p. 190): "All too often, generally praiseworthy authors have repeated that America is, in every sense of the word, a new continent. The luxuriance of the vegetation, the enormous amounts of flowing water, the disquiet of mighty volcanoes announce (so they say) that the continually quaking, as yet not completely dried out Earth is closer there to the chaotic, primordial state than it is in the Old Continent. Such ideas, long before the beginning of my trip, seemed to me to be as unphilosophical as they were at variance with generally accepted physical laws. Fanciful images of youth and unrest, of increasing dryness and inertia of the aging Earth can arise only among those who easily snatch up contrasts between the two hemispheres without making the effort to comprehend in a general way the construction of the planet. Should one presume that southern Italy is newer than its northern regions because it is almost continuously shaken by earthquakes and volcanic eruptions? Moreover, what trivial phenomena are our current volcanoes and earthquakes in comparison to the revolutions of Nature that the geognost must postulate when pondering the chaotic conditions of the Earth at the lifting, the solidification, and the fracturing of the mountain masses? In distant climates, differences in causes must engender different effects of the forces of Nature. In the New Continent, the volcanoes" (I now count more than 28 of them) "continue to burn longer because the high mountain combs, upon which they burst forth in rows following long faults, are closer to the ocean, and because with few exceptions this proximity, in a way that has yet to be explained, seems to modify the energy of the subterranean fire. Also, the activity of earthquakes and fire-spewing mountains is periodical. Now" (so I wrote 42 years ago!) "physical unrest and political calm prevail in the New Continent, while in the old, the devastating conflicts of the people disturb the enjoyment of a Nature at peace. Perhaps times will come when, in this curious conflict between the physical and moral powers, one part of the world will take over the role of the other. Volcanoes rest for centuries before they erupt anew; the idea that in the older country a certain peace in Nature must prevail is based upon a mere flight of our imagination. No reason exists to presume that an entire side of our planet is older or newer than the other. There are, of course, islands pushed up by volcanoes and raised gradually by coral organisms, such as the Azores and many flat islands of the Pacific. These are certainly newer than many plutonic formations of the Central European chain. A small section of the earth (such as Bohemia, Kashmir, and many so-called moon valleys) that is surrounded by a ring of mountains can, with partial flooding, lie covered like a lake for a long time, and upon the runoff of this inland water, one might figuratively maintain that the ground which the plants are gradually beginning to colonize is of newer origin. Islands have been connected to continental masses by the process of upheaval, and other parts of the land have disappeared through a sinking of the oscillating ground surface; but consistent with hydrostatic laws, general submersions may only be imagined on all continents and in all climates as existing simultaneously. The ocean cannot for any long duration flood the vast plains around the Orinoco and the Amazon without at the same time desolating our Baltic countries. Also, the order and identity of the sedimentary layers, like the organic remains of prehistoric plants and animals contained within them, show that many great geological depositions occurred almost simultaneously over the entire surface of the Earth." (Regarding vegetable materials in the coal formations of North America and Europe, see Adolph Brongniart, *Prodrome d'une Hist. des Végétaux fossiles*, p. 179, and Charles Lyell, *Travels in North America*, vol. II, p. 20.)

20. A hemisphere which . . . is cooler and more humid than is our northern hemisphere

Due to the increasing narrowness of the continent as it extends southward, Chile, Buenos Aires, and the southern portions of Brazil and Peru have a true island climate, with cool summers and mild winters. This advantage of the Southern Hemisphere is evident until 48° and 50° southern latitude; but farther down toward the ice-covered South Pole, South America grows, little by little, into a barren waste. The dissimilarity of the latitudes in which the southern extremities of Australia, including the island of Van Dieman's Land (Tasmania), of Africa, and of America come to their end gives to each of these continents a unique character. The Strait of Magellan lies between the 53rd and 54th degrees of latitude, and yet the thermometer sinks there, even in December and January when the sun shines for 18 hours a day, down to 4° Réamur. It snows nearly every day on the plains, and the highest air temperature observed by Churruca in December of 1788—the summer there—was not more than 9°. Cabo Pilar, whose towerlike cliff is only 218 toises high and yet all the same forms the southernmost tip of the Andes chain, lies at a southern latitude nearly the same distance from the equator as the northern latitude of Berlin (*Relacion de Viage al Estrecho de Magallanes*, appendix 1793, p. 76).

While in the Northern Hemisphere the continents, as they extend toward the pole, all share a limit that fairly regularly coincides with the parallel of 70°, the southern points of America (Tierra del Fuego, with its many inlets carved by the sea), of Australia, and of Africa are separated from the South Pole by 34°, 46½°, and 56° respectively. The temperature of the great and so dissimilarly sized sections of the ocean that separate the southern points of land from the ice-covered pole contributes substantially to the modification of climate. The surface area of the dry land in the two hemispheres separated by the equator is proportioned at a ratio of 3 to 1. But this scarcity of continental masses in the Southern Hemisphere applies more to the temperate zones than the hot zones. The former stand in a ratio between the two hemispheres of 13 to 1, while the latter occupy a ratio of 5 to 4. Such a great inequity in the distribution of dry land exercises a noticeable influence on the strength of the ascending air current that turns toward the South Pole, and indeed on the temperature of the entire Southern Hemisphere. The noblest plant species of the tropics, for example the arborescent ferns, will grow south of the equator as far as the region between the 46th and 53rd parallels, while north of the equator they do not thrive above the Tropic of Cancer (Robert Brown, *Appendix to Flinder's Voyage*, pp. 575 and 584; Humboldt, *De distributione geographica Plantarum*, pp. 81–85). The arborescent ferns (tree-ferns) thrive excellently at Hobart Town in Van Dieman's Land (42°53' lat.) at a mean annual temperature of 9°: i.e., at an isothermal latitude that is 1.6° lower than that of Toulon. Rome is almost an entire degree of latitude farther away from the equator than Hobart Town, yet Rome has an annual temperature of 12.3°, in winter 6.5°, in summer 24°, while these averages for Hobart Town are 8.9°, 4.5°, and 13.8°. In Dusty Bay, New Zealand, arborescent ferns thrive at 46°8', on Lord Auckland's and Campbell's islands as far as 53° (Jos. Hooker, *Flora antarct.*, 1844, p. 107).

On the archipelago of Tierra del Fuego, which lies at the southern latitude analogous to Dublin's northern one and has a mean temperature of 0.4° in winter and of only 8° (10° C.) in summer, Captain King found the ground covered with attractive plants ("vegetation thriving most luxuriantly in large woody stemmed trees of Fuchsia and Veronica"). Yet this vegetative potency, which Charles Darwin, especially on the west coast of America at 38° to 40° southern latitude, so picturesquely described, suddenly disappears south of Cape Horn, on the Rocks of the South Orkney and South Shetland Islands, and on the Sandwich archipelago. These islands, sparsely covered with grass, moss, and lichens—*Terres de Désolation*, as the French mariners called them—still lie quite far north of the Antarctic Circle, while in the northern hemisphere, at 70° latitude in the extreme reaches of Scandinavia, spruce trees rise up to 60 feet (compare Darwin, *Journal of Researches*, 1845, p. 244, and King, *Narrative of the Voyages of the "Adventure" and "Beagle,"* vol. I, p. 577). If one compares Tierra del Fuego, especially Port Famine in the Strait of Magellan (53°38' lat.) to Berlin, which lies about one degree closer to the equator, one will find the values

		$\dfrac{-0.5}{13.9}$					$\dfrac{1.2}{8.0}$
Berlin	6.5			Port Famine		4.7	

At the end of this note, I have compiled the few reliable temperature figures we presently possess for the temperate land zones of the Southern Hemisphere, for comparison with the northern temperatures and their unequal distributions of summer heat and winter cold. The convenient method of designation that I employ, in which the figure before the fraction designates the mean annual temperature, and the numerator and the denominator of the fraction signify the mean winter and summer temperatures, has already been explained above.

Place	Southern latitude	Mean annual, winter, and summer temperatures in Réaumur degrees	
Sidney and Parramatta (Australia)	33°50′	14.5	$\dfrac{10.0}{20.2}$
Cape Town (Africa)	33°55′	15.0	$\dfrac{11.8}{18.3}$
Buenos Aires	34°17′	13.5	$\dfrac{9.1}{18.2}$
Montevideo	34°54′	15.5	$\dfrac{11.3?}{20.2?}$
Hobart Town (Tasmania)	42°45′	9.1	$\dfrac{4.5}{13.8}$
Port Famine (Strait of Magellan)	53°38′	4.7	$\dfrac{1.2}{8.0}$

21. A single connected ocean of sand

Just as the socially growing *Erica* that form the heathlands from the mouth of the Scheldt to the Elbe, from the point of Jutland to the feet of the Harz Mountains, may be viewed as a connected body of plants, so too can one trace the ocean of sand through Africa and Asia, from Cabo Blanco to beyond the Indus, over a course of 1,400 geographical miles. Herodotus's sandy region, which the Arabs call the Sahara Desert, crosses, with the interruption of oases, the entirety of Africa like a desiccated arm of the sea. The Nile River Valley marks the eastern border of the Libyan Desert. Beyond the Isthmus of Suez, beyond the porphyry, syenite, and greenstone crags of the Sinai, begins the desert plateau of Nadj, which fills the entire interior of the Arabian Peninsula and is bordered on the west and south by the fertile and happy coastal countries of Hejaz and Hadramaut. The Euphrates marks the end of the Arabian and Syrian deserts. Tremendous sand dunes, *bejaban*, cut across the whole of Persia from the Caspian Sea to the Indian Ocean. Here are found the deserts, rich in salt and potash, of Kerman, Sistan, Balujistan, and Makran. This last is separated from the desert of Multan by the Indus River.

22. The western portion of the Atlas range

The question regarding the placement of the Atlas of the ancients has aroused spirited discussion in recent times. There are some who, in the course of this inquiry, intermingle the oldest Phoenician folk legends with the tales of the Atlas later put down by the Greeks and Romans. One man who combined profound knowledge of languages with a sound understanding of astronomy and mathematics, Professor Ideler (the father), was the first to cast a clear light upon this intermingling of ideas. I should like to introduce here, if I might, what this highly observant scholar imparted to me as regards this important subject:

"The Phoenicians, in a very early age of this world, ventured beyond the Strait of Gibraltar. On the Spanish coast they built Gades (Cadiz) and Tartessus, and on the Mauritanian coast of the Atlantic Ocean, Lixus and several other cities. They sailed forth from these coasts: northward to the Cassiterite Islands, from whence they took tin, and to the Prussian coast, where they procured amber; southward past Madera to the islands of Cape Verde. They visited, among others, the archipelago of the Canary Islands. Here the peak of Tenerife caught their eye, the height of which, already very significant in and of itself, appears all the higher in that it rises directly from the sea. Through the colonies that they sent to Greece, especially under Cadmus to Boeotia, the news of this mountain that reached above the region of the clouds, and of the happy islands upon which the mountain sat, islands bejeweled with fruits of all sorts, especially the golden oranges, spread into Greece. Here the tradition was propagated by the songs of the bards and thus came to Homer. Homer speaks of an Atlas that knows all the ocean depths, and that supports great pillars that separate the heavens from the Earth (*Odyssey* I, 52); he speaks of the Elysian Fields, which he presents as an enticing land in the West (*Iliad* IV, 561). Hesiod expresses himself in a similar way about Atlas and makes him a neighbor to the Hesperidean nymphs (*Theogony* V, 517). The Elysian Fields, which he relocates to the western edge of the world, he calls the 'Isles of the Blessed' (*Op. et dies* v. 167). The later poets embellished these myths of Atlas, of the Hesperides and their golden apples, and of the Isles of the Blessed, the home to which better people were directed after death, and connected to them the expeditions of Melicertes, the Tyrian god of trade, called Hercules by the Greeks.

"The Greeks did not rival the Phoenicians or the Carthaginians as mariners until quite late. While they did visit the coasts of the Atlantic Ocean, they seem never to have penetrated far into it. Whether they ever saw the Canary Islands or the peak of Tenerife strikes me as doubtful. They believed that the Atlas, described to them by their poets and folk legends as a very tall mountain standing at the western edge of the world, must be found on the west coast of Africa. There, then, was he displaced by their later geographers: Strabo, Ptolemy, and others. Since there is, incidentally, no single distinguishably high mountain situated in Northwest Africa, there was some confusion about the actual position of the Atlas, and so it was looked for: now on the coast, now in the interior; now close to the Mediterranean, now farther to the south. It then became usual (in the first century of our current reckoning, when the arms of the Roman Empire had penetrated into the interior of Mauretania and Numidia) to refer to the entire mountain range that runs west to east across Africa, almost parallel with the Mediterranean, as 'Atlas.' Pliny and Solinus felt quite sure, however, that the descriptions that the Greek and Roman poets gave of the Atlas did not fit this mountain range; they thus believed that the Atlas, which they describe in a picturesque manner after the fashion of the poetic tales, must be placed in the terra incognita of Central Africa. The Atlas of Homer and Hesiod can therefore be no other peak than that of Tenerife, just as the Atlas of the Greek and Roman geographers is to be found in Northern Africa."

I would like to add only the following remarks to Professor Ideler's instructive elucidation. In Pliny and Solinus, the Atlas rises from the sandy plain (*e medio arenarum*); elephants (which Tenerife has certainly never known) forage on its slopes. That which we now give the name Atlas is a long comb. How did the Romans come to recognize in this mountain comb the isolated cone described by Herodotus? Might not the cause of this lie in the optical illusion by which any mountain range, when viewed from the side, appears in the elongated field of view as a narrow cone? On the ocean, I have often mistaken long combs for isolated mountains. According to Höst, the Atlas near Morocco is covered with perpetual snow. It follows, then, that its height there must be over 1,800 toises. It is also worthy of note that, according to Pliny, the Barbarians, the ancient Mauretanians, gave the Atlas the name *Dyris*. Even now, the Atlas range is known among the Arabs as *Daran*, a word that has almost the same consonants as *Dyris*. Hornius, on the other hand (*De originibus Americanorum*, p. 195), believes he recognizes the word *Dyris* in the Guanche name of the peak of Tenerife, *Aya-dyrma*. For more on the connection of purely mythical ideas to legends of a geographical nature, and on the manner in which the Titan Atlas was caused to take the form of a mountain beyond the pillars of Hercules that supports the heavens, see Letronne, *Essai sur les idées cosmographiques qui se rattachent au nom d'Atlas*, in Férussac's *Bulletin universel des sciences*, mars 1831, p. 10.

If, with our current and admittedly very limited geological knowledge of the mountainous region of North Africa we can find no trace of volcanic eruptions there in historical times, it is even more conspicuous that we find among the ancients many indications of belief in such phenomena having occurred in the western Atlas and on the nearby western coast of the continent. The rivers of fire so often mentioned in the ship's log of Hanno could certainly have been signal fires or burning strips of grass, set by the wild inhabitants of the coast when danger threatened or at the first sight of enemy vessels. Alight with flame, the high summit of Mount Cameroon, the "Chariot of the Gods" (θεῶν ὄχημα), could be a dark reminder of the peak of Tenerife, but Hanno goes on to describe a peculiar ground formation. He finds in the gulf of the western horn a large island, and upon this island a salt sea; in this salt sea, there is another, smaller, island. This same configuration is repeated south of the Bay of Gorilla-apes. Are these coral reefs, lagoon islands (atolls), or volcanic crater lakes (*cratères-lacs*) in the center of which a cone has arisen? Lake Tritonis did not lie near little Sirte but on the oceanic west coast (*Asie centrale*, vol. I, p. 179). Earthquakes accompanied by great eruptions of fire caused the lake to disappear. Diodorus (lib. III, 53, 55) expressly states: πυρὸς ἐκφυτήματα μεγάλα (a great erupting of fire). But the most wondrous description is ascribed to the Atlas by a previously little-noted passage in one of the philosophical dissertations of Maximus of Tyre. This Platonic philosopher lived in Rome under Commodus. His

Atlas lies "upon the continent, there, where the Western Libyans inhabit a protruding peninsula." The mountain possesses upon the seaward side a deep, semicircular cavity with a cliff face so steep that it cannot be climbed. This abyss is filled with forest: "One looks upon the tops of the trees and the fruit they bear, as if one were looking down into a well" (Maximus Tyrius, VIII, 7, ed. Markland). The description is so uniquely picturesque that it quite probably presents the memory of an actual sight.

23. The Mountains of the Moon, Djebel al-Komr

Ptolemy's (lib. IV, cap. 9) Mountains of the Moon, σελήνης ὄρος, form on our older maps a colossal, uninterrupted mountainous parallel that cuts across the entirety of Africa. The existence of the mountains seems certain, but their full extent, their distance from the equator, and their general orientation are problematic. I have already pointed out in another work (*Kosmos*, vol. II, pp. 225 and 440) how a more accurate familiarity with Indian dialects and with Old Persian (the Zend) teaches us that part of the geographical nomenclature of Ptolemy is a historical preservation of the commercial connections between the Occident and the farthest regions of South Asia and East Africa. This train of thought may be discerned in a very recently initiated investigation. One wonders whether the great geographer and astronomer of Pelusium, in the naming of the Mountains of the Moon, merely wished to advance the Greek translation of the indigenous name of the mountains, as is the case with the "Island of Barley" (Jabadiu, Java); whether, as is most likely, al-Istakhri, Idrisi, Ibn al-Wardi, and other early Arabian geographers translated Ptolemaic nomenclature into their language; or whether similarity of word sounds and spelling misled them. In the notes to the translation of Abd al-Latif's famous description of Egypt, my great teacher Silvestre de Sacy (1810 edition, pp. 7 and 353) expressly states: "One normally translates the name of these mountains, which Leo Africanus considers as the sources of the Nile, as 'mountains of the moon,' and I have followed that tradition. I do not know if Arabs originally used this term of Ptolemy. One can believe that today they do indeed understand the word قمر meaning *moon*, while pronouncing it 'kamar': I do not believe, however, that was the opinion of the ancient Arab writers who pronounce it, as Makrizi proves, 'komr.' Abu'l-Fida completely rejects the idea of those who pronounce it 'kamar' and who derive this name from that of the moon. Since the word *komr*, considered the plural form of أقمر, means a *greenish* or off-white-*colored* object, according to the author of the *Kamous*, it seems that some writers believed that mountain got its name from its color."

The learned Reinaud, in his recently released and excellent translation of Abu'l-Fida (vol. II, part 1. pp. 81–82), believes it is probable that Ptolemy's interpretation of the name as "Mountains of the Moon" (ὄρη σεληναῖα) was the one originally accepted by the Arabs. He notes that in the *Al-Mushtarik* of Yakut and in Ibn Said the mountain group's name is written "al-Komr," and that Yakut writes the name of the Isle of Zendj (Zanzibar) in a similar manner. In his learned critical examination of the Nile and its tributaries (*Journal of the Royal Geographical Society of London*, vol. XVII, 1847, pp. 74–76) Beke, who traveled throughout Abyssinia, seeks to prove that Ptolemy, having learned from the tales that came to him thanks to the widespread commercial travel of the day, was merely imitating an indigenous name with his σελήνης ὄρος. "Ptolemy knew," he says, "that the Nile originates in the mountainous region of Moezi, and in the languages that are spread out over a great portion of Southern Africa (for example, in the idioms of Congo, Monjou, and Mozambique), the word *moezi* means moon. A large southwestern country was named *Mono-Muezi* or *Mani-Moezi*, i.e., the 'Land of the King of Moezi' (King of the Moon Country); for in the same linguistic family in which *moezi* or *muezi* signifies the moon, *mano* or *mani* means 'king.' Even Alvarez, in his *Viaggio nella Ethiopa* (Ramusio, vol. I, p. 249), speaks of the *regno di Manicongo*, the 'Empire of the King of Congo.'" Unlike Arnaud, Werne, and Beke, who look for the source of the White Nile (Bahr el-Abiad) near to or even south of the equator (at 29°0' Parisian longitude), Beke's detractor Mr. Ayrton, like d'Abbadie, seeks it far to the northeast in the Gojeb and Gibbe of Eneara (Iniara), that is, in the high mountains of Habesh at 7°20' northern latitude and 30°0' Parisian longitude. He assumes that, due to tonal similarity, the Arabs interpreted the indigenous name *Gamaro*, which comes to the Abyssinian mountain group in the southwest of Gaka that is the source of the Gojeb (or of the White Nile?), as "Mountains of the Moon" (Djebel al-*Komr*), such that Ptolemy himself, familiar with traffic between Abyssinia and the Indian Ocean, could have adopted the Semitic interpretation made by immigrant Arab settlers in prehistoric times. (Compare Ayrton in *Journal of the Royal Geogr. Soc.*, vol. XVIII, 1848, pp. 53, 55, and 59–63, with Ferd. Werne's instructive *Expedition zur Entdeckung der Nilquellen*, 1848, pp. 534–536.)

The lively interest that has recently been sparked in England as regards discovering the southernmost sources of the Nile has inspired the above-mentioned traveler to Abyssinia, Charles Beke (barely two months ago, at the gathering of the British Association for the Advancement of Science in Swansea), to develop in greater detail his ideas on the relationship of the Mountains of the Moon with those of Habesh. "The Abyssinian plateau, in most parts elevated to 8,000 feet, is growing longer," he asserts, "to the south, down to 9° and 10° northern latitude. The eastern slope of the highland looks, to the inhabitants of the coast, like a mountain range. The plateau drops considerably lower at its southern end and passes over into the Mountains of the Moon, which do not proceed from east to west, but parallel to the coast (from 10° N. to 5° S.), namely from NNE to SSW. The sources of the White Nile lie in the country of Mono-Moezi, probably lower than 2½° southern latitude, at the point where the Sabaki River, on the eastern slope of the Mountains of the Moon, falls into the Indian Ocean at Melindeh. The two missionaries to Abyssinia, Rebmann and Dr. Krapf, were still on the littoral at Mombasa last fall (1847). They have founded there, near to the Wakamba tribe, a mission called Rabbay Empie which also promises to be very useful for geographical discoveries. Families of the Wakamba tribe extend westward as much as five to six hundred English miles into the interior, to the upper course of the Lusidji, to the great Lake Nyassa or Zambezi (5° south lat.?), and to the nearby sources of the Nile. The expedition to these sources, for which"—with Beke's advice—"Mr. Friedrich Bialloblotski of Hannover is preparing, is supposed to be starting out in Mombasa. The Nile that flows from the west, as mentioned by the ancients, is probably the Bahr al Ghazal or Keilah which, below 9° northern latitude, flows from the west into the Nile above the mouth of the Gojeb, or Sobot."

Russegger's scientific expedition, motivated by Mehemed Ali's desire for the gold deposits of Fazokl on the Blue (Green) Nile or Bahr al Azrek (1837 and 1838), cast great doubt upon the existence of a "Mountains of the Moon" group. The Blue Nile, Ptolemy's Astapus, which has as its source Lake Coloe (now Lake Tsana), wends its way out of the colossal Abyssinian mountains, but an extensive lowland extends toward the southwest. The three expeditions sent off by the Egyptian government at Khartoum at the confluence of the Blue and White Niles (the first being in Nov. 1839, under the leadership of Selim Bimbashi; the next in the fall of 1840 in the company of the French engineers Arnaud, Sabatier, and Thibaut; and the final one in August 1841) were the first to uncover the high group of mountains which, running first from west to east and later from northwest to southeast, between the parallels of 6° to 4° (and perhaps farther south), approach the left bank of the Bahr al Abiad. The second expedition of Mehemed Ali, according to Werne's report, saw the mountain range for the first time at 11⅛° lat., where the Gebel Abul and Gebel Kutak ascend to 3,400 ft. The plateau continued and came nearer the river farther to the south, from 4¼° latitude down to the parallel of Tschenker Island at 4°4', the endpoint of the expedition of Selim Capitan and Feizulla Effendi. The shallow river pushes through the rocks, and the individual mountains in the region of Bari again ascend to as much as 3,000 feet in elevation. These are undoubtedly a part of the Mountains of the Moon as they appear on the most recent maps: clearly not a mountain group bedecked by perpetual snow, as Ptolemy (lib. IV, cap. 9) would have it. The snow line in these latitudes does not begin below 14,500 feet above sea level. Perhaps Ptolemy transferred the information he may have possessed regarding the high mountains of Habesh, in Upper Egypt near the Red Sea, to this source land of the White Nile. In Gojam, Kaffa, Miecha, and Simien, according to precise measurements (unlike those of Bruce, who puts Khartoum not at 1,430 but at 4,730 feet of elevation!), the Abyssinian mountain groups rise to 10,000 and 14,000 feet. Rüppell, one of the most precise observers of our time, finds the Abba Jarat at 13°10' to be only 66 feet lower than Mont Blanc (see Rüppell, *Reisen in Abyssinia*, vol. I, p. 414, vol. II, p. 443). Rüppell found a high plateau, lightly covered with fresh-fallen snow, to be touching on Mount Buahat and lying at an elevation of 13,080 feet above the Red Sea (Humboldt, *Asie centrale*, vol. III, p. 272). The famous inscription of Abdulis, which Niebuhr holds to be older than Juba and Augustus, also speaks of Abyssinian snow "that reaches up to the knees"—the oldest description of snow between the tropics, I believe, in antiquity (ibid., vol. III, p. 235), for the Paropanisus is still 12 degrees of latitude away from the edge of the tropics.

Zimmermann's map of the regions of the Upper Nile presents the dividing line that delineates the basin of the great river and separates the basin from the river drainages that belong to the Indian Ocean: from the Doara, whose mouth opens north of Magadoxho; from the Teb on the Amber Coast near Ogda; from the water-rich Goschop, which is formed by the confluence of the Gibu and the Zebi, and should not be confused with the Gojeb, which is well known since 1839, thanks to Antoine d'Abbadie, the missionary Krapf, and Beke. Immediately upon their appearance in 1843, I greeted these findings that Zimmermann had so concisely compiled from the recent journeys of Beke, Krapf, Isenberg, Russegger, Rüppell, Abbadie, and Werne with joy and enthusiasm in a letter to Carl Ritter: "When, over the course of a long life," I wrote, "some discomforts should arise for an aging man, and some also for those around him, what serves as a compensation is the intellectual pleasure of being able to compare earlier states of human knowledge with newer ones, to see before our very eyes how great things grow and develop just there, where everything has long slumbered, where one had often hypercritically endeavored again to repudiate that which one had already achieved. Such a beneficial enjoyment has come to you and me from time to time in the course of our geographical studies, in those very parts, in fact, where one could speak out only with a certain timorous diffidence. The inner form and structure of a continent is for the most part dependent upon certain malleable conditions which are usually the very ones that are the last to be unraveled. A new and excellent work by our friend Carl Zimmermann on the Upper Nile country and East-Central Africa has actively renewed my interest in these considerations. The new map very clearly shows by way of specific shading that which is as yet unknown, and that which, through the courage and endurance of travelers of all nations (among whom those of the Fatherland fortunately play an important part), has now been opened. It may be viewed as a fruitful enterprise that at certain points in time, the current state of human knowledge should be graphically depicted by men who are thoroughly familiar with the available and very disparate material, and who do not merely draft and compile but compare and select and, wherever possible, narrow down routes of travel through the application of astronomical determinations of place. Anyone who has contributed to our knowledge as richly as you have certainly has a special right to expect a good deal, for the number of points of connection has been increased by their combinations, yet I believe, nevertheless, that at the time of the editing of your great work on Africa in 1822, you could not have expected so many additions as have come to us." Admittedly, it is often only regarding the courses of rivers, with which we are familiar in terms of their direction, their branches, and the many synonyms by which they are known and the linguistic families thereof—but the courses of rivers reveal the formation of the earth's surface: they are the life-giving element that unifies mankind and indicates his future.

The northern course of the White Nile and the southern course of the great Goschop attest to the fact that a swelling of the ground separates the two river basins. Just how this swelling is directly connected to the high plain of Habesh as it continues southward far beyond the equator we admittedly understand only partially. Probably, and this is also the opinion of my friend Carl Ritter, the Lupata mountain group, which according to the observations of the admirable Wilhelm Peters extend to 26° southern latitude, is connected to that northern swelling of the earth's surface and to the Abyssinian plateau by the Mountains of the Moon. *Lupata*, according to the report of this African traveler, is used as an adjective meaning "closed" in the Tette language. The mountain range is also called "Locked" and "Barricaded" (penetrated only by single rivers). "The Lupata range of the Portuguese writers," says Peters, "lies some 90 *legoas* from the mouth of the Zambezi and is barely two thousand feet high. The wall-like range is oriented for the most part from north to south, but several times diverts, now to the east, now to the west. It is occasionally interrupted by plateaus. All along the coast of Zanzibar, those commercial travelers who have pressed into the interior give witness to this long, though not very high, comb of mountains, which stretches between 6° and 26° southern lat. down to the factory in Lourenzo-Marques on the Rio de Espirito Santo (in the Bai da Lagoa of the English). The farther the Lupata range pushes southward, the closer it comes to the coast; at Lourenzo-Marques it is only 15 *legoas* away."

24. An effect of the great rotational current

In the northern part of the Atlantic Ocean, between Europe, North Africa, and the New Continent, the waters are driven round and round into a true rotational current, forever turning back into itself. Below the tropics, the general current, which considering its cause one could also call a rotational current, is known to move, like the trade wind, from east to west. It accelerates the travel of ships that sail from the Canary Islands to South America. It makes it nearly impossible to navigate in a straight line from Cartagena de Indias to Cumana (against the current). This westward current, ascribed to the trade wind, is increased in the Sea of the Antilles, however, by a much stronger movement of waters that has a very distant cause, discovered in 1560 by Sir Humphrey Gilbert (Hakluyt, *Voyages*, vol. III, p. 14) and further clarified by Rennell in 1832. The Mozambique Current forces itself from north to south between Madagascar and the eastern coast of Africa, onto the Agulhas Bank and then to its north around the southern point of Africa; it follows with great force the African west coast up to the island of São Tomé just above the equator, at the same time giving a part of the waters of the South Atlantic a northwesterly direction, causing them to strike Cape St. Augustine and move on to follow the coast of Guyana past the mouth of the Orinoco, the *Boca del Drago*, and along the littoral of Paria (Rennell, *Investigation of the Currents of the Atlantic Ocean*, 1832, pp. 96 and 136). From the Isthmus of Panama up to the northern part of Mexico, the New Continent forms a dam that stands in opposition to this movement of the ocean. Thus the current is forced to take a northerly direction from Veragua on, and to follow the bends of the coasts of Costa Rica, Mosquito, Campeche, and Tabasco. The waters that flow into the Gulf of Mexico between Cape Catoche in Yucatan and Cape San Antonio de Cuba, after completing a great rotation between Veracruz, Tamiagua, and the mouths of the Rio Bravo del Norte and the Mississippi, press onward to the north through the Bahama Canal and back into the open ocean. Here they form what mariners call the Gulf Stream, a flow of warm, quickly moving water that flows in a diagonal direction ever farther from the North American coast. Because of this oblique direction of the current, ships bound for this coast from Europe, yet unsure of its geographical length, orient themselves through simple observations of latitude as soon as they reach the Gulf Stream. Its location and bearing were first accurately identified by Franklin, Williams, and Pownall.

Past the 41st parallel of latitude, this flow of warm water, which grows wider and wider as it gradually loses speed, turns suddenly to the east. It almost touches the southern edge of the Grand Banks of Newfoundland, where the temperature difference between the waters of the Gulf Stream and those upon the chilling Banks is greatest. Before the warm flow reaches the most westerly of the Azores, it divides into two arms, one of which, at least at certain times of year, turns toward Ireland and Norway, while the other turns toward the Canary Islands and the western coast of North Africa. Through this rotational current of the Atlantic, which I have described in detail elsewhere (in the 1st volume of my journey to the tropics), one can explain how it can be that despite the trade winds, the trunks of South American and West Indian dicotyledons have washed up on the coasts of the Canary Islands. Around the Banks of Newfoundland, I have done many experiments on the temperature of the Gulf Stream. With great speed, it brings the warmer waters of the lower latitudes into the northern regions. Thus the temperature of the stream is two to three degrees Réaumur warmer than the bordering still waters that form the bank of this warm, oceanic river.

The flying fish of the equinoctial zone (*Exocetus volitans*), loving the warmth of the waters, migrates in the bed of the Gulf Stream far northward into the temperate zone. Floating seaweed (*Fucus natans*), which the stream picks up primarily in the Gulf of Mexico, makes the entry into the Gulf Stream easy for mariners to recognize. The position of the floating blades of the seaweed indicates the direction of the water's motion. The great mast of the English warship *Tilbury*, which during the Seven Years' Ocean War went up in flames on the coast of Santo Domingo, was carried by the Gulf Stream to the coast of Northern Scotland; indeed, barrels full of palm oil, flotsam of an English ship that had been crushed on the rocks of Cape Lopez in Africa, also made it to Scotland, after having twice crossed the Atlantic Ocean: once from east to west, following the Equinoctial Current between 2° and 12° latitude, and then from west to east, between 45° and 55°, with the help of the Gulf Stream. Rennell relates (*Investigation of Currents*, p. 347) the journey of a floating bottle that was thrown, complete with message, from the English ship *Newcastle* on the 20th of January, 1819, at 38°52′ lat. and 66°20′ long., and was found on the 2nd of June, 1820, at the Rosses in Northwest Ireland near Aran Island. Shortly before my arrival at Tenerife, the ocean at the harbor of Santa Cruz washed ashore the trunk, encased in its lichen-covered bark, of a South American cedar (*Cedrela odorata*).

The effects of the Gulf Stream (the washing ashore on the Azorean islands of Fayal, Flores, and Corvo of bamboo stems, wood cut by artificial means, trunks of Mexican or Antillean varieties of pine unknown to the area, even of the corpses of a particular race of broad-faced men) undoubtedly contributed to the discovery of America, for they reinforced for Columbus the presumption of the existence of Asiatic countries and islands lying near to the west. The great explorer even learned straight from the mouths of the settlers of the Azorean Cap de la Verga: "On a westward voyage, they met with covered barks piloted by strange-looking people, and seemingly built as if they could not sink— *almadias con casa movediza, que nunca se hunden*." As much as the fact has long been doubted, there exists the most reliable evidence of a true crossing by Native Americans (probably Eskimos from Greenland or Labrador), coming from the northwest through currents and storms to our continent. In his *Account of the Islands of Orkney* (1700, p. 60), James Wallace relates that in 1682, a Greenlander in his small boat was seen by several people off the southern point of Eday Island. The attempts to catch him were unsuccessful. Also, a Greenland fisherman appeared in 1684 near Westray Island. An Eskimo boat that had been carried away by storm and current hung in the church on Burray Island. The inhabitants of the Orkney Islands refer to the Greenlanders who have appeared there as "Finnmen."

In Cardinal Bembo's history of Venice, I find the report that in the year 1508, near the coast of England, a small boat carrying seven people of foreign appearance was captured by a French ship. The description exactly fits that of the Eskimos ("homines errant septem, mediocre statura, colore suboscuro, lato et patente vultu, cicatriceque una violacea signato"). No one understood their language. Their clothes were of fish-skins, sewn together. On their heads they wore "coronam e culmo pictam, septum quasi auriculis intextam." They ate raw meat and drank blood as we drink wine. Six of these men died on the trip; the seventh, a youth, was presented to the king of France, who was at that time in Orleans (Bembo, *Historiae Venetae*, ed. 1718, lib.VII, p. 257).

The appearance of so-called Indians on the western Germanic shores during the Ottonian reigns in the 10th century and under Frederick Barbarossa in the 12th, and even those reported by Cornelius Nepos in the Fragments (ed. Van Staveren, cur. Bardili, vol. II, 1820, p. 356), by Pomponius Mela (lib. III, cap. 5, §8), and by Pliny (*Hist. Nat.* II, 67), when Quintus Metellus Celer was proconsul in Gaul, may find their explanation in the effects of the ocean currents and long-prevailing northwest winds. A king of the

Boyars (or, others maintain, of the Suebi) presented the stranded, swarthy people as a gift to Metellus Celer. In his *Historia general de las Indias* (Saragossa, 1553 fol. VII), Gómara already proposes that the "Indians" of the Boyar king are actually natives of Labrador. "Si ya no fuesen," he says, "de Tierra del Labrador, y los tuviesen los Romanos por Indianos, engañados en el color." One can believe that the appearance of Eskimos on the Northern European coasts can have happened more frequently in earlier times, for the reason that this branch of humanity, as we know from the research of Rasmus Rask and Finn Magnussen, was, in the 11th and 12th centuries, spread in great numbers from Labrador far southward to Good Vinland (i.e., to the littoral of Massachusetts and Connecticut) under the name Skralingers (*Kosmos*, vol. II, p. 270; *Examen critique de l'hist. de la Géographie*, vol. II, p. 247–278).

Just as the winter cold of the northern part of Scandinavia is rendered milder by the returning Gulf Stream, which carries up, past the 62nd parallel of latitude, fruits of the American tropics (fruits of the coconut palm, of the *Mimosa scandens* and the *Anacardium occidentale*), in the same way, Iceland enjoys from time to time the beneficial influence of a dispersion of the warm water of the Gulf Stream far into the north. On the Icelandic coasts, as on those of the Faroe Islands, a great number of American tree trunks are washed ashore. In the past, this driftwood that arrived in great amounts was used as building material. It was cut into planks and boards, while the fruits of tropical plants, which are collected on Icelandic shores especially between Raufarhöfn and Vopnafjordur, confirm the direction of the waters coming from the south (Sartorius von Waltershausen, *Physisch-geographische Skizze von Island*, 1847, pp. 22–35).

25. Neither lecidea nor other lichens
In the northern countries, the areas of ground empty of plants cover themselves with *Baeomyces roseus, Cenomyce rangiferinus, Lecidea muscorum, Lecidea icmadophila*, and other cryptogams, which may be said to prepare the soil for the growth of grasses and herbs. In the tropics, where mosses and lichens are found in large numbers only in shaded areas, certain oily plants provide a place for the ground lichens.

26. The care of milk-giving animals . . . ruins of the Aztec Castle
Two oxlike creatures that we have mentioned above, *Bos americanus* and *Bos moschatus*, are indigenous to the northern part of the New Continent. But the natives,

> *Queis neque mos, neque cultus erat, nec jungere tauros* (Virg., *Aen.*, I, 316)
> (Who had neither laws nor policy, knew neither to yoke the laboring steer),

drank the fresh blood rather than the milk of these animals. Individual exceptions have been found, however, but among tribes who also cultivated maize. I have mentioned above how Gómara tells of a people in Northwestern Mexico who kept herds of tamed bison and used these animals for clothing, food, and drink. The drink was perhaps blood (Prescott, *Conquest of Mexico*, vol. III, p. 416), for, as I have often heard, the distaste for milk, or at least the absence of its being used, seems to have been, prior to the arrival of Europeans, a common trait of all natives of the New Continent along with the inhabitants of China and Cochinchina, even though they were closely surrounded by true shepherd folk. The herds of domesticated llamas that were to be found in the highlands of Quito, Peru, and Chile belonged not to nomads but to settled agrarian tribes. As one certainly very rare exception to this way of life, Pedro de Cieza de León (*Chronica del Peru*, Sevilla, 1553, cap. 110, p. 264) seems to indicate that in the Peruvian high plain of Callao, llamas were used to pull the plow (cf. Gay, *Zoologia de Chile, Mamiferos*, 1847, p. 154). Generally in Peru, however, the plowing was done by humans alone (see Garcilaso [El Inca] de la Vega, *Comentários reales*, part I, lib. V, cap. 2, p. 133, and Prescott, *History of the Conquest of Peru*, 1847, vol. I, p. 136). Mr. Barton has shown it to be probable that among certain West Canadian tribes, the American buffalo has nearly forever been an object of cattle husbandry for its meat and its leather (*Fragments of the Natural History of Pennsylvania*, part I, p. 4). In Peru and Quito, the llama is no longer to be found in its original wild state. As the natives explained to me, the llamas on the western slope of Chimborazo reverted to a wild state when Lican, the old residence of the ruler of Quito, burned to the ground. Similarly, there are cattle in Central Peru, in the Ceja de la Montaña, that have turned completely wild: a small and doughty breed that is often hunted by the Indians. The natives call these cattle *vacas del monte* or *vacas cimarronas* (Tschudi, *Fauna Peruana*, p. 256). Cuvier's assertion that the llama descends from the still wild guanaco has been (unfortunately!) spread widely by the commendable Meyen, but thoroughly refuted by Herr von Tschudi.

Llama, paco or alpaca, and guanaco are three originally distinct species of animal (Tschudi, pp. 228 and 237). The guanaco (*Huanacu* in the Quichua language) is the largest of the three; the alpaca, measured from the ground to the top of the head, is the smallest. The llama is second in height to the guanaco. Llama herds, in such great numbers as I saw on the high plains between Quito and Riobamba, adorn the landscape charmingly. The moro-moro of Chile seems to be merely a variation of the llama. Camelids currently living at elevations of 13,000 to 16,000 feet above sea level include the vicuña, the guanaco, and the alpaca. These last two can be domesticated, although it is uncommon with the guanaco. The alpaca is not as well suited for the hot climate as the llama. Since the introduction of the more versatile horses, mules, and donkeys (the last two being especially fine and lively in the tropics), the breeding and use of the llama and the alpaca as pack animals have declined greatly in the mining regions. The wool, so different in its fineness, remains, however, an important product to the industry of the mountain people. In Chile, they differentiate between the wild and the domestic guanaco; the first is called *luan*, the second *chilehueque*. Important to the wide distribution of the wild guanaco from the Peruvian cordilleras to Tierra del Fuego, sometimes in herds of 500, is the fact that this animal is able to swim from island to island with great ease and they are thus not hindered in their migrations by the arms of the sea (fjords) in Patagonia (see the graceful representations in Darwin, *Journal*, 1845, p. 66).

South of the Gila River, which together with the Rio Colorado empties into the Gulf of California (*Mar de Cortes*), the puzzling ruins of the Aztec Palace, named *las Casas Grandes* by the Spanish, lie lonely upon the steppe. When the Aztecs came forth from the unknown land of Aztlan and appeared in Anahuac around the year 1160, they settled for a time on the Gila River. The Franciscan monks Garces and Font are the last travelers to have visited the *Casas Grandes* (1773). They affirm that the ruins cover an area of a square quarter-mile. The entire plain there is covered with shards of artfully painted earthenware. The main palace (if a house built of unfired clay can earn such a name) is 420 feet long and 260 feet wide. (See the rare *Cronica seráfica y apostólica del Colegio de Propaganda Fide de la Santa Cruz de Querétaro por Fr. Juan Domingo Arricivita*, printed in Mexico in 1792.)

The *tayé* of California, as described by Father Venegas, seems to differ little from the mouflon (*Ovis musimon*) of the Old Continent. The same animal has been seen in the northern Rocky Mountains near the sources of the Peace River. Quite different, on the other hand, is the small white-and-black dappled creature that grazes on the meadows near the Missouri and Arkansas Rivers. The synonymy of *Antilope furcifer*, *Antilope tememazama* Smith, and *Ovis Montana* remains very uncertain.

27. Cultivation of grasses that yield meal or flour

The place of origin of grasses that yield flour, like that of the domesticated animals that have accompanied humans since our earliest migrations, is cloaked in the same obscurity. Jacob Grimm wisely derives the German word *Getreide* (grain; cereal) from the Old German *gitragidi* or *gitregidi* (*v*: to carry; *n*: yield, produce): "It is that which is tamed, the fruit (*fruges, frumentum*) which has come into the hands of humans, like tamed animals, as opposed to wild ones" (Jacob Grimm, *Geschichte der deutschen Sprache*, 1848, part I, p. 62). It is certainly a striking phenomenon that there are people on one side of our planet for whom dairy foods and the making of flour from narrow-eared grasses (*Hordeacea* and *Avenacea*) was originally completely unknown, while in the other hemisphere there were nations virtually everywhere who cultivated cereals and raised animals for milk. The cultivation of various grasses characterizes both of these parts of the world. In the New Continent, we see only one grass, maize, being cultivated between the latitudes of 52° north and 46° south. On the Old Continent, however, we discover everywhere, since the earliest days of recorded history, the fruits of Ceres: the cultivation of wheat, barley, spelt, and oats. A belief of some ancients, mentioned even by Diodorus Siculus (lib. V, pp. 199 and 232; Wessel.), was that wheat grew wild in the Leontine Fields, as well as several other places in Sicily. Ceres was also found in the alpine meadow of Enna, and Diodorus tells the tale that the Atlanteans "knew not of the fruits of Ceres, for they had separated from the other human races before those fruits had been shown to mortals." Sprengel collected several interesting passages, through which it came to seem probable to him that the greater portion of our European grains originally grew wild in Northern Persia and India, namely summer wheat in the land of the Musicani, a province in Northern India (Strabo, XV, 1017); barley (*antiquissimum frumentum*, as it is called by Pliny, and the only grain known to the Guanches of the Canary Islands) on the Araxes or Kura River, according to Moses of Chorene, and in Balasham in Northern India, according to Marco Polo (Ramusio, vol. II, p.10); spelt in Hamadan. But these passages, as my perceptive friend and teacher Link showed in a substantial essay (*Abhandlungen der Berliner Akademie 1816*, p. 123), leave behind a great deal of uncertainty. I too (*Essai sur la Géographie des Plantes*, 1805, p. 28) early on doubted the existence of wild-growing grains in Asia and held them to have become wild. From Reinhold Forster, who before his travels with Captain Cook had taken by order of Empress Catherine a natural history expedition into Southern Russia, came the information that near the point where the Samara joins the Volga, two-rowed summer barley (*Hordeum distichon*) grew wild. At the end of September 1829, on the journey from Orenburg and Uralsk to Saratov and the Caspian Sea, Ehrenberg and I also camped on the Samara. The number of stands (turned wild) of wheat and rye upon uncultivated ground was indeed conspicuous in this area, but the plants seemed no different from the usual cultivated varieties. Ehrenberg received from Mr. Carelin a sample of a rye, *Secale fragile*, from the Kyrgyz steppe that Marschall von Bieberstein had for a long time held to be the mother plant of our cultivated rye, *Secale cereale*. That Olivier and Michaux maintained that spelt grew wild near Hamadan in Persia is, as Achilles Richard reports, not proved by Michaux's herbarium. More reliable are the most recent reports that we receive thanks to the tireless efforts of a very knowledgeable traveler, Professor Carl Koch. He found several ryes (*Secale cereale* var. β *pectinata*) in the Pontic Mountains between five and six thousand feet high, at places where this grain had never been cultivated in the memory of the inhabitants. "The appearance," he says, "is all the more important, as this grain, in our part of the world, has never been known to propagate elsewhere on its own." In the Shirvan region of the Caucasus, Koch collects a variety of barley that he calls *Hordeum spontaneum* and takes to be the original wild *Hordeum zeocriton* Linn. (Carl Koch, *Beiträge zur Flora des Orients*, issue I, pp. 139 and 142).

A Negro slave of the great Cortes was the first to cultivate wheat in New Spain. He found three kernels of it beneath the rice that had been brought from Spain as a provision for the army. In a Franciscan cloister in Quito, I saw preserved as a relic the earthenware pot that held the first wheat planted by the Franciscan monk Fra Iodoco Rixi de Gante de Quito. Rixi was born in Ghent (Gante), Flanders. The first grain was cultivated in front of the cloister that had been built, after the encroaching forest at the foot of the Pinchincha volcano had been cleared, upon the *Plazuela de San Francisco*. The monks, whom I often visited during my time in Quito, asked me to explain to them the inscription on the pot, in which they suspected there would be some mysterious connection to the wheat. I read the motto in its Old German dialect: "Whoever drinks from me, let him not forget his God." I too sensed something venerable about the Old German drinking vessel! If only it were so, that instead of the names of those who lay waste to the land in bloody conflict, there were enshrined all over the New Continent the names of those who first bestowed upon it the fruits of Ceres!

In terms of the old linguistic interrelations in general "this is more seldom found among varieties of grains and the implements of farming than it is in animal husbandry. Migrating herdsmen still shared much in common, for which farmers who came later had to choose specialized vocabulary. But the fact that when compared to Sanskrit, Romans and Greeks are in a similar position to Germans and Slavs speaks to a very early shared migration of the last two. Even so, the Indian *java* (*Frumentum hordeum*), compared to the Lithuanian *jawai* and the Finnish *jywa*, offers a rare exception" (Jacob Grimm, *Geschichte der deutschen Sprache*, vol. I, p. 69).

28. Preferring the cold, pursued the ridge of the Andes

In all of Mexico and Peru can be found the traces of a great human culture that remained only in the high mountain regions. We found ruins of palaces and baths on the spine of the Andes chain at elevations from 1,600 to 1,800 toises. Only northern peoples in a stream of migration from north to south toward the equator could have enjoyed such a climate.

29. The history of the populating of Japan

I believe I have demonstrated the probability that the western peoples of the New Continent were associated with East Asia long before the arrival of the Spaniards in my works on the monuments of the primitive Americans (*Vues des Cordillères et monumens des peuples indigènes de l'Amérique*, 2 vols.) through the comparison of the Mexican and Tibetan/Japanese calendars, the well-oriented stepped pyramids, and the ancient myths of the four ages of time or the destruction of the world, and of the spread of the human race after a great flood. Since the appearance of my work on the extraordinary structures in the ruins of Guatemala and Yucatan, which are in a style nearly like that of India, the things that have been published in England, France, and the United States give to these analogies a still greater value. (Compare Antonio del Rio, *Description of the Ruins of an Ancient City, discovered near Palenque*, 1822 [translated from the original manuscript report by Cabrera; del Rio's investigation took place in 1787], p. 9, tbl. 12–14, with Stephens, *Incidents of Travel in Yucatan*, 1843, vol. I, pp. 391 and 429–434, vol. II, pp. 21, 54, 56, 317, and 323; also with the impressive major work by Catherwood, *Views of Ancient Monuments in Central America, Chiapas, and Yucatan*, 1844; and finally, with Prescott, *The Conquest of Mexico*, vol. III, append., p. 360.)

The ancient structures on the Yucatan Peninsula, more even than Palenque, speak of a culture that arouses astonishment. They lie between Valladolid, Merida, and Campeche, mostly in the western part of the territory. But the structures on the island of Cozumel (actually Cuzamil), east of Yucatan, were the first ones seen by the members of the Spanish expeditions of Juan de Grijalva in 1518 and Cortes in 1519. It was they who spread the idea of the great progress of ancient Mexican civilization throughout Europe. The important ruins of the Yucatan Peninsula, which have as yet not been properly measured and drawn up by architects, are the *Casa del Gobernador* of Uxmal, the teo-callis and vaultlike constructions at Kabah, the ruins of Labna with coupled columns, those of Zayi with pillars approaching the Doric style, and those of Chichen with great, ornamented pilasters. An old manuscript, put down in the Mayan language by a Christian Indian, and which is currently in the hands of the *jefe politico* of Peto, don Juan Pio Perez, presents the different epochs (in *katuns* of 52 years) in which the Toltecs settled the various parts of the peninsula. From this information, Perez extrapolates that the structures in Chichen go back to the end of the fourth century of our reckoning, while those of Uxmal belong to the tenth century. The accuracy of these historical conclusions is, however, subject to much doubt (Stephens, *Incidents of Travel in Yucatan*, vol. I, p. 439, and vol. II, p. 278).

I consider an ancient association of Western Americans and Eastern Asians as more than probable, but upon which routes of travel and with which Asiatic tribes the connection existed can at the moment not be determined. A small number of individuals from the educated caste of priests could have perhaps sufficed to bring about great changes in the social conditions in Western America. The yarns once spun about Chinese expeditions to the New Continent really refer merely to voyages to Fusang or Japan. On the other hand, it is possible that Japanese mariners or Sian Pi from Korea, battered by storms, may have landed on the American coast. We know historically that bonzes (Japanese *bōzu*, "Buddhist monk") and other adventurers sailed the East China Sea in hopes of finding an elixir of immortality. Thus was a cohort of 300 couples of young men and women sent to Japan under Qin Shi Huang-ti in 209 BC. Instead of returning to China, however, they settled upon Nippon (Klaproth, *Tableaus historiques de l'Asie*, 1824, p. 79; *Nouveau Journal asiatique*, vol. X, 1832, p. 335; Humboldt, *Examen critique*, vol. II, pp. 62–67). Might not chance have sent similar missions to the Fox Islands, to Alaska, or to New California? Since the western coasts of the American continent are oriented from NW to SE, while the eastern coasts of Asia run from NE to SW, it seems that the distance between those zones of the two continents that lie below 45° latitude and are milder and more conducive to intellectual development is simply too great to support the idea of an accidental Asiatic migration there. One is forced to assume that the first landing was in the inhospitable region between 55° and 65°, and that the gathering of knowledge was done at successive stages as the general American migration of peoples moved from north to south (Humboldt, *Relat. hist.*, vol. III, pp. 155–160). On the coasts of Northern Dorado (also called Quivire or Cibora) at the beginning of the 16th century, there were supposedly found the wrecks of ships from *Catayo*, that is, Japan or China (Gómara, *Hist. general de las Indias*, p. 117).

Up to now we remain insufficiently familiar with American languages to be able to give up completely, in the face of their tremendous variety, the hope that a dialect might one day be discovered that, with certain modifications, might have been spoken simultaneously in the interiors of South America and Asia, or that at least indicates an old interrelation. Such a discovery would certainly be among the most brilliant that one could hope for in the history of the human race! Linguistic analogies, however, first gain credibility when, instead of dwelling only on tonal similarities of root words, they press on into the organic structure, into the wealth of grammatical forms, into that which in languages reveals itself to be the product of the intellectual power of humanity.

30. Many other creatures

On the steppes of Caracas, whole herds of the so-named *Cervus mexicanus* run about. The young deer has light spots similar to those of the roe fawn. We also saw many pure white specimens among them, which is remarkable for such a hot climate. In the Andes chain, the *Cervus mexicanus* does not climb above 700 or 800 toises up the mountain sides. But above as much as 2,000 toises elevation, a large deer, also frequently white, may be found which I could hardly differentiate by any specific characteristic from the European. The *Cavia capybara*, called *Chiguire* in Caracas Province, is the unfortunate animal that is hunted in the water by crocodiles and on the plain by tigers (jaguars). It is such a poor runner that we were often able to catch it by hand. The extremities are smoked like hams, having a taste that is very unpleasant due to the musky smell, and to which we greatly preferred the smoked monkey flesh near the Orinoco. The prettily striped but odiferous animals are *Viverra mapurito*, *Viverra zorilla*, and *Viverra vittata*.

31. The Guarani and the fan palms, Mauritia

The small coastal tribe of the Guarani (called, in British Guyana, the Warraws or Guaranos, and by the Caribs the *U-ara-u*) does not live only in the swampy delta and the greater river system of the Orinoco, concentrated on the banks of the *Manamo grande* and the Caño Macareo; the Guarani are also spread out, with little change to their lifestyle, over the littoral region between the mouth of the Essequibo and Boca de Navios of the Orinoco. (Compare my *Relation historique*, vol. I, p. 492, vol. II, pp. 653 and 703, with Richard Schomburgk, *Reisen in Britisch Guiana*, vol. I, 1847, pp. 62, 120, 173, and 194.) According to the report of this last-mentioned and excellent natural explorer, there are still around 1,700 Warraws or Guarani living in the region surrounding Cumana and along the Barima River, which empties into the Gulf of Boca de Navios. The customs of the tribes living in the Orinoco delta were already familiar to the great historian Cardinal Bembo, the contemporary of Columbus, Amerigo Vespucci, and Alonso de Hojeda.

Bembo writes, "Quibusdam in locis propter paludes incolae domus in arboribus ae-
dificant" (*Historiae Venetae*, 1551, p. 88). It is unlikely that Bembo is referring to the
natives at the opening of the Gulf of Maracaibo rather than to the Guarani at the mouth
of the Orinoco; it was at the Gulf's opening in August of 1499 that Alonso de Hojeda,
accompanied then by Vespucci and Juan de la Cosa, also "found a population, *fondata
sopra l'acqua come Venezia*" (text by Ricardi in my *Examen crit.*, vol. IV, p. 496). In
Vespucci's travel report—in which we find the first trace of the etymology of the phrase
"Province of Venezuela" (Little Venice) for the Province of Caracas—there is mention
only of houses built on pillars, not of homes in trees.

A later, indisputable account comes from Sir Walter Raleigh. In his description of
Guyana, he expressly states that on his second trip into the estuary of the Orinoco in
1595, he saw the fires of the Tivitivis and Oua-rau-etes (as he calls the Guarani) high up
in the trees (Raleigh, *Discovery of Guyana*, 1596, p. 90). The drawing of the fire may
be found in the Latin edition: *Brevis et admiranda Descriptio regni Guianae* (Norib.,
1599), tbl. 4. Raleigh also was the first to bring the fruit of the Mauritia palm back to
England, which he quite correctly compared to the cones of the fir because of its scales.
Padre José Gumilla, who twice visited the Guarani as a missionary, even says that this
tribe lives in the *palmares* (palm shrubs) of the marshes; he only mentions, however,
certain lofty homes built on poles, but not the singular scaffolds attached to still-living
trees (Gumilla, *Historia natural, civil y geografica de las Naciones situadas en las riv-
eras del Rio Orinoco*, nueva impr. 1791, pp. 143, 145, and 163). Hilhouse and Sir Robert
Schomburgk (*Journal of the Royal Geographical Society*, vol. XII, 1842, p. 175, and *Description of the Murichi or Ita Palm, read in the meeting of the British Association held
at Cambridge, June 1845*, reprinted in Simonds, *Colonial Magazine*) are of the opinion
that Bembo, from others' accounts, and Raleigh, as an eyewitness, were deceived by the
fact that fires lying farther inland illuminated the high palms at night, such that those
on the passing ships believed that the homes of the Guarani were attached to the trees
themselves. "We do not deny that, in order to escape the attacks of the mosquitos, the
Indian sometimes suspends his hammock from the tops of trees; but on such occasions
no fires are made under the hammock" (compare also Sir Robert Schomburgk's new
edition of Raleigh, *Discovery of Guyana*, 1848, p. 50).

According to Martius, the lovely palm called *Moriche, Mauritia flexuosa, Quiteve*, or
Ita palm (Bernau, *Missionary Labours in British Guiana*, 1847, pp. 34 and 44) belongs,
along with *Calamus*, to the group of *Lepidocarea* or *Coryphinea*. Linnaeus described it
very incompletely, for he incorrectly held it to be leafless. The trunk is up to 25 feet tall,
but attains this height only after 120 to 150 years. The Mauritia climbs high up the slope
of Duida, north of the Mission Esmeralda, where I found very beautiful examples. In
damp areas it forms splendid groups of a fresh, gleaming green, reminiscent of the green
of our alders. Through their shade, the trees preserve the moisture of the ground, causing the Indians to believe that the Mauritia draws the water to its roots by means of some
mystical attraction. Based on a similar theory, they advise against the killing of snakes,
for the eradication of the snakes will cause the water holes (*lagunas*) to dry up. This
is how the rough Man of Nature confuses cause and effect. Gumilla calls the *Mauritia
flexuosa* of the Guarani the Tree of Life, *arbol de la vida*. In the Roraima mountain group
east of the sources of the Orinoco, it grows at elevations as high as 4,000 feet.

On the rarely visited banks of the Rio Atabapo in the interior of Guyana we discovered a new species of Mauritia with thorny trunks (boles), our *Mauritia aculeate* (Humboldt, Bonpland, and Kunth, *Nova genera et species Plantarum*, vol. I, p. 310).

32. An American Stylite

The founder of the Stylite sect, the fanatic pillar-saint Simeon Sisanites, whose father was a Syrian shepherd, is said to have spent 37 years in holy contemplation upon five pillars of successively greater height. He died in the year 461. The last pillar upon which he lived was 40 ells high. For seven hundred years there were people who imitated this way of life and were called *sancti columnares* (pillar-saints). Even in Germany, in the small city of Trier, there were attempts to found an elevated cloister, but the bishops opposed the dangerous undertaking (Mosheim, *Institut. Hist. Eccles.*, 1775, p. 215).

33. Cities on the rivers of the steppes

Families that live from raising livestock rather than farming have gravitated together in the middle of the steppe into small cities that would barely be considered villages in the cultivated portions of Europe, cities like Calabozo (lying at 8°56′14″ north latitude and 4°40′20″ west longitude, by my astronomical observations), Villa del Pao (8°38′1″ lat., 4°27′47″ long.), San Sebastian, etc.

34. In the form of funnel clouds

The strange phenomenon of these sandspouts, an analogy of which can be seen at any crossroads in Europe, is especially common to the Peruvian sand desert between Amotape and Coquimbo. Such a thick cloud of dust can be dangerous to the traveler who does not with foresight avoid it. It is curious that these separate, opposing air currents occur only at times of relative calm. The "ocean of the air" is in this respect similar to the sea; for in the sea too there are small currents where the water often ripples audibly as it moves (*filets de courant*), but it is noticeable only at dead calm (*calme plat*).

35. The smothering heat of the air

In the Llanos de Apure, at the Guadalupe dairy farm, I have observed that the Réaumur thermometer climbed from 27° to 29° whenever the hot wind began to blow from the nearby desert, which was covered only by sand and parched turf. In the middle of the dust cloud, the temperature was for several minutes 35°. The dry sand in the village of San Fernando de Apure had a heat of 42°.

36. The illusion of the waving surface of water

The well-known occurrence of light refraction, the mirage, is called "thirst of the gazelle" in Sanskrit (see my *Relation historique*, vol. I, pp. 296 and 625, and vol. II, p. 161). All objects appear to be floating and seem to be reflected in the lower layer of air. The entire desert then looks like an immense lake, its surface in wavelike motion. Palm trunks, cattle, and camels occasionally appear upside down on the horizon. During the French expedition in Egypt, this optical illusion often brought the thirsty soldiers to despair. This phenomenon can be observed in all parts of the world. Even the ancients were familiar with this peculiar breaking of light rays in the Libyan Desert. In Diodorus Siculus, I have found mention of wondrous illusions, an African *Fata Morgana*, with even more outlandish explanations regarding the clustering conglomeration of air particles (lib. III, p. 184 Rhod.; p. 219 Wessel.).

37. The melon cactus

The *Cactus melocactus*, which often has a diameter of 10 to 12 inches and generally has 14 ribs. The natural group of cactus species, the entire family of Jussieu's nopalea, was originally exclusive to the New Continent. The cactus appears in many forms: ribbed and melonlike (*Melocacti*), paddle-formed (*Opuntia*), upright and pillarlike (*Cerei*), creeping, snakelike (*Rhipsalides*), or those with leaves (*Pereskia*). Many will climb high up on the mountain slopes. Near the foot of Chimborazo, in the high sandy plain around Riobamba, I found a new variety of pitahaya at elevations up to 10,000 feet (Humboldt, Bonpland, and Kunth, *Synopsis Plantarum aequinoct. Orbis Novi*, vol. III, p. 370).

38. The scene on the steppe suddenly changes

I have attempted to represent the arrival of the rainy season and the symptoms that announce it. The deep and dark blue of the sky is the result of the complete dissolution of vapors in the tropical air. The cyanometer shows a lighter blueness as soon as the vapors begin to condense. The black flecks in the Southern Cross grow less distinct to the extent that the clarity of the atmosphere decreases, and this change announces the approaching rain. In the same way, the bright gleam of the Magellanic Clouds (Nubecula Major and Minor) grows dim. The fixed stars, which previously shone with a steady and unwavering light like that of the planets, now twinkle even at the zenith. (Compare to Arago in my *Relation hist.*, vol. I, p. 623.) All of these occurrences are the effects of the increasing water vapor suspended in the atmosphere.

39. One sees the clay slowly lift itself in clods

Drought brings out in plants and animals the same appearances as does the removal of stimulating warmth. During drought, many tropical plants drop their leaves. The crocodiles and other amphibians hide themselves in the clay. They lie, seemingly dead, just as they do when the cold sends them into hibernation. (See my *Rel. hist.*, vol. II, pp. 192 and 626.)

40. A vast inland sea

Nowhere are the floods more widespread than in the network of rivers formed by the Apure, the Arachuna, the Pajara, the Arauca, and the Cabuliare. Large vessels sail across the land here, for 10 to 12 miles over the steppe.

41. To the mountainous plain of Antisana

The mountainous plain that surrounds the Antisana volcano has a height of 2,107 toises (12,642 ft) above sea level. The air pressure there is so low that the wild bulls, when the dogs are set on them, lose blood out of the nose and mouth.

42. Bera and Rastro

I have described this method of trapping the gymnotids in other works in detail (*Observations de Zoologie et d'Anatomie comparée*, vol. I, pp. 83–87, and *Relation historique*, vol. II, pp. 173–190). In trials without a chain conducted on a living gymnotid that was still quite powerful upon its arrival in Paris, Mr Gay-Lussac and I were completely successful. The discharge depends solely upon the will of the animal. We did not observe light flashing, but other physical scientists often have.

43. Awakened by the contact of moist and dissimilar parts

In all living organisms, dissimilar parts lie in contact with one another. In all of them are paired the fixed and the fluid. Wherever are found organism and life, electrical tension or the play of voltaic piles appears, as is shown by the experiments of Nobili and Mateucci, and even more by the recent admirable work of Emil Dubois. This last physical scientist has succeeded in "substantiating the presence of electrical muscle currents in living and completely unharmed bodies of animals." He shows "how the human body, by means of a copper wire, can at will move a magnetic needle back and forth at a distance" (*Untersuchungen über thierische Electricität*, by Emil du Bois-Reymond, 1848, vol. I, p. XV). I was a witness to these movements brought forth at will, and I see a great, unexpected light cast upon phenomena to which I arduously and hopefully devoted so many youthful hours.

44. Osiris and Typhon

For information on the battle of the two races of humanity, the Arabian shepherd people of Lower Egypt and the civilized agrarian tribes of Upper Egypt, and about the blond-haired Prince Baby or Typhon, the founder of Pelusium, and the dark-skinned Dionysos or Osiris, see Zoëga's older, now mostly abandoned views in his masterwork *De origine et usu Obeliscorum*, p. 577.

45. The region of European semi-culture

In the *Capitania general de Caracas*, as in the entirety of the eastern part of America, the culture as introduced by Europeans is restricted to the narrow strip of land along the coast. In Mexico, New Granada, and Quito, however, European customs penetrate far into the country, as far as the spine of the cordilleras. In this latter region, namely, there already existed in the 15th century an early civilization among the people who lived there. Wherever the Spanish found this civilization, they continued with it, regardless of whether its place of origin lay near to or far from the seacoast. The old cities were expanded, and the Indian names with their old meanings were sometimes shortened, sometimes replaced by the names of Christian saints.

46. Lead-colored granite massifs

In the Orinoco, especially in the cataracts of Maypures and Atures (not in the Black River, the Rio Negro), all of the blocks of granite, even the pieces of white quartz, take on wherever the water of the Orinoco touches them a grayish-black coating that penetrates less than 0.01 of a line into the stone. It seems as though one is looking at basalt or at fossils colored by graphite. And this coating indeed seems in fact to contain manganese and carbon. I say "seems," for this phenomenon has not yet been investigated with sufficient rigor. Rozier observed something quite similar on the syenite rocks of the Nile (near Syene and Phila), as did the unfortunate Captain Tuckey on the rocky banks of the Zaire River, and Sir Robert Schomburgk on the Berbice (*Reisen in Guiana und am Orinoko*, p. 212). On the Orinoco, these lead-colored rocks give off harmful vapors when wetted. Their proximity is considered to be a cause of fever (*Rel. hist.*, vol. II, pp. 299–304). Notably, the rivers in South America with black water (*aguas negras*), like those with coffee-brown or yellow-green water, do not discolor the granite rocks; that is, they do not bring forth upon the stone the effect of creating from its components a black or lead-colored coating.

47. The howls of the bearded monkeys, heralding the rain

Some hours before the rain begins, the melancholy howling of the monkeys is heard: the *Simia seniculus*, *Simia beelzebub*, et al. One believes he can hear the storm raging in the distance. The intensity of the noise coming from such small animals can be explained only by the fact that a tree often plays host to a troop of 70 to 80 monkeys. Regarding the vocal cords and bony larynx of these animals, see my anatomical treatise in the first issue of my *Recueil d'Observations de Zoologie*, vol. I, p. 18.

48. Often covered with birds

The crocodiles lie so motionlessly that I have seen flamingos (*Phoenicopterus*) resting on their heads. Meanwhile, the entire body lay there like a log, covered with waterfowl.

49. Down its expanding throat

The saliva with which the boa covers its prey promotes quick decomposition. It causes the muscle tissue to become gelatinous, so that the snake is able to pass entire limbs of its kill down its expanding throat. The Creoles are thus persuaded to name the giant snake *Tragavenado*, "deer swallower," so to speak. They tell tales of snakes in whose jaws have been spied antlers that could not be swallowed. On numerous occasions I have seen boas swimming in the Orinoco and in the smaller forest rivers, the Tuamini, the Temi, and the Atabapo. It lifts its head out of the water like a dog. Its skin is covered in gorgeous markings. It is claimed that it reaches 45 feet in length, but the biggest snakeskins that anyone has been able to measure carefully in Europe have not exceeded 20–22 feet. The South American boa (a python) differs from the East Indian variety. (On the Ethiopean boa, see Diodorus, Lib. III, p. 204, ed. Wesseling.)

50. Who eat ants, rubber, and earth

Along the coasts of Cumana, New Barcelona, and Caracas, which are visited by the Franciscan monks of Guyana as they return from their missions, the legend of people on the Orinoco who eat earth is widespread. On our return trip from the Rio Negro (on June 6, 1800), after we had navigated down the Orinoco for 36 days, we spent a day at the mission where the earth-eating Otomacs reside. The little village is called La Concepcion de Uruana and is picturesquely set against a granite cliff. I calculated its geographical position to be 7°8′3″ north latitiude, and by a chronometric reading to be 4ʰ38′38″ west Parisian longitude. The earth that the Otomacs eat is a mild, oily loam, a true potter's clay of a yellow-gray hue, colored with a bit of iron oxide. They choose it carefully, seeking it in certain banks on the edge of the Orinoco and the Meta. They differentiate the taste of different types of earth, and not all loams appeal equally to their taste. They knead this earth into balls of 4 to 6 inches diameter and burn the outside with a weak flame until the crust turns reddish. The ball is moistened again before it is eaten. These Indians are for the most part wild people who abhor cultivating plants. An expression common to the most remote nations of the Orinoco, when describing something as particularly dirty, is "so dirty that the Otomac will eat it."

For as long as the Orinoco and the Meta are at low water, the people live on fish and turtles. The former are taken with arrows when they come to the surface of the water, a hunt whereby we often marveled at the great skill of the Indians. When the streams periodically swelled, the fishing ended, for it is as hard to fish in deep river water as in a deep ocean. During this interval, which lasts from 2 to 3 months, the Otomacs are seen to consume tremendous quantities of earth. We found large stockpiles of it in their huts, pyramidal heaps into which the clay balls were stacked. The knowledgeable monk Fra Ramon Bueno, a native of Madrid who has lived among these Indians for 12 years, assures us that an Indian consumes ¾ to ⅚ of a pound per day. According to the statements of the Otomacs themselves, this earth is their primary source of nutrition during the rainy season. Meanwhile, they will occasionally eat a lizard (when they can get it), a small fish, and a fern root. Yet they have such an appetite for their clay that even in the dry season, when they have plenty of fish to eat, they will still have a bit of earth for dessert after a meal.

These people have a complexion of dark copper-brown. They possess unpleasant Tartaric facial features; they are husky but not big-bellied. The Franciscan monk who lives among them as a missionary assures us that he has noticed no change in the health of the Otomacs during the times when they eat this earth. The simple facts then are these: the Indians consume great quantities of loam with no harm to their health; they themselves consider the loam to be a foodstuff—that is, that upon eating it, they feel full for a long while. They ascribe this fullness to the loam and not to the other sorts of sparse nourishment that they are able to gather here and there along with it. Were one to ask the Otomac about his provisions for winter (winter is usually called "the rainy season" in the hot parts of South America), he would show them the heaps of clay in his hut. But these simple facts still do not answer the questions: Can this loam truly be a foodstuff? Can earth be assimilated? Or does it only serve as ballast in the stomach? Do they merely stretch its walls and thus stave off hunger? To all of these questions I have no answer (*Relation hist.*, vol. I, pp. 618–620). It is notable that the usually credulous and uncritical Father Gumilla flatly denies the eating of earth as such (*Historia del Rio Orinoco*, nueva impr. 1791, vol. I, p. 179). He maintains that the balls of clay are thoroughly mixed with maize meal and crocodile fat. But the missionary Fra Ramon Bueno and our friend and traveling companion, the lay brother Fra Juan Gonzales (who was swallowed by the ocean off the coast of Africa, along with part of our collection), have both assured us that the Otomacs never mix the loam with crocodile fat. In Uruana we heard nothing at all about mixing meal into it.

The earth that we brought back with us and that Vauquelin chemically analyzed is completely pure and unadulterated. Could it be that Gumilla, confusing heterogenous facts, is alluding to the the preparation of bread from the long pod of a certain variety of *Inga*? This fruit is indeed buried in the ground, so that it will rot more quickly. It does seem remarkable to me that the Otomacs can consume so much earth without becoming sick. Has this people been accustomed to this practice for many generations?

In all tropical countries, the people have an astonishing, nearly irresistible desire to eat earth, and not just the so-called alkaline soil (lime) in hopes perhaps of neutralizing acids, but rather, oily, strong-smelling loams. Children must often be kept inside to stop them running outside after a fresh-fallen rain and eating earth. As I watched with astonishment, the Indian women who turn pots in the small village of Banco on the Magdalena River put large pieces of clay in their mouths as they worked. Gilij cites this very thing in *Saggio di Storia Americana*, vol. II, p. 311. The wolves, too, eat earth in the winter, especially loam. It would be very much worthwhile to examine the excrement of all earth-eating humans and animals. Except for the Otomacs, individuals of all other tribes who give in to this peculiar inclination to consume loam for any extended time will grow sick. In the Mission San Borja we found a child who, according to the child's Indian mother, enjoyed eating almost nothing other than earth but who was skeletally emaciated because of it.

Why is it that in the temperate and frigid zones, this morbid desire for earth-eating is so much less common and almost completely restricted to children and pregnant women? In this light, one may maintain that the practice of earth-eating is indigenous to the tropic zones of all parts of the world. In Guinea, the Negroes eat a yellowish earth that they call *Caouac*. When they are brought to the West Indies as slaves, they seek to procure something similar there. Meanwhile, they assert that earth-eating was quite harmless in their African homeland. But the *Caouac* of the American islands makes the slaves sick. For this reason, the eating of earth was long forbidden in the Antilles, even though there was earth (*un tuf rouge, jaunâtre*) being sold secretly at the markets in Martinique in 1751. "The Negroes of Guinea say that, in their country, they normally eat a certain dirt, the taste of which pleases them, without ill effects. Those who overeat of the Caouac are so fond of it that there is no punishment that could stop them from de-vouring this dirt" (Thibault de Chanvalon, *Voyage à la Martinique*, p. 85). On the island of Java between Surabaya and Samarang, Labillardière saw little square reddish cakes being sold in the villages. The natives call them *tana ampo* (*tanah* means "earth" in Ma-layan and Javanese). Upon examining them more closely, he found that they were cakes of reddish clay, made to be eaten (*Voyage à la Recherche de la Pérouse*, vol. II, p. 322). The edible clay of Samarang was very recently (1847) sent by Mohnike to Berlin in the shape of sticks similar to cinnamon and was investigated by Ehrenberg. It is a freshwa-ter formation set down upon on tertiary limestone, consisting of microscopic bacteria (*Gallionella, Navicula*) and phytolites (*Bericht über die Verhandl. der Akad. D. Wiss. zu Berlin aus dem J. 1848*, pp. 222–225). In order to satisfy their hunger, the inhabitants of New Caledonia eat fist-sized pieces of easily pulverized soapstone, in which Vauquelin also found a not inconsiderable copper content (*Voy. à la Rech. de la Pérouse*, vol. II, p. 205). In Popayan and in several parts of Peru, lime soil is offered for sale on the street as a food for the Indians. This lime is enjoyed with coca (the leaves of the *Erythroxylon peruvianum*). Thus do we find the eating of earth spread throughout the torrid zone amongst the lethargic peoples who inhabit the most magnificent and fertile parts of the world. But from the North too, reports have come from Berzelius and Retzius, accord-ing to which hundreds of wagonloads of infusorial earth are eaten yearly by the rural people of the remotest parts of Sweden, more as a hobby (like the smoking of tobacco) than from need. Here and there in Finland, similar soil is mixed into bread. It consists of the empty shells of tiny animals, so small and delicate that one doesn't notice them when biting down, that are filling, but without any actual nutritional value. In times of war, the chronicles and archival documents often mention the consuming of infusorial earth under the uncertain and common name "mountain meal." So it was in the Thirty Years' War in Pomerania (near Cammin), in Lusatia (near Moscow), in the Dessau region (near Klieken), and later, in 1719 and 1733, at the fortress of Wittenberg. (See Ehrenberg, *Über das unsichtbar wirkende organische Leben*, 1842, p. 41.)

51. Pictures carved in stone

In the South American interior, between the 2nd and 3rd degrees of northern lati-tude, there lies a forested plain surrounded by four rivers: the Orinoco, the Atabapo, the Rio Negro, and the Cassiquiare. There are granite and syenite cliffs here that, like those of Caicara and Uruana, are covered with symbolic pictures (colossal figures of crocodiles, tigers, household items, moon- and sun-signs). Meanwhile, this remote corner of the Earth which comprises over 500 square geographical miles is nearly devoid of people. The neighboring tribes live at the lowest level of human civilization, a rabble running about naked and far from having attained the capacity to carve hieroglyphics into stone. It is possible in South America to follow a whole zone of such pictorially adorned cliffs and rocks from the Rupununi, the Essequibo, and the Pacaraima Mountains all the way to the banks of the Orinoco and the Japurá across more than eight degrees of longitude. The carved images may well belong to widely differing epochs, for Sir Robert Schomburgk even found depictions of a Spanish galleon on the Rio Negro (*Reisen in Guiana und am Orinoko*, translated by Otto Schomburgk, 1841, p. 500): thus an origin later than the beginning of the 16th century, and in a wilderness where the natives were then undoubtedly just as coarse as they are now. What I said in another work should not be forgotten here, namely, that peoples of greatly differing derivation, driven by an innermost spiritual disposition toward the rhythmic repetition and arrangement of the pictures, can bring forth similar signs and symbols with the same coarseness and the same tendency toward simplification and generalization of shapes. (Compare *Relation historique*, vol. II, p. 589, with Martius, *Über die Physiognomie des Pflanzenreichs in Brasilien*, 1824, p. 14.)

At the meeting of the Society of Antiquaries in London on November 17th, 1836, a memoir by Mr. Robert Schomburgk was read, concerning the religious myths of the Macushi Indians who inhabit the Upper Mahu and a portion of the Pacaraima Mountains, a nation that for an entire century (since the journey of the bold Hortsmann) has not changed its place of residence. "The Macushis," says Mr. Schomburgk, "believe that the single person who survived a Great Flood repopulated the Earth by turning the stones into people." If this myth, the fruit of the living fantasy of these people, is reminiscent of Deucalion and Pyrrha, it also springs up in a somewhat altered form among the Tamanacs of the Orinoco. If asked how the human race survived this Great Flood (the Mexicans' "Age of Water"), the Tamanacs reply without hesitation, "A man and his wife saved themselves on the peak of the high Tamanacu mountain on the banks of the Asiveru, then threw the fruit of the Mauritia palm over their heads behind them, and from its seeds sprang up men and women; this repopulated the Earth." Some miles from Encaramada, a rock called Tepu-Mereme, "painted rock," rises from the middle of the savanna. Upon it are several figures of animals and symbolic markings that have a great similarity to the ones we saw some distance above Encaramada at Caycara (7°5' to 7°40' lat., 68°50' to 69°45' long.). The same carved rocks can be found between the Cassaquiare and the Atabapo (2°5' to 3°20' lat.), and what is most remarkable, once again in the lonely Sierra Parime, 140 miles to the east. I put this last fact beyond all doubt after seeing a copy, straight from the hand of the famous d'Anville, of the journal of Nicholas Hortsmann of Hildesheim. This artless and humble traveler wrote down for every day and place that which seemed to him worth noting; he deserves to be believed even more, insofar as he looks down upon all that he encounters with a certain contempt and much dissatisfaction, having failed to attain the goal of his explorations: the Dorado Sea, the gold nuggets, and a diamond mine, which turned out to produce nothing more than very clean quartz. On the 16th of April, 1749, before he reaches the region around Lake Amucu, he finds on the banks of the Rupununi, where the river, full of little cascades, winds through the Macarena Mountains, "rocks covered with figures," or, as he puts it in Portuguese, "de varias letras." On the Culimacare rock on the banks of the Cassiquiare we were also shown markings that were said to have been measured out by use of a cord; they were, however, nothing more than poorly formed depictions of celestial bodies, crocodiles, boa snakes, and the implements used for preparing manioc flour. In these decorated rocks (*piedras pintadas*) I found no symmetrical order or regular, proportionally measured characters. The word *letras* in the German surgeon's journal may thus be understood, it seems to me, not in its strictest sense.

Mr. Schomburgk was not so fortunate as to have again found the rocks seen by Hortsmann, but he has described others that lie on the banks of the Essequibo near the Waraputa Fall. "This cascade," he says, "is not famous only for its height; it is also for the great number of figures carved into the stone nearby, which are very similar to the ones that I saw on St. John in the Virgin Islands, and which I consider to be indisputably the work of the Caribs who in earlier times populated this part of the Antilles. I attempted to do the impossible and cut away one of these stones bearing these inscriptions that I might take it with me, but the stone was too hard and the fever had weakened me. Neither threats nor promises could induce the Indians to strike a single hammer's blow to these massive stones, the venerable monuments to the civilization and superiority of their ancestors. They hold the stones to be the work of the Great Spirit, and the various tribes that we met are, despite great distances, still familiar with them. Horror showed on the faces of my Indian companions, who seemed to expect that the fire of heaven could fall upon my head at any moment. I now saw that my efforts were futile, and had to be satisfied with being able to take away a complete drawing of these monuments." This last decision was without a doubt the best one, and to my great pleasure, the publisher of the English journal added the note "It is to be hoped that others have no more luck than Mr. Schomburgk, and that no traveler from a civilized nation will again seek to destroy these monuments of the defenseless Indians."

The symbolic markings that Robert Schomburgk found carved in stone in the Essequibo River valley near the rapids (small cataracts) of Waraputa (Richard Schomburgk, *Reisen in Britisch Guiana*, vol. I, p. 320) are identical, he remarks, to the true Carib markings on one of the smaller Virgin Islands (St. John), but despite the broad dispersal that the incursions of the Carib tribes achieved, and despite the former might of this attractive people, I cannot believe that this enormous belt of carved rocks and cliffs which cuts from west to east across a large piece of South America can be the work of the Caribs. They are more likely the traces of an ancient civilization that belongs, perhaps, to an epoch when the races that we differentiate today were still unknown by name or interrelation. Even the widespread awe in which these rough carvings of the ancients are held indicates that the Indians of today have no concept of how to execute such works. Moreover, between Encaramada and Caycara on the banks of the Orinoco, these hieroglyphic figures frequently are found on cliff faces at such considerable heights as would be attainable only through the use of extraordinarily high scaffolds. If one asks the natives how these figures can have been carved, they answer with a smile, as though relating something that only a white man would be capable of not knowing: "In the days of the great waters, our fathers floated that high in their canoes." This is a geological dream that serves to solve the problem of a long extinct civilization.

I would like to introduce here a comment borrowed from a letter I received from the distinguished traveler Robert Schomburgk: "The hieroglyphic figures are much more widespread than you may have assumed. During my expedition, the objective of which was the investigation of the Corentyne River, I did not only observe some gigantic figures on the Timeri rock (4½° N. lat., 57½° W. long. from Greenwich), but I also discovered similar ones not far from the great cataract of the Corentyne at 4°21′30″ N. lat. and 57°55′30″ W. long. from Greenwich. These figures were completed with much greater industry than any others I have discovered in Guyana. Their height is around 10 feet, and they seem to depict human figures. The headdress is most curious; it encompasses the entire head, spreads out considerably, and is not dissimilar to a halo. I have left drawings of these pictures in the colony and will probably be in a position, once they are collected, to place them before the public. I saw less sophisticated figures at the Cuyuni, which pours into the Essequibo from the northwest at 2°16′ N. lat., and later came upon similar figures on the Essequibo itself, at 1°40′ N. lat. These figures extend, following actual sightings, from 7°10′ to 1° 0′ N. lat. and from 57°30′ to 66°30′ W. long. from Greenwich. The zone of these pictorial cliffs and rocks, as far as it has currently been investigated, spreads over an area of 12,000 square miles (according to the formula of 15 linear miles per degree) and includes the basins of the Corentyne, the Essequibo, and the Orinoco, a circumstance from which one may draw conclusions concerning the former population of this part of the continent."

Other peculiar remnants of bygone culture are the granite vessels, decorated with delicate labyrinths, and the earthen masks, similar to those of the Romans, that have been found among the wild Indians of the Mosquito Coast (*Archaelogia Britannica*, vol. V, 1779, pp. 318–324, and vol. VI, 1782, p. 107). I had engravings of them included in the pictorial atlas that accompanies the historical portion of my trip. Antiquarians are astounded by the similarity that these *à la grecs* have to those that adorn the Palace at Mitla (near Oaxaca in New Spain). I have never seen in Peruvian carvings the large-nosed race that is so frequently depicted in Aztec pictures, as well as in the reliefs on the Palenque of Guatemala. Klaproth remembered having found such oversized noses among the Chalcas, a Northern Mongolian horde. It is generally recognized that many of the tribes of the North American and Canadian copper-colored natives feature handsome aquiline noses, and it is an essential physiognomical mark of differentiation between them and the current inhabitants of Mexico, New Granada, Quito, and Peru. Do the large-eyed, fair-skinned people on America's northwest coast, to whom Marchand refers as being between 54° and 58° latitude, stem from the Uysyns in Inner Asia, an Alano-Gothic race?

52. And yet equipped for murder

The Otomacs often poison their thumbnail with curare. A mere pricking with this nail will become deadly if the curare mixes with the blood. We possess one of the twining plants from whose juices the curare is extracted in the Esmeralda region on the Upper Orinoco. Unfortunately, we did not find the plant in bloom. In its physiognomy, it is related to *Strychnos* (*Rel. hist.*, vol. II, pp. 547–556).

Since I wrote down these notes on the curare (or urare, as both plant and poison were named by Raleigh), the two brothers Robert and Richard Schomburgk have earned themselves great merit for having arrived at an accurate understanding of the nature and preparation of this substance, of which a sizable quantity was first brought to Europe by me. Richard Schomburgk found the climbing plant in blossom in Guyana near the banks of the Pomeroon and Sururu in the region of the Caribs, who are, however, not versed in the method of making the poison. His instructive work (*Reisen in Britisch Guiana*, vol. I, pp. 441–461) contains the chemical analysis of the juice of the *Strychnos toxifera*, which despite its name and its organic structure contains no trace of strychnine, according to Boussingault. Virchow and Münter's interesting physiological experiments confirm that the curare or urare poison does not seem to kill through resorption from without, but primarily only when it is resorbed by living animal tissue after the separation of the cohesion of said tissue; that curare does not belong to the group of tetanus poisons, and that it especially causes paralysis, that is, failing of voluntary muscular movement accompanied by continuing function of the involuntary muscles (heart, gastrointestinal). Compare also the earlier chemical analyses of Boussingault in the *Annales de Chimie et de Physique*, vol. XXXIX, 1828, pp. 24–37.

2

Concerning the Waterfalls of the Orinoco near Atures and Maypures

In the previous chapter, which I used as the subject of an academic lecture, I described the immeasurable flatlands, the natural characteristics of which are so diversely modified by climatic conditions, appearing in one instance as deserts devoid of vegetation and in another as steppes or as far-reaching grassy plains. In contrast to the Llanos in the southern portion of the New Continent are the terrible oceans of sand that lie in the African interior; in contrast to these are the steppes of Central Asia, home to invading shepherd peoples who, driven from the East, once spread barbarism and desolation over the face of the Earth.

While I dared at that time (1806) to unify great land masses into a single portrait of Nature and to discourse before the public upon subjects of a complexion that bespoke the somber disposition of our minds, I will now, restricting myself to a narrower variety of phenomena, attempt to sketch the friendlier picture of luxuriant vegetation and effervescent river valleys. I refer to two nature scenes from the wilderness of Guyana: Atures and Maypures, the waterfalls of the Orinoco—of wide renown, and yet visited by few Europeans before me.

The impression that the sight of Nature leaves within us is determined less by the properties of the region than by the light in which mountain and meadow appear—now in ethereal sky-blue, now in the shadow of low-hanging clouds. In the same way, descriptions of Nature more strongly or weakly affect us depending upon the greater or lesser extent to which they correspond to the needs of our feelings. For in the innermost receptive mind, the physical world is reflected, living and true. That which designates the character of a landscape—the profile of the mountains that border the horizon in the hazy distance, the darkness of the fir forests, the roaring forest river that plummets between overhanging cliffs—all of it stands in an ancient and mysterious association with the disposition of human temperament.

Upon this association rests the nobler part of the enjoyment that Nature provides. Nowhere does she more completely fill us with the sense of her greatness, nowhere does she address us more mightily, than in the world of the tropics—under "the Indian sky," as the climate of the torrid zone was called in the early Middle Ages. If I may thus dare entertain this audience anew with a description of those regions, so might I hope that their inherent charm will not remain unfelt. The memory of a distant and richly endowed land, the sight of a free and powerful growth of vegetation, refreshes and fortifies the mind, much as the upwardly striving spirit, embattled by the present, gladly takes joy from the early age of humanity and its simple grandeur.

Westward currents and tropical winds facilitate the voyage across the peaceful arm of the ocean[1] that fills the broad valley between the New Continent and West Africa. Even before the coast rises from the vaulted ocean bed, one notices a churning of foaming waves cutting across and through one another. Mariners unfamiliar with the region would assume the proximity of shallows or an astonishing eruption of freshwater springs, as may be found in the ocean between the Antillean Islands.[2]

Nearer to the granite coast of Guyana appears the wide estuary of a mighty river that breaks forth like a shoreless sea and covers the ocean around it with freshwater. The river's waves of green, or milky-white in the shallows, contrast with the ocean's indigo blue, which forms a sharp perimeter around the river waves.

The name Orinoco, which was given to the river by its first discoverers and which probably owes its existence to a linguistic confusion, is unknown deep in the country's interior. Primitive peoples distinguish with particular geographical names only such objects as can be confused with others. The Orinoco, the Amazon, and the Magdalena Rivers are simply called "the river," or at best "the great river" or "the great water," while the inhabitants of the banks differentiate the smallest streams with individual names.

The current that is created by the Orinoco between the South American continent and the bitumen-rich island of Trinidad is so powerful that ships sailing into it with a fresh west wind and sails unfurled can hardly overcome it. This bleak and fearsome region is called the "Gulf of Sadness" (*Golfo Triste*). The entrance is the "Dragon's Mouth" (*Boca del Drago*). Here individual cliffs rise up like towers amidst the raging flood, indicating the ancient dam of rock[3] that, now penetrated by the current, once connected the island of Trinidad with the Paria Coast.

The view of this region first convinced the bold explorer Columbus of the existence of an American continent. "Such a gigantic amount of freshwater," concluded this man learned in the ways of Nature, "could come together only in a river of great length. The land that gives forth these waters would have to be a continent, not an island." Just as the companions of Alexander on crossing the snow-covered Parapanisus[4] believed, according to Arrian, that they were seeing in the crocodile-filled Indus a part of the Nile, so did Columbus, unaware of the physiognomic similarities of all products of tropical climates, imagine the coast of the New Continent to be the eastern coast of far-reaching Asia. The mild coolness of the evening air, the ethereal purity of the starry firmament, the balsam scent of blossoms carried on the land wind: all led him to suspect (so says Herrera in the *Décadas*)[5] that he was approaching here the Garden of Eden, the sacred home of the first generation of humanity. To him, the Orinoco seemed to be one of the four rivers that, according to the venerable myth of the world's origin, flowed down from Paradise to divide and water the Earth, now newly adorned with plant life. This poetic passage from Columbus's travel report, or rather, a letter to Ferdinand and Isabella, sent from Haiti (October 1498), is of singular psychological interest. It instructs us once more that the creative imagination of the poet expresses itself in the explorers of the world, as in any of humanity's great characters.

When one considers the amount of water that the Orinoco carries to the Atlantic Ocean, the question arises: which of the South American rivers, the Orinoco, the Amazon, or the Plata, is the largest? The question is ambiguous, as is the concept of size. The broadest estuary is that of the Rio de la Plata, with a width that covers 23 geographical miles. But this river, like those of England, is relatively short. Its insufficient depth already hinders shipping even at the city of Buenos Aires. The Amazon is the longest of all rivers. From its source at Lauricocha Lake to its mouth, its course covers 720 geographical miles. On the other hand, its breadth at the cataract of Rentama in the province of Jaen de Bracamoros, where I measured it beneath the picturesque mountains of Patachuma, is barely equal to the breadth of our Rhine at Mainz.

The Orinoco is narrower at its mouth than either the Plata or Amazon River, and its length, according to my astronomical observations, is only 280 geographical miles. Deep in the interior of Guyana, however, 140 geographical miles from its mouth, I found the river at high water to be over 16,200 feet wide. Its periodical rising lifts the water level annually 28 to 34 feet above the low-water mark. There is as yet insufficient information for an exact comparison of the tremendous rivers that cut across the South American continent. To accomplish this, one would have to become familiar with the profile of each river's bed and with its speed, which can vary so greatly in every area.

The delta that is embraced by the Orinoco's many separate and as yet unexplored arms shows manifold similarities with the Nile in terms of the regularity of its rising and falling and the number and size of its crocodiles. The two rivers are also analogous insofar as both wind their way for a long distance between granite and syenite mountains as rushing sylvan streams and then flow slowly forth, contained now by treeless banks, onto virtually level plains. From the celebrated mountain lake at Gondar in the alpine mountains of Gojam in Abyssinia, an arm of the Green Nile (*Bahr el Azraq*) rolls through the mountains of Shangalla and Sennar down to Syene and Elephantine. In the same way, the Orinoco originates on the southern slope of the mountain range that, stretching along the 4th and 5th degrees of northern latitude, reaches westward from French Guyana toward the Andes of New Granada. The sources of the Orinoco[6] have never been seen by a European; indeed, they have never been seen by a native who has had dealings with Europeans.

When we paddled up the Upper Orinoco in the summer of 1800, we reached, beyond the Esmeralda Mission, the mouths of the Sodomoni and Guapo. Jutting above the clouds here is the summit of Yeonnamari or Duida: a mountain that rises, according to my trigonometric measurements, to 8,278 feet above sea level, and whose aspect is one of the most superb nature scenes that the tropical world has to offer. Its southern slope is a treeless, grassy meadow. There the scent of pineapple fills the humid evening air. Amongst the low-growing meadow herbs rise the bromeliads, their stalks bursting with juices. Beneath the blue-green crown of leaves, the golden-yellow fruit may be seen glowing from afar. Where the mountain waters break forth from the covering grasses, there stand isolated groups of fan palms. In this hot region, their foliage is never moved by cooling currents of air.

East of Duida begins a thicket of wild cacao plants that surround the admirable almond tree *Bertholletia excelsa*, the mightiest product of the tropical world.[7] The Indians collect the materials for their blowguns here—colossal grass stalks, which have segments of more than 17 feet from knot to knot.[8] Some Franciscan monks have penetrated as far as the mouth of the Chiguire, where the river is already so narrow that the natives have woven a bridge of tendrillar plants across it near the waterfall of the Guaharibes. The Guaicas, a people with whitish skin but small in stature, arm themselves with poisoned arrows to defend against further eastward incursions.

Everything that has been proposed about the Orinoco originating in a lake is therefore fantasy.[9] In vain does one seek in Nature the *Laguna Dorada*, which Arrowsmith's maps still show as an inland sea 20 geographical miles long. Could the small, reed-covered Amucu Lake, where the Pirara (a branch of the Mayu) originates, have given rise to this myth? This swamp, however, lies 4 degrees farther east than the region in which it is presumed the source of the Orinoco lies. Erroneously placed within it was the Island of Pumacena, a rock of mica schist, the glittering of which has played since the 16th century a memorable and, to a misguided humanity, often ruinous role in the legend of El Dorado.

According to the legends of many natives, the Magellanic Clouds of the southern sky (the patches of fog that accompany the ship-constellation Argo) are a reflection of the metallic gleam of those silver mountains of the Parima hills. It is also an ancient custom of dogmatic geographers to have every river of considerable size on the planet originate in an inland sea.

The Orinoco belongs among those remarkable rivers that, after a great number of turns to the west and the north, finally runs back to the east in such a manner that its mouth lies on almost the same parallel as its source. From the Chiguire and Gehette to the Guaviare, the course of the Orinoco runs westward, as though it wishes to take its waters to the Pacific. In this stretch it puts forth the Cassiquiare, an unusual arm little known in Europe which joins the Rio Negro (as the natives call it) or Guainia. This is the only example of a bifurcation in the deepest interior of a continent, a natural connection between the two great river valleys, the Orinoco and the Amazon. The nature of the ground surface and the influx of the Guaviare and the Atabapo into the Orinoco cause the latter to turn suddenly to the north. Due to ignorance of the geography, it was for a long time erroneously believed that the Guaviare, flowing from the west, was the true source of the Orinoco. The doubts regarding the possibility of a connection with the Amazon, which were engendered since 1797 by a famous geographer, Mr. Buache, have, I hope, been completely refuted by my expedition. On an uninterrupted boat trip of 230 geographical miles, by way of an extraordinary network of rivers, I succeeded in traveling across the interior of the continent—from the Rio Negro via the Cassiquiare to the Orinoco—from the Brazilian border to the coast of Caracas.

In this upper portion of the river region between the 3rd and 4th degrees of northern latitude, Nature has several times repeated the curious phenomenon of the so-called black water. The Atabapo, with its banks bejeweled by Caro-linias and arborescent Melastomas, the Temi, the Tuamini, and the Guainia are rivers of a coffee-brown color. In the shade of the palms, this color changes to a nearly inky black. In a transparent container, the water is golden-yellow. With marvelous clarity the southern constellations are reflected in these black rivers. Where the water flows smoothly, they offer the astronomer who observes with reflective instruments a most excellent artificial horizon.

Scarcity of crocodiles, and fish as well, greater cooling, fewer plagues of the biting mosquitos, and salubrity of the air characterize the region of the black rivers. They probably owe their unusual color to a solution of carbonized hydrogen, the luxuriance of the tropical vegetation, and the abundance of plants in the ground over which they flow. Indeed, I have noted that on the western slope of Chimborazo, inclining toward the Pacific coast, the waters that spill over from the Rio de Guayaquil gradually take on a golden-yellow or almost coffee-brown color after they have covered the meadows for weeks.

Not far from the shared mouth of the Guaviare and the Atabapo can be found one of the noblest forms of all palm plants, the piriguao,[10] whose 60-foot trunk is adorned with tender, reedy foliage with rippled edges. I know of no palms that bear fruit as large and as beautifully colored. These fruits are similar to peaches, yellow mixed with purplish-red. Seventy to eighty of them form bunches like enormous grapes, of which each trunk will produce three per year. One might call this magnificent plant a "peach palm." The fleshy fruits are for the most part seedless, thanks to the great luxuriance of the vegetation. They thus provide the natives with a nourishing and farinaceous food that, like pisangs and potatoes, may be prepared in a great many ways.

Up to this point, that is, up to the mouth of the Guaviare, the Orinoco runs parallel to the southern slope of the Parima range; but from its left bank southward to far beyond the equator approaching the 15th degree of southern latitude, stretches the tree-covered basin of the Amazon River. The Orinoco now suddenly turns northward at San Fernando de Atabapo, piercing a part of the range itself. Here are found the great waterfalls of Atures and Maypures. The riverbed here is narrowed everywhere by colossal rocks and at the same time broken into separate reservoirs by natural dams.

Before the mouth of the Meta there stands in a mighty whirlpool an isolated cliff, which the natives very aptly refer to as the "Stone of Patience," for at times of low water, it can cost those attempting to ship upstream a delay of two full days. Pressing deep into the countryside, the Orinoco here forms scenic, rocky bays. Across from the Indian mission of Carichana, the traveler is surprised by a remarkable sight. The eye is drawn irresistibly to a craggy granite cliff, el Mogote de Cocuyza, a great block that thrusts 200 vertical feet upward and has upon its flat top a forest of deciduous trees. Like a cyclopean monument of simple grandeur, this cliff rises far above the tops of the surrounding palms. It stands out in sharp relief against the blue of the sky: a forest above the forest.

Upon navigating farther downstream from Carichana, one comes to the point where the river has cut a way through the narrow pass of Baraguan. Here one can discern signs of chaotic devastation all around. Farther to the north toward Uruana and Encaramada rise granite masses of a grotesque appearance. Broken into jagged points and of a brilliant white, they blaze upward from the forest.

In this region, from the mouth of the Apure on, the river leaves the granite range. Moving eastward all the way to the Atlantic Ocean, it separates the impenetrable forests from the grasslands, upon which, at an inconceivable distance, the vault of the heavens rests. Thus does the Orinoco surround on three sides—to the south, the west, and the north—the high Parima mountain range that fills the broad region between the sources of the Jao and the Caura. The river also remains free of cliffs and whirlpools from Carichana all the way to its mouth, with the exception of Hell's Mouth (*Boca del Infierno*) near Muitaco, a whirlpool brought about by rocks that, however, do not dam the entire riverbed as they do at Atures and Mayapures. In this region near the sea, the boatmen know of no other danger than that of the natural rafts, upon which their canoes are often dashed, especially at night. These rafts consist of forest trees from the banks that are pulled up at the roots and borne away by the swelling current. Covered like meadows with blooming water plants, these rafts are reminiscent of the floating gardens of the Mexican lakes.

After this quick overview of the course of the Orinoco and the general circumstances around it, I turn to the description of the waterfalls of Maypures and Atures.

From the high mountain mass of Cunavami, between the sources of the rivers Sipapo and Ventuari, a granite spine thrusts westward toward the Uniama Mountains. From this spine, four streams flow down, confining the cataracts of Maypures: on the east bank of the Orinoco the Sipapo and the Sanariapo, on the west bank, the Cameji and the Toparo. Where the mission town of Maypures lies, the mountains form a sort of wide bay that opens on the southwest.

The river now flows foaming down the eastern slope of the mountain. Far to the west, one can recognize the old, abandoned bank. A wide grassland stretches out between the two ranges of hills. Upon this the Jesuits built a small church of palm trunks. The plain is raised barely 30 feet above the high-water mark of the river.

The geognostic appearance of this region, the insular form of the cliffs Keri and Oco, the hollows that the flood washed out in the first of these hills and that lie at exactly the same height as the holes in the island of Uivitari standing opposite: all of these phenomena indicate that the Orinoco once filled this entire, now dry, bay. The waters probably formed a wide lake for as long as the northern dam provided resistance. Upon the penetration of this dam, the grassy plain now inhabited by the Guarequena Indians emerged, first as an island. It may be that the river for a long time enclosed the Keri and Oco cliffs, which, rising like mountain castles from the old riverbed, provide a splendid sight. As the waters gradually receded, they withdrew completely in the direction of the eastern mountain range.

This assumption is supported by many circumstances. For example, the Orinoco possesses the remarkable characteristic, like the Nile at Philae and Syene, of changing the color of the granite masses around which it has flowed for thousands of years from reddish-white to black. As far as the waters reach, there appears on the rocky banks a lead-colored coating containing manganese and perhaps carbon penetrating the rock surface to a depth of barely one-tenth of a line. This blackening and the hollowing mentioned above are indications of the old water level of the Orinoco. In the Keri cliff, in the islands of the cataracts, in the gneisslike Cumadaminari hill chain that runs above the Island of Tomo, and at the mouth of the Jao, those black cavities may be found 150 to 180 feet above the current water level. Their existence teaches us (as may also be observed, by the way, in all riverbeds in Europe) that the watercourses whose size arouses our admiration today are but weak remnants of the tremendous bodies of water of ages past.

Even the primitive natives of Guyana did not fail to make these simple observations. Everywhere, the Indians brought to our attention the signs of the ancient presence of water. Indeed, in the midst of a grassy plain near Uruana there lies an isolated granite rock into which (according to the narratives of credible men) pictures of the sun, the moon, and many various animals, especially crocodiles and boa snakes, are carved almost in rows at a height of 80 feet. Without scaffolding, no one today can climb up this vertical wall, which deserves the most attentive investigation of future travelers. This is the remarkable position in which are found the hieroglyphic stone-carvings in the mountain regions of Uruana and Encaramada.

If one asks the natives how these pictures can have been carved, they answer: It happened in the time of the high water, for their ancestors then piloted boats at that height. Such a water level was thus contemporaneous with these rude memorials of human artistic endeavor. It is indicative of a former time of very different distribution of water and land, of an earlier stage of the Earth's surface; it is a stage, however, that may not be confused with that time during which the first adorning plants of our planet, the gigantic bodies of extinct land animals, and the pelagic creatures of a chaotic prehistoric world found their graves in the then-hardening crust of the Earth.

The northernmost outflow of the cataracts draws attention to itself through the so-called natural pictures of the sun and the moon. The Keri cliff, to which I have already referred several times, gets its name from a white spot that can be seen from far away, in which the Indians believe they see a conspicuous similarity to the disc of the full moon. I was not able to climb this sheer cliff face myself, but the white spot is probably a very large chunk of quartz formed by converging veins in the gray-black granite.

Across from the Keri cliff on the basaltlike twin mountain of the island of Uivitari, the Indians indicate with mysterious awe a similar disc which they revere as the image of the sun, *Camosi*. Perhaps the geographical juxtaposition of the two cliffs contributed to this appellation, for I did, in fact, find Keri oriented toward the evening and Camosi toward morning. Etymologizing language scholars have thought to discern in the American word *Camosi* some similarity to Camosh, the sun name in one of the Phoenician dialects, along with Apollo Chomeus, or Beelphegor, and Ammon.

Unlike the 140-foot Niagara Falls, the cataracts of Maypures do not consist of a single plunge of a great mass of water. They are also not narrows—passes through which the stream is forced with accelerated speed—like the Pongo de Manseriche in the Amazon River. The cataracts of Maypures appear as a multitude of small cascades that follow one another in a series of steps. The *raudal* (as the Spanish call this sort of cataract) is formed by an archipelago of islands and cliffs that narrow the 8,000-foot-wide riverbed to such an extent that often an opening of barely 20 feet is left for the water to pass through. The eastern side is at the present far more impassable and dangerous than the western.

At the mouth of the Cameji, goods are unloaded so that the empty canoe, or as they say here, the *piragua*, may be piloted by the Indians familiar with the *raudal* to the mouth of the Toparo, where the danger is considered to be past. If the individual cliffs or steps (each of which has been given its own name) are not more than 2 to 3 feet in height, the Indians venture to pilot the canoe down over them; should the direction be upstream, however, they swim ahead and, after many fruitless efforts, manage to get a line around the points of rock that jut out of the swirling waters and then use this line to pull the craft upward. In the course of this strenuous work, the canoe is often filled with water or capsized.

Occasionally, and only in this case do the natives show concern, the canoe is shattered upon the rocks. With bloodied body, the pilots attempt to escape the maelstrom and swim to the riverbank. Where the steps are high, or where the dam of rock has crossed the entire riverbed, the light boat is brought to land and pulled along the near bank on top of cut tree trunks like rollers.

The most celebrated and difficult steps are Purimarimi and Manimi. They have a drop of 9 feet. I was astonished to find through barometric measurements (it is impossible to conduct a geodetic leveling because of the inaccessibility of the location and the pestilential and mosquito-filled air) that the entire vertical drop of the *raudal*, from the mouth of the Cameji to that of the Toparo, amounts to only 28 to 30 feet. I say "astonished" for it becomes clear from this that the terrible roaring and wild foaming of the river are the result of the narrowing of the bed by innumerable rocks and islands, the result of the back current brought about by the form and positioning of the masses of stone. The most convincing way to see the truth of this assertion of the small drop of the entire system of falls is to climb down from the village of Maypures over the Manimi cliff to the riverbed.

This is the spot where one may enjoy a marvelous view. All at once, a roiling mile-long surface offers itself to the eye. From this surface, iron-black cliffs tower like ruins and fortresses. Each island, each stone is adorned by lushly thriving forest trees. Thick fog drifts eternally over the water's surface. Through the steaming cloud from the foam the tops of the tall palms emerge. When a ray of the glowing evening sun penetrates the damp vapors, there begins an optical magic. Colorful rainbows disappear and then return. The ethereal image fluctuates in the play of the airs.

Here and there on the naked rock, the trickling waters of the long rainy season have heaped up islands of topsoil. Bedecked with *Melastoma* and *Drosera*, with small, silver-leaved mimosas and ferns, they form flowerbeds on the desolate stone. To the European, they call to mind the memory of those plant groups that the people of the Alps call *Courtils*: blocks of granite that protrude, lonely and adorned with flowers, out of the Savoyan glaciers.

In the blue distance, the eye comes to rest upon the Cunavami mountain range, a long mountain ridge that ends in a steep and truncated cone. This last (which has the Indian name *Calitamini*) we saw glowing like red fire in the setting sun. This spectacle returns daily. No one has ever been close to this mountain. Perhaps its gleam arises from a reflective decomposition of talc or mica slate.

During the five days that we spent in the area around the cataracts, it was striking how one perceived the roaring of the rushing stream to be three times louder by night than by day. The same phenomenon may be noted at all European waterfalls. What might the cause for this be in a remote area where nothing disturbs the peace of Nature? It is probably the currents of warm rising air which, through the disparate admixture of the elastic medium, impede the propagation of sound, continually breaking up the sound waves, and which then cease this action during the nightly cooling of the Earth's crust.

The Indians showed us traces of wagon tracks. They speak with amazement of the horned animals (oxen) that pulled the canoes on wagons along the left bank of the Orinoco from the mouth of the Cameji to that of the Toparo in the days when the Jesuits were pursuing their business of proselytizing. In those days, the canoes remained loaded and, unlike today, were not worn away by the continual beaching and dragging along on the rough cliffs.

The plan of the surrounding area that I drew up shows that a canal could be opened up from the Cameji to the Toparo. The valley in which those water-rich streams flow is smoothly flat. The canal, the building of which I have recommended to the governor general of Venezuela, would serve as a navigable adjacent arm of the river and would thus render the old, dangerous riverbed unnecessary.

The *raudal* of Atures is quite similar to the *raudal* of Maypures: again, an island world through which the river presses on for a distance of three to four thousand toises, a grove of palms rising right from the middle of the foaming water's surface. The best-known steps of the cataract lie between the islands of Avaguri and Javariveni, between Suripamana and Uirapuri.

When we, Mr. Bonpland and I, returned from the banks of the Rio Negro, we decided to risk navigating the last or lower half of the *raudal* of Atures in the laden canoe. Several times we climbed out onto the rocks, which, like causeways, join island to island. Sometimes the water crashes over these rocks; other times it falls into the hollows in the rocks with a dull roar. Thus whole stretches of riverbed are often dry, for the stream now makes its way through subterranean canals. The golden-yellow cliff hens (*Pipra rupicola*) nest here—one of the most beautiful birds of the tropics, with a double-rowed crest of movable feathers, and as aggressive as the East Indian rooster.

In the Canucari *raudal*, piles of granite boulders form the rock dam. There we crawled into the interior of a cavity, the damp walls of which were covered with conferva and luminescent *Byssus*. With a dreadful roaring, the river rushed on above us. By chance we were presented with the opportunity to observe this great nature scene longer than we could have wished. The Indians left us there in the middle of the cataract. The canoe was supposed to navigate in a long detour around a narrow island, in order to pick us up again. For one and a half hours we waited under a fearful rainstorm. Night fell, and in vain we sought shelter between the cloven masses of granite. The plaintive cries of the little monkeys in woven cages that we had carried with us for months enticed the crocodiles, whose size and lead-gray color indicated great age. I would not mention this occurrence, so common in the Orinoco, had the Indians not assured us that a crocodile had never been seen in the cataracts; indeed, trusting in their assertion, we had even dared several times to bathe in this section of the river.

Meanwhile, the worry increased with each passing moment that we, soaked to the skin and deafened by the thundering of the crashing water, would have to wait through a sleepless tropical night in the middle of the *raudal*—until the Indians finally appeared with our canoe. They had found the step down which they had intended to ride impassable due to the water's being far too low. The pilots were thus compelled to seek a more navigable passage through the labyrinth of channels.

At the southern entrance to the *raudal* of Atures, on the right bank of the river, is the cavern of Ataruipe, famed far and wide amongst the Indians. The surrounding region is Nature of a great and solemn character, making it a suitable place for a national cemetery. One must laboriously scale a sheer wall of granite, not without danger of a great fall. It would hardly be possible to gain a firm foothold on the flat surface were it not for the large feldspar crystals that protrude an inch or more from the stone, defying the erosive effects of weather.

Immediately upon gaining the summit, one is surprised by a wide vista of the surrounding region. Rising from the foaming riverbed are hills ornamented with forest. On the opposite side of the river, beyond the western bank, one's view falls upon the immense grassland of Meta. Looming on the horizon like a bank of growing, threatening clouds are the Uniama mountain group. Such is the distant view; nearer at hand, all is bleak and closed in. In the deeply furrowed valley soar the lonely vulture and the cawing Caprimulgiformes. Their retreating shadows glide across the bare rock walls.

This kettle-shaped valley is surrounded by mountains whose rounded summits support granite boulders of monstrous size. The diameter of these boulders is around 40 to 50 feet. They appear to touch the ground below them with but a single point, as if they would come rolling down upon even the smallest tremor of the Earth.

The rear side of the rocky valley is covered in dense deciduous forest. In this shady spot is the opening of the cavern of Ataruipe—actually not a cavern but a sort of vault, a cliff with a very large overhang, a bight of sorts that the waters wore out in the days when they reached this height. This place is the tomb of an extinct tribe.[11] We counted approximately 600 well-preserved skeletons, each in a basket woven from the stalks of palm fronds. These baskets, which the natives call *mapires*, are in the shape of a sort of four-cornered sack, in different sizes depending upon the age of the deceased. Even newborn children have their own *mapire*. The skeletons are so complete that neither a rib nor a phalange is missing.

The bones are prepared in three ways: some are bleached, some are colored red with *Onoto*, the pigment of the *Bixorellana*, and some, in the manner of mummies, are rubbed with aromatic resins in pisang leaves. The Indians affirm that the fresh corpse is buried for some months in moist earth that gradually absorbs the muscle tissue; it is then exhumed, and any remaining tissue is scraped from the bones with sharp stones. This is still the practice of some tribes in Guyana. Next to the *mapires* or baskets, there are also half-fired clay urns that seem to contain the bones of entire families.

The largest of these urns are 3 feet high and 5½ feet long, of an attractive oval shape, greenish, with handles in the shapes of crocodiles and snakes, and decorated around the rim with twining or labyrinthine designs. These decorations are quite similar to those that cover the walls of the Mexican palace of Mitla. Indeed, they are to be found in all zones, and at the most differing stages of human civilization: among the Greeks and Romans, as well as on the shields of the Tahitians and other Pacific Islanders—everywhere, the rhythmic repetition of regular forms is pleasing to the eye. The causes of these similarities, as I develop further in another work, rest on psychological bases, upon the inner nature of our mental faculties, more than being indicative of shared ancestry and ancient intercourse between the different peoples.

Our interpreters were unable to give us any reliable information regarding the age of these vessels. But the majority of skeletons appeared to be no more than one hundred years old. There is a legend among the Guarequena Indians that relates that the courageous people of Atures, pursued by cannibalistic Caribs, saved themselves on the cliffs of the cataracts—a sad place to settle, where the persecuted tribe and with them their language died out.[12] In the most inaccessible parts of the *raudal* there are similar tombs; it is indeed possible that the last of the Atures people did not die out until recently. For in Maypures (a curious fact), there lives an old parrot, of which the natives maintain that no one can understand him because he is speaking the language of the Atures people.

We left the cavern at nightfall, after collecting several skulls and the complete skeleton of an older man, much to the great irritation of our Indian guides. One of these skulls has been copied by Blumenbach and included in his excellent work on craniology. The skeleton, however, like a great portion of our collection of natural specimens (especially the entomological ones), was lost in a shipwreck on the African coast that also took the life of our friend and former traveling companion, the young Franciscan monk Juan Gonzalez.

As though with a premonition of this painful loss, it was in a somber mood that we left behind us this tomb of an extinct tribe. It was one of those serene and cool nights so common in the tropics. Surrounded by colorful rings, the disc of the moon stood high at the zenith. It illuminated the edge of the fog that, in sharp outline, covered the foaming river like clouds. Innumerable insects poured out their reddish phosphorescent light over the verdant earth. The ground glowed with this living fire, as though the starry roof of the heavens had laid itself down upon the grassland. Twining begonias, fragrant vanilla, and yellow-blossomed Banisteria adorned the entrance to the cavern. Over the graves, the tops of the palm trees rustled.

Thus do the races of men die away. The admirable lore of the different peoples fades away. But with the wilting of each blossom of the spirit, whenever, in the storm of the times, the works of creative art are scattered, so forever will new life sprout forth from the womb of the Earth. Restlessly, procreative Nature opens her buds: unconcerned whether outrageous humanity (a forever discordant race) should trample the ripening fruit.

Annotations and Additions

1. Across the peaceful arm of the ocean

Between the 23rd degree of southern and the 70th degree of northern latitude, the Atlantic Ocean has the form of a long, groove-shaped valley in which the jutting and receding profiles on opposite sides fit into one another. I first developed this idea in my *Essai d'un Tableau géologique de l'Amérique méridionale*, which is printed in the *Journal de Physique*, vol. LIII, p. 61 (*Geognostische Skizze von Südamerika* in Gilbert's *Annalen der Physik*, vol. XVI, 1804, pp. 394–449). From the Canary Islands, especially from the 21st degree of northern latitude and the 25th degree of western longitude to the northeast coast of South America, the surface of the ocean is so peaceful and moves with waves so gentle and low that an open boat could surely navigate it easily.

2. Freshwater springs between the Antillean Islands

On the southern coast of the island of Cuba, southwest of Puerto Batabano in Xagua Bay, but at a distance of about two to three nautical miles from dry land and in the middle of salty ocean water, springs of freshwater burst forth from the seabed, probably as a result of hydrostatic pressure. This eruption occurs with such power that it is only with great caution that canoes will approach the place, famous for the height and crisscrossing motion of the waves there. Trade ships that sail along the coast and do not wish to land sometimes visit these springs in order to take on a supply of freshwater while still remaining out at sea. The deeper one scoops down to get the water, the more potable it is. Also, the "river cow" (*Trichecus manati*) is often taken by hunters there; this animal does not thrive in saltwater. This incredible phenomenon, of which mention has never been made until now, has been most carefully investigated by one of my friends, don Francisco Lemaur, who trigonometrically measured the Bahi de Xagua. I was farther south, in the so-called Gardens of the King on the Jardines del Rey island group in order to determine their positions through astronomical observations; I was not in Xagua itself.

3. The ancient dam of rock

Christopher Columbus, whose restlessly observant mind was directed toward all things, presents in his letters to the Spanish monarchs a geognostic hypothesis regarding the conformation of the Greater Antilles. Intensely occupied with the strength of the often westerly equinoctial current, he ascribes to it the fragmentation of the Lesser Antilles group and the remarkable latitudinally extended configuration of the southern coasts of Puerto Rico, Haiti, Cuba, and Jamaica, which follow the circles of latitude almost exactly. On the third voyage (from the end of May 1498 to the end of November 1500), upon which, in traveling from the *Boca del Drago* to the island of Margarita, and later, from this island to Haiti, he felt the full power of the equinoctial current, the motion of the waters "in agreement with the motion of the heavens" (*movimiento de los cielos*), he expressly states that the violence of the current tore the island of Trinidad from the continent. He refers the monarchs to a map that he presents to them, a *pintura de la tierra* of his own devising, which will later be referred to frequently during the famous legal proceedings against don Diego Colon in regards to the first admiral's rights. "Es la carta de marear y figura que hizo el Almirante señalando los rumbos y vientos por los quales vino á Paria, que dicen parte del Asia" (It is the navigational chart that the admiral did indicating the course and winds by which he came to Paria, said to be part of Asia) (Navarrete, *Viages y Descubrimientos, que hiciéron por mar los Españoles*, vol. I, pp. 253 and 260, vol. III, pp. 539 and 587).

4. On crossing the snow-covered Parapanisus

There are some who see in Diodorus's description of the Parapanisus (Diodorus Siculus, lib. XVII, p. 553, Rhodom.) a depiction of the Peruvian Andes chain. The army marched through settled areas where snow fell daily!

5. Herrera in the Décadas

Historia general de las Indias occidentales, Dec. I, lib. III, cap. 12 (ed. 1601, p. 106); Juan Bautista Muñoz, *Historia del Nuevo Mundo*, lib. VI, c. 31, p. 301; Humboldt, *Examen crit.*, vol. III, p. 111.

6. The sources of the Orinoco

I wrote this about these sources in the year 1807 in the first edition of *Views of Nature*, and I can repeat the assertion today with the same validity, 41 years later. The journeys of the brothers Robert and Richard Schomburgk, journeys so important to all areas of natural science and geographical knowledge, have brought to light other (and more interesting) facts, but the problem of the location of the Orinoco sources has only been approximately determined by Sir Robert Schomburgk. Approaching from the west, Mr. Bonpland and I had pressed on as far as Esmeralda, or the confluence of the Orinoco and the Guapo. Through careful investigation, I was able to describe the upper course of the Orinoco to beyond the mouth of the Gehette, to the small waterfall Raudalde los Guaharibos. Approaching from the east, Robert Schomburgk, having come from the Majonkong Indians' mountainous territory, which, judging by the boiling point of water, he estimated to lie at 3,300 feet of elevation in its inhabited portion, succeeded in reaching the Orinoco via the Padamo, which the Majonkongs and the Guinaos call the Paramu (*Reisen in Guiana*, 1841, p. 448). In my Atlas, I had estimated the position of this confluence of the Padamo and the Orinoco to be 3°12′ lat., 68° 8′ long.; by immediate observation, Robert Schomburgk finds 2°53′ lat., 68°10′ long. The primary objective of this traveler's undertaking was in the discipline of natural history. It was to find the solution to a problem presented in a prize competition, which was sponsored by the Royal Geographical Society in London in 1834: to connect the eastern littoral of British Guyana to the easternmost point that I had reached on the Upper Orinoco. After many tribulations, the attempt to find this solution met with complete success. On the 22nd of February, 1839, Robert Schomburgk arrived, with his instruments, in Esmeralda. His calculations of the longitude and latitude of the place agreed with mine more exactly than I had expected (pp. XVIII and 471). Let us allow the observer himself to speak here: "I cannot find the words to describe the feelings that overwhelmed me as I sprang out onto the bank. My goal was attained, and my observations, which had begun on the coast of Guyana, were now brought to a connection with those of Humboldt at Esmeralda, and I freely admit that at a time when nearly all of my physical powers had left me, when I was surrounded by dangers and difficulties that were of no typical nature, it was only through the hope of recognition from him that I was encouraged to unshakable perseverance in pursuing the goal that I had now achieved. The emaciated appearance of my Indians and faithful guides indicated more clearly than any words could ever describe the difficulties there were to overcome, and that we had overcome." After these words, so kind to me, I must be allowed to insert here the opinions concerning the great undertaking that had been initiated by the London Geographical Society that I expressed in 1841 in the preface to the German edition of Robert Schomburgk's travel writings: "Immediately upon my return from Mexico, I made suggestions as to the direction and ways by which the unknown portions of the South American continent between the sources of the Orinoco, the Pacaraima mountain range, and the seacoast near Essequibo might be opened up. These wishes, which I expressed so avidly in my historical travel reports, have now, after nearly half a century, been to a large extent fulfilled. I have had the joy of experiencing such an important expansion of our geographical knowledge; also the joy of seeing that such a bold and well-led undertaking, demanding the most self-sacrificing endurance, should be accomplished by a young man with whom I feel myself connected as much by a similarity of our efforts as by the connection of a common fatherland. These conditions alone could move me to overcome the aversion and disinclination that I, perhaps unfairly, feel toward introductory prefaces from a strange hand. I needed to express publicly my sincere respect for a talented traveler who was led by one idea: the intention of pushing his way, from east to west, from the valley of the Essequibo to Esmeralda; and who, after five years of effort and suffering (the excessive extent of which I, to some degree, know from my own experience), reached the goal set before him. It is easier to find courage for the daring act of a moment, and it requires less inner strength, than does the long patience of enduring physical suffering, deeply driven by a spiritual interest to move forward, disregarding the certainty of encountering again, but with weakened faculties, the same deprivations on the return trip. A cheerful temperament, possibly the first requirement for an undertaking in inhospitable regions, a passionate love for all classes of scientific work (be they of a natural historical, astronomical, hypsometric, or magnetic nature), a pure sense of the enjoyment that open Nature provides: these are the elements that, when they meet within one individual, will assure the success of a great and important journey."

I will begin with my own assumptions regarding the location of the sources of the Orinoco. The dangerous path taken in 1739 by the surgeon Nicolas Hortsmann of Hildesheim, in 1775 by the Spaniards don Antonio Santos and his friend Nicolas Rodriguez, in 1793 by Lt. Col. don Francisco José Rodrigues Barata of the First Line Regiment of Para, and, according to manuscript maps (for which I am grateful to the former Portuguese ambassador to Paris, Chevalier de Brito), by several English and Dutch colonists who arrived in Para in 1811, by way of the portage of Rupunuri and the Rio Branco of Suriname—this dangerous path divides the terra incognita of the Parima into two uneven halves; in so doing it sets boundaries for a place that is very important to the geography of this region, the place of the sources of the Orinoco: boundaries that one may no longer possibly push back eastward into the blue without intersecting the Rio Branco, which flows from north to south through the river system of the Upper Orinoco, while the Upper Orinoco itself follows an east-west direction. Since the beginning of the 19th century, the Brazilians have, for political reasons, shown a lively interest in the broad plains east of the Rio Branco. See the memoir that I prepared in 1817 at the request of the Portuguese court: *Sur la fixation des limites des Guyanes française et portugaise* (Schoell, *Archives historiques et politiques, ou Recueil de Pièces officielles, Mémoires etc.*, vol. I, 1818, pp. 48–58). Due to the position of Santa Rosa on the Uraricapara, whose course seems to have been quite accurately plotted by the Portuguese engineers, the sources of the Orinoco cannot lie east of the meridian of 65½°. This is the eastern limit beyond which they may not be presumed to be, and considering the condition of the river at the *raudal* of the Guaharibos (above Caño Chiguire, in the land of the unusually white-skinned Guaica Indians, 52′ east of the great Cerro Duida), it strikes me as probable that the upper course of the Orinoco reaches at most the meridian of 66½°. By my calculations, this point is 4°12′ farther west than little Amucu Lake, which was reached by Mr. Schomburgk.

I will now follow my own, older assumptions with those of this gentleman. According to Mr. Schomburgk, the course of the Upper Orinoco east of Esmeralda is oriented from southeast to northwest, as my estimates of the mouths of the Padamo and the Gehette seem to be 19′ and 36′ short in terms of latitude. Robert Schomburgk assumes that the sources of the Orinoco lie at 2°30′ latitude (p. 460), and the fine map *Map of Guyana, to illustrate the route of R. H. Schomburgk*, which is included in the magnificent English work *Views in the Interior of Giana*, sets the geographical position of the sources at 67°18′, i.e., 1°6′ west of Esmeralda, and only 0°48′ Parisian longitude farther west than I believed myself allowed to push against the Atlantic littoral. Through astronomical calculations, Robert Schomburgk placed the nine- to ten-thousand-foot Maravaca mountain at 3°41′ lat. and 68°10′ long. The width of the Orinoco at the mouth of the Padamo, or Paramu, is barely 300 yards, and where it expands west of there to four to six hundred yards, it is so shallow and full of sandbars that the expedition had to dig canals, for the riverbed itself had a depth of barely 15 inches. The freshwater dolphins still showed themselves in abundance: an occurrence in the Orinoco and the Ganges for which zoologists of the 18th century would not have been prepared.

7. The mightiest product of the tropical world

The *Bertholletia excelsa* (Juvia; Brazil-nut tree), of the family *Myrtaceae*, and in the subcategory described by Richard Schomburgk of *Lecythidaceae*, was first described by us in *Plantes équinoxiales*, vol. I, 1808, p. 122, tbl. 36. The enormous, majestic tree, through its coconutlike round fruit with its dense, woody shell that encases the three-sided, again woody, seed pods, presents a most remarkable example of advanced organic development. The Bertholletia grows in the forests of the Upper Orinoco between the Padamo and the Ocamu, not far from the mountain Mapaya, and also between the Amaguaca and Gehette rivers (*Relation historique*, vol. II, pp. 471, 496, 558–562).

8. Grass stalks, which have segments of more than 17 feet from knot to knot

Robert Schomburgk, when he visited the small mountain country of the Majonkongs on his way to La Esmeralda, was so fortunate as to designate the species *Arundinaria*, which provides the material for their blowguns. He says of the plant: "It grows in large clumps, like the *Bambusa*; the first segment rises without knots from the old reed for 15 or 16 feet, and only then produces leaves. The full height of the *Arundinaria* at the foot of the large Maravaca mountain reaches 30 to 40 feet, with a diameter thickness of barely one-half inch. The top is always bowed, and this variety of grass is native only to the sandstone mountains between the Ventuari, Paramu, and Mavaca Rivers. The Indian name is *curata*; because of the excellence of these widely celebrated long blowguns, the Majonkongs and Guinaus of this region have been given the name 'Curata People'" (*Reisen in Guiana und am Orinoko*, p. 451).

9. The Orinoco originating in a lake is therefore fantasy

The lakes of this region, of which theorizing geographers have invented some and exaggerated others, may be divided into two groups. The first of these groups comprises those that have been determined to lie between Esmeralda, the easternmost mission on the Upper Orinoco, and the Rio Branco; to the second group belong the lakes that have been assumed to lie in the tract of land between the Rio Branco and the French, Dutch, and British Guyanas. This overview, which travelers should always bear in mind, indicates that the question of whether there is a Lake Parime east of the Rio Branco other than Lake Amucu (which has been seen by Hortsmann, Santos, Colonel Barata, and Mr. Schomburgk), has nothing to do with the problem of the Orinoco sources. As the name of my famous friend don Felipe Bauza, the former director of the Hydrographic Bureau in Madrid, is so important in the field of geography, I am obligated by the objectivity that should dominate every scientific debate to mention that this learned man was inclined to believe that west of the Rio Branco, rather near to the sources of the Orinoco, there must be lakes. Shortly before his death, he wrote me from London: "I wish that you were here, that I might discuss with you the geography of the Upper Orinoco, which has occupied you so much. I was so lucky as to have saved from complete destruction the documents belonging to Marine General don José Solano, the father of the Solano who perished so sadly at Cadiz. These documents pertain to the drawing up of borders by the Spanish and the Portuguese, to which Solano, in connection with *Chef d'escadre* Yturriaga and don Vicente Doz, had been assigned since 1754. On all of these plans and drawings I see a Laguna Parime, designated sometimes as a source of the Orinoco, other times as being totally separate from these sources. But may one admit, above all, that east and northeast of Esmeralda, some sort of lake exists?"

As the botanist to this last-mentioned expedition, Linné's famous pupil Löffling came to Cumana. He died, after having been to the missions on the Piritu and the Caroni, on the 22nd of February, 1756, at the mission of Santa Eulalia de Murucuri, somewhat south of the confluence of the Orinoco and the Caroni. The documents of which Bauza speaks are the same ones that form the basis of the great map of de la Cruz Olmedilla. They became the model for all maps of South America that appeared before the end of the last century in England, France, and Germany; they also contributed to the two maps drawn in 1756 by Father Caulin, historiographer of the Solano expedition, and by Mr. de Surville, archivist of the State Secretariat of Madrid and an unskilled compiler. The contradiction that these maps present evinces the unreliability of the mapping surveys that arise from that expedition. Moreover, Father Caulin, the expedition's historiographer, unveils with acuity the circumstances that gave rise to the legend of Lake

Parime, and Surville's map, which accompanies his work, not only reinserts this lake (under the name of White Sea and *Mar Dorado*) but also includes another, smaller one, from which arise, fed partly by side branches, the rivers Orinoco, Siapa, and Ocamo. I was able to convince myself on site of the fact that is well known in the missions: that don José Solano had merely passed the cataracts of Atures and Maypures but had not reached the confluence of the Guaviare and the Orinoco at 4°3′ lat. and 70°31′ long., and that the astronomical instruments of the border expedition were not carried as far as the isthmus of Pimichin and the Rio Negro, nor to the Cassiquiare, and, on the Upper Orinoco, not even past the mouth of the Atabapo. This immense country, in which no exact observations had been attempted before my journey, was crossed since Solano's time only by a few soldiers who had been sent out on discovery missions; meanwhile, don Apolinario de la Fuente, whose journals I obtained from the archives of the Province of Quixos, collected without critique everything from the tall tales of the Indians that could flatter the gullibility of Governor Centurion. No member of the expedition saw a lake, and don Apolinario could come no farther than the Cerro Yumariquin and the Gehette.

Now that a dividing line that forms the basin of the Rio Negro has been established within the expanse of this land (to which it would be desirable to direct the inquisitive enthusiasm of explorers), it remains only to remark that for the last century our geographical knowledge of the country west of this valley, between 64° and 68° longitude, has in no way advanced. The attempts that the government of Spanish Guyana has repeatedly made since the expeditions of Iturria and Solano to reach and cross over the Pacaraima Mountains has been crowned only with truly insignificant success. The Spanish, on the way to the missions of the Catalonian Capuchins of Barceloneta at the confluence of the Caroni and the Paragua, in sailing southward up the latter river to its intersection with the Paraguamusi founded at this intersection the Guirion mission, which at the beginning was given the ostentatious name "Ciudad de Guirion." I place it at approximately 4½° northern lat. From there, Governor Centurion, who was incited to seek El Dorado by the exaggerated tales of Paranacare and Arimuicaipi, two Indian chiefs of the mighty Ipurucoto people, extended what they then called the "spiritual conquests" to beyond the Pacaraima Mountains, where he founded the two villages Santa Rosa and San Bautista de Caudacala: the first on the high eastern bank of the Uraricapara, a tributary of the Uraricuera, which I find under the name Rio Curaricara in the travel reports of Rodriguez; the second, six to seven miles farther east-southeast. The astronomer-geographer of the Portuguese Border Commission, Frigate Captain don Antonio Pires de Sylva Pontes Lema, and the captain of engineers don Ricardo Franco d'Almeida de Serra, who from 1787 to 1804 mapped out with the utmost care the entire run of the Rio Branco and its upper branches, call the western portion of the Uraricapara the "Valley of Inundation." They situate the Spanish mission of Santa Rosa at 3°46′ northern lat. and indicate the path that leads north over the mountain range to the Caño Anocapra: a tributary of the Paraguamusi by way of which one may travel from the basin of the Rio Branco to that of the Caroni. Two maps by these Portuguese officers, which contain all details of the trigonometric data concerning the windings of the Rio Branco, the Uraricuera, the Tacutu, and the Mahu, were most helpfully shared with me and Colonel Lapie by Count von Linhares. These valuable, unprinted documents which I used remain in the hands of the learned geographer who long ago had engravings begun at his own cost. The Portuguese sometimes call the entire Rio Branco "Rio Parime"; sometimes they restrict this name to the single tributary, the Uraricuera, somewhat below Caño Mayare and above the old Mission San Antonio. Since the words Paragua and Parime can at once mean "water," "big water," "lake," and "ocean," it should surprise no one to find the words so often repeated among the Omaguas on the Upper Marañon, among the western Guarani, and among the Caribs—in short, among peoples who live very far from one another. In all zones, as I mention above, the greatest rivers, among the people who inhabit their banks, are called "the River," with no other special designations. Paragua, a branch of the Caroni, is also the name that the natives give to the Upper Orinoco. The name *Orinucu* is Tamanac, and Diego de Ordaz heard it spoken for the first time in 1531, when he sailed to the mouth of the Meta. Beside the above-mentioned Valley of Inundation, one may find other large lakes between the Rio Xumuru and the Parime. One of these bays is a tributary to the Tacutu and the other to the Uraricuera. Even at the foot of the Pacaraima Mountains the rivers are susceptible to periodic massive flooding, and Lake Amucu, of which more will be said later, brings this very characteristic to its position at the beginning of the plains. The Spanish missions of Santa Rosa and San Bautista de Caudacacla or Cayacaya, founded in 1770 and 1773 by the governor don Manuel Centurion, were destroyed before the end of the last century, and since that time, no new attempt has been made to penetrate from the basin of the Caroni to the southern slope of the Pacaraima Mountains.

The terrain spreading to the east of the valley of the Rio Branco has in recent years been the object of successful exploration. Mr. Hillhouse has navigated the Massaruni as far as the Bay of Caranang, from which a path, he says, is supposed to have led travelers in two days to the source of the Massaruni and in three days to the feeders of the Rio Branco. In reference to the windings of the large river Massaruni, which Mr. Hillhouse described, he remarks in a letter to me (Demerary, January 1st, 1831): "The Massaruni, measured from its sources on, flows first to the west, then for a degree of latitude to the north, thereafter almost 200 English miles to the east, and finally north and north-north-east to join the Essequibo." Since Mr. Hillhouse was unable to reach the southern slope of the Pacaraima range, he is not familiar with Lake Amucu; in his published report, he himself relates that he, "upon the advice received from the Akawais, who continually travel across the land that lies between the Amazon and the shore . . . became convinced that there is no lake in these regions." This assurance surprised me somewhat; it directly contradicted the ideas that I had gained regarding Lake Amucu—from which, according to the travel reports of Hortsmann, Santos, and Rodriguez (which gave me even more confidence when they agreed completely with the new Portuguese manuscript map), the Caño Pirara was said to flow. Finally, after five years of anticipation, Mr. Schomburgk's journey removed all doubt.

"It is hard to believe," says Mr. Hillhouse, in his interesting memoir about the Massaruni, "that the legend of a great inland sea should have absolutely no basis. In my opinion, the following circumstances can perhaps have contributed to the origination of the belief in the existence of the fabled Lake Parime. At a rather great distance from the rockfall of Teboco, the waters of the Massaruni present to the eye no stronger movement than the quiet surface of a lake. When, in a more or less distant epoch, the horizontal granite beds of Teboco were fully compact and without fissures, the waters must have risen at least 50 feet over their present level, which would then have formed a gigantic lake of 10 to 12 English miles width and 1,500 to 2,000 English miles length" (*Nouvelles Annales des Voyages*, 1836, Sept., p. 316). It is not the extensive length of this flooded region alone that hinders me from supporting this explanation. I have seen plains (Llanos) where the rainy season annually gives the flooding of the feeder streams of the Orinoco a surface area of 400 square geographical miles. The labyrinth of branches between the Apure, Arauca, Capanaparo, and Cinaruco (see maps 17 and 18 of my geographical and physical atlas) disappears then completely; the shape of the riverbeds is obliterated, and everything has the appearance of a tremendous lake. But the locality of the myth of El Dorado and Lake Parime belongs historically to a completely different region of Guyana, that is, to the southern end of the Pacaraima Mountains. As I believe I established in another work (already 30 years ago), the glittering rocks of Mount Ucucuamo, the name of the Parime River (the Rio Branco), the flooding of its tributaries, and especially the existence of Lake Amucu, which lies near the Rio Rupunuwini (Rupunuri) and is connected via the Pirara with the Rio Parime—these have given rise to the legend of the White Sea and El Dorado of Parime.

I have seen with pleasure how the journey of Mr. Schomburgk perfectly confirms these views. The part of his map that shows the course of the Essequibo and the Rupunuri is all new and of great importance to geography. It depicts the Pacaraima range at 3°52′ to the 4th degree of latitude; I had given it a position of 4° to 4°10′. The range reaches the confluence of the Essequibo and the Rupunuri at 3°57′ north latitude and 60°23′ west longitude (calculated, as always, from the Paris meridian); I had placed this point one-half of a degree too far north. Schomburgk gives the name of the latter river as "Rupununi," following the accent of the Macushis; as synonyms, he offers "Rupunuri," "Rupunuwini," and "Opununy," the Cariban tribes here having some difficulty pronouncing the letter *r*. The orientation of Lake Amucu and its relationship to the Mahu (Maou) and the Tacutu (Tacoto) consistently agree with my 1825 map of Colombia. Similarly, we are in agreement regarding the latitude of Lake Amucu. The explorer finds it to be 3°33′, while I believed the point at which I had to stop was 3°35′; but the Caño Pirara (Pirarara), which connects Lake Amucu with the Rio Branco, flows northward out of the lake, not westward. The Sibarana of my map, which Hortsmann has rising near a beautiful mine of rock crystal somewhat north of Cerro Ucucuamo, is the same as the Sibarana of Schomburgk's map. His "Waa-Ekuru" is the same as the "Tavaricuru" of the Portuguese geographer Pontes Leme; it is the tributary of the Rupunuri that most closely approaches Lake Amucu.

The following remarks from the reports of Robert Schomburgk shed some light upon this object that had occupied our attention. "Lake Amucu," says the traveler, "is without dispute the nucleus of Lake Parime and the alleged White Sea. In December and January, when we visited it, it was barely one English mile long, and half-covered with rushes." (This expression was already to be found on d'Anville's map of 1748.) "The Pirara flows from the lake west-northwest of the Indian village of Pirara and falls into the Maou or Mahu. This latter river, according to information I gathered, springs up north of the swell of the Pacaraima Mountains, which in their eastern portions ascend to only 1,500 feet. The sources are located on a plateau upon which the river makes a beautiful waterfall called Corona. We were in the very act of going to visit this place when, on the third day of this excursion, up in the mountains, the illness of one of my companions obliged me to return to the station at Lake Amucu. The Mahu has black (coffee-colored) water, and its current is more rapid than that of the Rupunuri. In the mountains through which it forges its way, it has a width of about 60 yards, and its surroundings are uncommonly picturesque. This valley, like the banks of the Buro Buro, which flows to the Ciparuni, is inhabited by the Macushis. In April, the savannas are entirely flooded and present the peculiar phenomenon of the waters of two different river systems intermingling. The enormous extent of this intermittent flooding has probably given rise to the Myth of Lake Parime. During the rainy season, the interior of the country features a water connection of the Essequibo with the Rio Branco and the Gran Para. Some copses of trees rise like oases from the sand hills of the savanna, and during the time of flooding give the appearance of being islands scattered in a sea; these are without a doubt the Ipomucena Islands of don Antonio Santos."

In the manuscripts of d'Anville, which his heirs most generously permitted me to peruse, I found that the surgeon Hortsmann of Hildesheim, who described these regions with great care, saw a second alpine lake, which he places two days' travel above the confluence of the Mahu and the Parime (Tacutu?) Rivers. It is a black-water lake on the summit of a mountain. He definitively differentiates it from Lake Amucu, which he describes as "covered in rushes." The travel reports of Hortsmann and Santos allow as little for the idea of an enduring connection between the Rupunuri and Lake Amucu as the Portuguese manuscript maps at the Marine Bureau in Rio de Janeiro. In this respect, the drawings of the rivers on d'Anville's maps in the first edition of the *Amerique Meridionale* of 1748 are better than those of the more widely distributed 1760 edition. Schomburgk's journey completely confirms the independence of the basin of the Rupunuri and the Essequibo, but points out that "during the rainy season, the Rio Waa-Ekuru, a tributary of the Rupunuri, becomes connected with the Caño Pirara." This is the situation of these basins of rivers that are not highly developed and are almost completely devoid of separating ridges (combs).

The Rupunuri and the village of Annai (3°56′ lat., 60°56′ long.) are currently recognized as the political border between the British and Brazilian territories of these desolate regions. Mr. Schomburgk, being very ill, was compelled to remain for an extended period in Annai; he bases the chronometric positioning of Lake Amucu on the mean of several lunar distance readings which he measured (from east to west) during his stay in Annai. The longitudes of this traveler for these points on the Parime are in general almost one degree farther east than the longitudes on my map of Colombia. Far from casting doubt upon the lunar readings made in Annai, I wish only to point out that the calculation of these distances becomes important if one wishes to carry over the time reading of Lake Amucu to Esmeralda, which I found to be at 68°23′19″ longitude.

Thus then do we see through recent research how the great *Mar de la Parima*, which had been so difficult to remove from our maps, and which at the time of my return from America was still ascribed a length of 40 miles, has been traced back to Lake Amucu, at a length of two to three English miles! The delusions that were nourished throughout nearly two centuries (the last Spanish expedition to seek El Dorado in 1775 cost more than one hundred people their lives) have ended in such a manner that the study of geography gained some fruits from it. In 1512, thousands of soldiers died on the expedition undertaken by Ponce de Leon to discover the Fountain of Youth on an island in the Bahamas named Bimini, which can hardly be found on our maps. This expedition led to the conquest of Florida and to the recognition of the great ocean current, the Gulf

Stream, that flows through the Bahama Channel. The thirst for treasures and the desire for regained youth: El Dorado and the Fountain of Youth have incited the passions of humanity as though in competition with one another.

10. One of the noblest forms of all palm plants, the piriguao

Compare Humboldt, Bonpland, and Kunth, *Nova Genera Plant. Aequinoct.*, vol. I, p. 315.

11. The tomb of an extinct tribe

While I sojourned in the forests of the Orinoco, investigations of these bone-filled caves were initiated by royal decree. The missionary of the cataracts was falsely accused of having discovered treasure in these caverns, supposedly hidden there by the Jesuits before they fled the region.

12. And with them their language died out

The Atures Parrot has been made the object of a charming poem for which I am indebted to my friend Professor Ernst Curtius, tutor to the young and promising Prince Friedrich Wilhelm of Prussia. He will forgive me if I insert his poem, which was not intended for publication and was merely sent to me in a letter, here at the end of the first volume of *Views of Nature.*

> In the Orinoco forest
> An old parrot sits alone,
> Never stirring, like the poorest
> Little statue carved in stone.
>
> Its course through rock-dams laying
> Foams the river's wild flow,
> While above the palms are swaying
> In the sun's quiescent glow.
>
> How the waves strive on, all acting
> Like their race may yet be won;
> In the water's mist refracting,
> Flash the colors of the sun.
>
> Down below where swells are breaking,
> There a tribe speaks nevermore;
> As the foe their lands were taking,
> Fled to cliffs along the shore.
>
> And the bold Atures perished
> As they lived, both free and brave;
> And the last things that they cherished
> Now lie hidden in a cave.
>
> For the last, now absent members
> Of the tribe the parrot grieves,
> Hones his beak, and he remembers,
> And his cry sounds through the leaves.
>
> Oh, all the boys who trained him
> In the phrases they thought best,

And the women who sustained him
With good food and cozy nest.

Now they all lie dead and broken,
Stretched out on the rocky shore;
Despite every word he's spoken,
He can't wake them anymore.

And now no one comprehends him
When he calls; alone he is.
Hears the water, but it sends him
Not a soul to comfort his.

And the savage who, unwilling,
Spies him, paddles fast to go;
All who see it find it chilling:
The Atures Parrot's woe.

3
The Nocturnal Wildlife of the Primeval Forest

One may assert that the many forms of active appreciation of Nature among different peoples, and the characteristics of the countries these peoples have inhabited or transmigrated, have to varying degrees enriched language. They have enriched it with sharply indicative words for the shapes of mountains, the properties of vegetation, aspects of atmospheric motion, and the form and grouping of clouds. It is also true, however, that many of these descriptive terms, through long use and the vagaries of literature, have abandoned their original meaning. Things that should remain distinct are gradually considered synonymous, and language loses some of the grace and power with which it was able to depict, in its descriptions of Nature, the physiognomic character of the landscape. To show the linguistic richness that intimacy with Nature and the privations of the strenuous nomadic lifestyle can bring forth, I cite the innumerable characteristic terms by which plains, steppes, and deserts are differentiated in Arabic and Persian,[1] according to whether they are completely bare, or covered with sand, or interrupted by cliffs, whether they have isolated pasture areas or feature long stretches of socially growing plants. Almost equally remarkable are the many expressions in Old Castilian dialects[2] for the physiognomy of those mountainous masses whose formations appear everywhere under the sun and indicate, even at great distances, the nature of the stone of which they are composed. Since peoples of Spanish extraction inhabit the slopes of the Andes chain, the mountainous portion of the Canary Islands, the Antilles, and the Philippines, and because the contours of the ground in these places influence the lives of the inhabitants to a greater degree than anywhere else on Earth (with the possible exception of the Himalayas and the Tibetan Plateau), the different terms for mountains in the trachyte, basalt, and porphyry regions, as well as in the slate, limestone, and sandstone mountains, have fortunately remained in daily use. New forms, too, enter the common treasury of language. The speech of humans is enlivened by everything indicative of natural truth, be it in the representation of sensory impressions received from the outer world or of profoundly stirred thought and inner feelings.

The goal of all descriptions of Nature is the ceaseless striving after this truth, both in understanding phenomena and in choosing the descriptive expression. It is most easily achieved by simple narration of what has been observed and experienced directly, through the limiting individualization of the situation on which the narrative hangs. Generalization of physical appearances and enumeration of results belong to the study of the Cosmos, which admittedly remains for us an inductive science. But the living description of organisms, of animals and of plants, within the context of their natural local relationship to the many-faceted surface of the Earth (as a small part of the Earth's collective life) presents the material of that study. Wherever this description can examine great natural phenomena in an aesthetic manner, it excites the mind.

Included among these great natural phenomena is surely the immeasurable forest region in the tropical zone of South America that fills the conjoined river systems of the Orinoco and the Amazon. In the strictest sense of the word, this region earns the name of "primeval" forest, a term that has of late suffered considerable misuse. Phrases using *primeval*, whether describing a forest, a period, or a people, are inexact and for the most part subjective. If every wild forest of densely growing trees upon which man has not yet laid his destructive hand is called a primeval forest, then the phenomenon is native to many parts of the temperate and frigid zones. But if the character lies in impenetrability, in the impossibility, over long stretches, of cutting a path with an ax through trees with a diameter of 8 to 12 feet, then the primeval forest belongs exclusively to the tropics. And it is by no means only the ropelike, tendrillar climbing vines or *lianas* that are, as in the tales told in Europe, the cause of this impenetrability. The lianas make up only a small part of the total mass of the undergrowth. The primary hindrance is created by the bushy growth that fills every open space—in a zone where everything that covers the ground becomes woody. When travelers who have just landed in a tropical region, even perhaps on an island, already think (while still near the coast) that they have pushed their way into a primeval forest, this misconception probably lies in their longing for the fulfillment of a long-cherished wish. Not every tropical forest is a primeval forest. I have almost never used this latter term in my travel works—yet I believe myself to be among those living explorers of Nature, like Bonpland, Martius, Pöppig, and Robert and Richard Schomburgk, who have lived the longest in the primeval forests of the deepest interior of a great continent.

In spite of the conspicuous wealth of terms descriptive of Nature in the

Spanish language, which I mention above, one and the same word, *monte*, is employed for both mountain and forest, for *cerro* (*montaña*) and *selva*. In a work on the true breadth and the longest eastward extension of the Andes chain, I showed how that dual meaning of the word *monte* was the reason behind a beautiful and widely distributed English map of South America having displayed rows of high mountains, rather than forests, standing on the plains. For whereas the Spanish map of La Cruz Olmedilla, which has served as the basis for so many others, had depicted cacao forest, *montes de cacao*,[3] cordilleras sprang up on the English map, even though cacao trees seek only the hottest depressions.

If one looks with a general overview upon the forested region that constitutes all of South America between the grass steppes of Venezuela (*los Llanos de Caracas*) and the Pampas of Buenos Aires, i.e., between 8° north and 19° south latitude, one will recognize that this continuous hylaea forest of the tropical zone is equaled in scope by no other on Earth. It possesses a surface area approximately 12 times that of Germany. Crisscrossed in all directions by rivers and streams, whose branches and tributaries of primary and secondary order occasionally surpass our Danube and Rhine in water volume, this forest owes the extraordinarily luxuriant growth of its trees to the dual beneficial effects of great humidity and heat. In the temperate zone, especially in Europe and Northern Asia, forests can be named after the species of trees that form them by growing together as social plants (*plantae sociales*). In the oak, fir, and birch forests of the North, in the linden forests of the East, a single species of Amentaceae, conifer, or Tiliaceae usually predominates; occasionally one coniferous species will be mixed together with a hardwood. Such homogeneity of species is unknown in the tropical forests. The immense diversity of blossoming forest flora forbids the question "What makes up the primeval forest?" An inconceivable number of families grow side by side here; even in small spaces, few species are found exclusively among their own kind. With each day, with each change of stopping-place, the traveler is met with new forms; often he sees blossoms that, though his attention is drawn by the shape of their leaves and their manner of branching, are simply beyond his reach.

The rivers, with their innumerable smaller arms, are the only paths through this country. Astronomical observations, or, lacking these, compass readings of the river bends, have on many occasions shown, in the region between the Orinoco, the Cassiquiare, and the Rio Negro, how there can be two lonely mission villages lying within the space of some few miles whose monks, following the windings of small streams in canoes fashioned from hollowed-out tree trunks, require one and a half days to go and visit one another. But the most conspicuous evidence of the impenetrability of certain parts of the forest is illustrated by the habits of the great American tiger, the pantherlike jaguar. These predators, thanks to the introduction of European cattle, horses, and mules, have been able to find bountiful nourishment in the Llanos and Pampas, those vast, treeless grasslands of Varinas, of the Meta, and of Buenos Aires, and thanks to this uneven conflict with the cattle herds they have, since the discovery of America, greatly increased their numbers there. But some individuals of this same breed lead a demanding life within the thicket of the forests, near to the sources of the Orinoco. The painful loss of a large dog of the German mastiff variety (our most faithful and friendly travel companion) at a bivouac near where the Cassiquiare flows into the Orinoco would later compel us, as we were returning from the insect swarms to the Esmeralda Mission, to spend a second night there, searching for the dog in vain and uncertain as to whether he had been savaged by a tiger. Quite nearby, we heard again the cry of the jaguar, probably the very one to which we could attribute the dreadful deed. Since the cloudy skies hindered astronomical observation, we had the interpreter (*lenguaraz*) repeat to us what the natives, our oarsmen, told about the tigers of the area.

Not uncommon among these is the so-called black jaguar, the largest and most bloodthirsty variation, with black, barely visible spots on a dark brown coat. It lives at the foot of the mountains Maraguaca and Unturan. "The jaguars," an Indian of the Durimund tribe related, "through their desire to wander and hunt, lose themselves in such impenetrable parts of the forest that they cannot hunt on the ground, and so live long in the trees, a terror to the monkey families and the Kinkajou with the curling tail [*Cercoleptes*]."

My German journals, from which I take this information, were not completely exhausted in the French travelogue that I published. They also contain a detailed depiction of the nocturnal wildlife—I could say the nocturnal animal voices—of the tropical forests. I consider this depiction especially well suited for a book that will have the title *Views of Nature*. Words that are written down in the presence of the phenomenon, or shortly afterward, can lay claim to more freshness of life than the echoes of later remembrance.

Traveling from west to east by way of the Rio Apure, whose flooding I discussed in the essay concerning steppes and deserts, we were able to reach the bed of the Orinoco. It was at the time of low water. The Apure's average width was barely 1,200 feet, while I found that of the Orinoco at its confluence with the Apure (not far from the granite Curiquima cliff, where I was able to take a line of bearing) to still be over 11,430 feet. But the Curiquima cliff is still, if measured in a straight line, a hundred geographical miles from the sea and from the delta of the Orinoco. One part of the plains, through which the Apure and the Payara flow, is inhabited by tribes of Yaruros and Achaguas. In the mission towns of the monks they are called savages because they desire to live independently. As for the degree to which their morality is primitive, they are quite on the same level as those who, while baptized and living "under the bell (*baxo la campana*)," nevertheless remain strangers to all instruction and to any doctrine.

Onward from the isle *del Diamante*, upon which the Spanish-speaking Zambos cultivate sugarcane, one enters a great and wild Nature. The air was filled with countless flamingos (*Phoenicopterus*) and other waterfowl that, like a dark cloud with an ever-changing outline, lifted themselves into the blue vault of the heavens. The riverbed narrows to a width of 900 feet and forms a perfectly straight canal that is hemmed in on both sides by thick forestation. The edge of the forest presents an unusual sight: before the nearly impenetrable wall of the gigantic trunks of *Caesalpinia*, *Cedrela*, and *Desmanthus*, there arises with great regularity from the sandy bank of the river a low *sauso* hedge. It stands only 4 feet high and consists of a small shrub, *Hermesia castaneifolia*, which constitutes a new species[4] of the family of Euphorbiaceae. Next to the hedge stand a few slender, thorny palms (varieties, perhaps, of the Martinezeia or Bactris), called Piritu and Corozo by the Spaniards. The whole resembles a trimmed garden hedge with gatelike openings at great distances from one another. The large quadrupeds of the forest undoubtedly created these openings for easy access to the stream. One may observe emerging from them, especially in the early morning and at sundown, the American tiger, the tapir, and the peccary (*Pecari*, *Dicotyles*) taking their young to water. Should they wish, upon being disturbed by the passing canoe of an Indian, to withdraw again to the forest, they do not attempt to penetrate the hedge by mere force; instead, one is treated to the sight of the wild animal running some four to five hundred paces between the riverbank and the hedge before disappearing into the next opening. While making our 74-day voyage, during which we were confined, with but few interruptions, to a narrow canoe for 380 geographical miles upon the Orinoco, the Casiquiare, and the Rio Negro, the same spectacle repeated itself for us at many points, each time, I dare say, renewing our delight. There appeared in groups, whether to drink, bathe, or fish, creatures of the most disparate classes: along with the large mammals were multicolored herons, *Palamedeae*, and the proudly strutting *Cracidae* (*Crax alector*, *C. pauxi*). Our pilot, an Indian who had been raised in the home of a clergyman, uttered with a pious expression, "It is like being in Paradise here [*es como en el paraiso*]." But the sweet peace of the primeval golden age does not reign in the Paradise of the American animal world. Instead, the creatures watch for and avoid one another. The capybara, the 3- to 4-foot "water swine"—a colossal version of the common Brazilian guinea pig

(*Cavia aguti*)—is eaten in the water by the crocodile and on land by the tiger. And yet it runs so poorly that we were able several times to overtake and capture individuals from the numerous herds.

We camped one night below the Mission of Santa Barbara de Arichuna, under the open sky as usual, lying on a stretch of sand on the bank of the Apure. The area was closely surrounded by the impenetrable forest. We had difficulty finding dry wood for the fires with which, according to local custom, every bivouac is surrounded to discourage jaguar attacks. The night was mildly humid with a bright moon. Several crocodiles approached the bank. What I observed, I believe, is that the sight of the fire actually attracts them, much as it does our crabs and other water creatures. The oars of our small boats were carefully planted into the ground so that we might attach our hammocks to them. Quiet reigned; one heard only the occasional snorting of the freshwater dolphins,[5] which are native both to the Orinoco river system and (according to Colebrooke) to the Ganges as far as Benares, following one another in long processions.

After 11 o'clock there arose in the forest nearby such a clamor that we were forced to abandon all hope of sleep for the rest of the night. The cries of wild beasts thundered through the woods. Among the many voices that simultaneously gave cry, the Indians could identify only those that might be heard singly after a short pause. There were the monotonous, plaintive howls of the alouattae (howler monkeys), the whining, finely piping tone of the little sapajous, the quavering grumble of the striped night monkeys[6] (*Nyctipithecus trivirgatus*, which I first described), the sporadic cries of the great tiger, the cougar or maneless American lion, the peccary, the sloth, and a host of parrots, parraquas (*Ortelida*), and other pheasantlike birds. Whenever the tigers came near to the forest edge, the dog (that we later lost), who had been barking without interruption, would whiningly seek refuge under our hammocks. Occasionally, the tiger's cry would come down from the top of a tree. In these instances it would always be accompanied by the piping tones of the monkeys, who sought to escape this unusual pursuit.

Should one ask the Indians why this incessant noise should arise on certain nights, they would answer with a smile, "The animals are enjoying the beautiful moonlight; they are celebrating the full moon." To me, the scene appeared to originate merely by chance, developing into a long-extended and ever-amplifying battle of the animals. The jaguar pursues the peccaries and tapirs, who, driven together, crash through the arborescent shrubbery, which hinders their flight. Frightened by this, the monkeys in the treetops add their cries to those of the larger beasts below. Together, they awaken the various breeds of fowl roosting together, and so, gradually, the entire animal world joins in the uproar. Long experience has taught us that in no way is it always the "celebrated moonlight" that disturbs the quiet of the forests. The voices were loudest during times of heavy rainfall, or when, with cracks of thunder, the lightning illuminated the forest interior. Good-natured despite several months of fever-sickness, the Franciscan monk who accompanied us past the cataracts of Atures and Maypures to San Carlos of the Rio Negro and onward to the Brazilian border used to say at nightfall, whenever he feared a storm was coming, "May Heaven grant a quiet night to us, and likewise to the wild beasts of the forest!"

In marvelous contrast to the Nature scenes that I describe here, which were played out for us time and again, stands the midday silence that reigns on unusually hot tropical days. From the same journal I now borrow a memory of the Baraguan strait. Here the Orinoco makes its way through the Parima mountain group. That which is referred to in this peculiar pass as a strait (*Angostura del Baraguan*) is really a water basin, the width of which is still 890 toises (5,340 feet). Aside from an old dry stem of the *aubletia* (*Apeibati bourbou*) and a new apocyne, *Allamanda salicifolia*, there was nothing to be found on the bare cliffs but a very few silvery croton shrubs. A thermometer, observed in the shade but within a few inches of the towering granite cliffs, climbed to over 40° Réaumur. All distant objects had wavelike, shimmering outlines, a result of reflection or optical displacement (*mirage*). Not a breath of air moved the dusty sand. The sun stood at zenith, and the tremendous light that it poured down upon the river, which the river in turn reflected back, sparkling in the gentle motion of its waves, intensified the hazy reddish blush that lay over the distance. All the blocks of stone and naked boulders were covered with innumerable large, thick-scaled iguanas, geckos, and colorfully speckled salamanders. Immobile, their heads lifted and mouths opened wide, they seem to inhale the hot air with delight. The larger animals are hiding now in the thickets of the forest, the birds under the foliage of the trees or within the clefts of the cliffs; if one were to listen now, however, for the quietest tones that come to us in this apparent stillness of Nature, then one perceives close to the ground and in the lower layers of the atmosphere a muffled sound, a whirring and buzzing of insects. Everything announces a world of active, organic powers. In every shrub, in the cracked bark of the trees, in the loose earth where live the hymenoptera, Life audibly stirs. It is one of the many voices of Nature, discernible to the solemn, receptive mind of humanity.

Annotations and Additions

1. In Arabic and Persian

One could cite more than 20 words by which the Arabs designate the steppe (*tanu-fah*) and the desert, whether completely bare or covered with gravelly sand interspersed with pastureland (*sahara, kafr, mikfar, tih, mehme*). *Sahl* is a low-lying plain, while *dakkah* is a desolate high plain. In Persian, *beyaban* is the arid sand-desert (such as the Mongolian *gobi* and the Chinese *han-hai* and *sha-mo*), and *yaila* better describes a steppe covered with grass than one with herbage (like the Mongolian *küdah*, the Turkish *tala* or *tschol*, the Chinese *huang*). *Deshti-reft* is a bare, high plain. (Humboldt, *Relation hist.* vol. II, p. 158.)

2. In Old Castilian dialects

Pico, picacho, mogote, cucurucho, espigon, lomatendida, mesa, panecillo, farallon, tablon, peña, peñon, peñasco, peñoleria, rocapartida, laxa, cerro, sierra, serrania, cordillera, monte, montaña, montañuela, cadena de montes, los altos, malpais, reventazon, bufa, etc.

3. Where the map presented montes de cacao

Compare to my *Rel. hist.*, vol. III, p. 238, regarding the row of hills that some have amplified to the high *Andes de Cuchao*.

4. Hermesia

The genus *Hermesia*, of the *Sauso*, has been described and depicted by Bonpland in our *Plantes équinoxiales*, vol. I, p. 162, tbl. XLVI.

5. Of the freshwater dolphins

These are not ocean dolphins that, like some varieties of *Pleuronectes* (plaice, which always have both eyes on one side of the body), have penetrated high upriver, such as the limande, or common dab (*Pleuronectes limanda*), which have been found to have reached Orleans. In the great rivers of both continents are versions of some ocean species, including dolphins and rays (*Raja*). The freshwater dolphin of the Apure and the Orinoco differs in terms of species from the *Delphinus gangeticus*, as from all ocean dolphins. See my *Relation historique*, vol. II, p. 223, 239, 406–413.

6. The striped night monkeys

This is the *Duruculi* or *Cusi-cusi* of the Casiquiare, which I first described in my *Recueil d'Observations de Zoologie et d'Anatomie comparée*, vol. I, p. 306–311, tbl. XXVIII, according to a drawing that I made after a live model. We later kept this live monkey in the menagerie of the Jardin des Plantes in Paris (op. cit. vol. II, p. 340). Spix also found this unusual little creature on the Amazon River and named it *Nyctipithecus vociferans*.

Potsdam, June 1849

4
Hypsometric Addenda

I am indebted to Mr. Pentland, whose scientific efforts have shed so much light upon the geography and geognostic characteristics of Bolivia, for the following positional determinations, which he shared with me in a letter from Paris after the publication (October 1848) of my large map.

Nevado of Sorata or Ancohuma

	Southern Latitude	Greenwich Longitude	Height (Engl. Ft)
Southern Peak	15°51′33″	68°33′55″	21,286
Northern Peak	15°49′18″	68°33′52″	21,043

Illimani

Southern Peak	16°38′52″	67°49′18″	21,145
Middle Peak	16°38′26″	67°49′17″	21,094
North Peak	16°37′50″	67°49′39″	21,060

The elevation values are, with the exception of the negligible difference of a few feet for the height of Illimani's southern peak, the same as those given on the map of Lake Titicaca. Reduced to the old French units, the highest summit of the Sorata is 19,974 Parisian feet, or 3,329 toises (21,286 English ft); the highest summit of the Illimani 19,843 Parisian feet, or 3,307 toises (21,145 English ft). Mr. Pentland had already presented a sketch of this last mountain, which is visible in all its majesty from La Paz, in the *Journal of the Royal Geographical Society*, vol. V (1835), p. 77, five years after the publication of the first measurements in the *Annuaire du Bureau des Longitudes* (1830), p. 323, the results of which I myself rushed to disseminate throughout Germany (see Hertha, *Zeitschrift für Erd- und Völkerkunde*, Berghaus, vol. XIII, 1829, pp. 3–29). In the Ymarra language, according to Pentland, the Nevado de Sorata, east of the village of Sorata or Esquibel, has the names Ancomani,

Itampu, and Illhampu. In the name "Illimani" can be recognized the Ymarra word *illi*, "snow."

Even while the elevations in the eastern Bolivian range were long overestimated (the Sorata by 3,718 and the Illimani by 2,675 Parisian feet), in Bolivia's western range, according to Pentland's map of Titicaca (1848), there are four peaks east of Arica between the latitudes of 18°7′ and 18°25′ which all surpass Chimborazo's height of 21,422 English or 20,100 Parisian feet. The four peaks are these:

Pomarape	21,700 English	or	20,360 Parisian ft
Gualateiri	21,960		20,604
Parinacota	22,030		20,670
Sahama	22,350		20,971

The investigation I introduced concerning the relationship of a comb of mountains (the average height of the passes) to the highest peaks (the culmination points), which varies so greatly from one chain to another (*Annales des Sciences naturelles*, vol. IV, 1825, pp. 225–253), was applied by Berghaus to the Andes chain in Bolivia. Using Pentland's map, he finds the average elevation of the passes in the eastern range to be 12,672, and in the western range 13,602 Parisian feet. The culmination points have the respective elevations of 19,972 and 20,971 Parisian feet; thus the ratio of pass elevation to summit elevation in the east is 1:1.57, in the west, 1:1.54 (Berghaus, *Zeitschrift für Erdkunde*, vol. IX, pp. 322–326). This ratio, also the measure of subterranean upward thrust, is similar to that of the Pyrenees, yet differs greatly from the outward formation of our Alps, the average pass elevation of which in comparison to the height of Mont Blanc is not as high. This measured ratio in the Pyrenees is 1:1.43; in the Alps it is 1:2.09.

According to FitzRoy and Darwin, however, the elevation of Sahama is exceeded by that of the volcano Aconcagua (southern lat. 32°39′) in the northeast of Valparaiso in Chile by 796 Parisian feet. In August of 1835, the officers of the expedition of the *Adventure* and *Beagle* found Aconcagua to rise to between 23,000 and 23,400 English feet. If one estimates Aconcagua to be 23,200 English feet (21,767 Par. ft), this means it stands 1,667 Parisian feet higher than Chimborazo (FitzRoy, *Voyages of the "Adventure" and Beagle*," 1839, vol. II, p. 481; Darwin, *Journal of Researches*, 1845, pp. 253 and 291). By more recent reckoning (Mary Somerville, *Phys. Geogr.*, 1849, vol. II, p. 425), Aconcagua is listed at 22,431 Par. ft high.

Thanks to the superb work of Charles Frémont (*Geographical Memoir on Upper California, an Illustration of his Map of Oregon and California*, 1848), of Dr. Wislizenus (*Memoir of a Tour to Northern Mexico connected with Col. Doniphan's Expedition*, 1848), of Lts. Abert and Peck (*Expedition on the Upper Arkansas*, 1845, and *Examination of New Mexico in 1846 and 1847*), the body of knowledge concerning the mountain systems lying north of the parallels of 30° and 31°, which are known as the Rocky Mountains and the Sierra Nevada of California, has in very recent times increased considerably in all areas: astrogeographical, hypsometric, geognostic, and botanical. There rules in these North American works a scientific spirit worthy of the most enthusiastic recognition. The remarkable high plain between the Rocky Mountains and the Sierra Nevada of California, the uninterrupted, four to five thousand feet high Great Basin, which I mention above, possesses an enclosed inner system of rivers, hot springs, and salt lakes. None of the rivers, not the Bear, the Carson, or the Humboldt, finds a way to the sea. Depicted by me (led by inferences and combinations) as Lake Timpanogos on the large map of Mexico that I completed in 1804, the "Great Salt Lake" of Frémont's map, 15 geographical miles long from north to south, and 10 wide, is connected with the freshwater, but higher-lying, Utah Lake, into which the Timpanogos or Timpanaozu River flows from the east (lat. 40°13'). If the Timpanogos Lake on my map is positioned not sufficiently to the north or west, the cause lies with the lack at that time of any astronomical positioning of Santa Fe in Nuevo Mexico. The error displaces the western shoreline of the lake by almost 50 arc minutes: a difference of absolute longitude that is less conspicuous when one remembers that my itinerary map of Guanajuato, over a distance of 15 degrees of latitude, could be based only upon resorting to the compass bearings (magnetic surveying) of don Pedro de Rivera (Humboldt, *Essai pol. sur la Nouvelle-Espagne*, vol. I, pp. 127–136). To my talented collaborator Mr. Friesen, who passed away so young, these directions indicated 107°58' for Santa Fe; to me, by other combinations, 107°13'. According to actual astronomical positioning, the true longitude seems to be 108°22'. The relative position of the rock-salt seam "in thick strata of red clay," southeast of the "Great Salt Lake" (Laguna de Timpanogos) with its many islands, and not far from Utah Lake and what is now Fort Mormon, is presented with perfect accuracy on my large Mexican map. I believe I may cite the quite recent report of a traveler who made the first reliable positional determinations in this area: "The mineral or rock salt, of which a specimen is placed in Congress Library, was found in the place marked by Humboldt in his map of New Spain (northern half) as derived from the Journal of the missionary Father Escalante, who attempted (1777) to penetrate the unknown country from Santa Fe of New Mexico to Monterey of the Pacific Ocean. Southeast of the Lake Timpanogos is the chain of the Wha-satch Mountains, and in this at the place

where Humboldt has written *Montagnes de selgemme*, this mineral is found."
(Frémont, *Geogr. Mem. of Upper California*, 1848, pp. 8 and 67; cf. Humboldt, *Essai politique*, vol. II, p. 261.)

This part of the highland, especially the region around Lake Timpanogos, which may be identical to Lake Teguayo, the birthplace of the Aztecs, is of great historical interest. For this people, over the course of their immigration from Aztlan to Tula and the valley of Tenochtitlan (Mexico), set up three stations, where ruins of the *casas grandes* can still be found. The first inhabitation of the Aztecs was at Lake Teguayo, south of Quivira, the second on the Gila River, the third not far from the Presidio de Llanos. Scattered across large areas on the banks of the Gila, Lieutenant Abert again found the same abundance of delicately painted shards of faience and earthenware that had in those same places already astonished the missionaries Francisco Garces and Pedro Fonte. The shards are considered artifacts that point to a time of elevated civilization in the now desolate area. Far to the east of the Rio Grande del Norte, in Taos for example, one may even now find recurrences of the peculiar architecture of the Aztecs and their houses of seven stories. (Compare Abert's *Examination of New Mexico* in the *Doc. of Congress*, no. 41, pp. 489 and 581–605, with *Essai pol.*, vol. II, pp. 241–244.) The Sierra Nevada of California runs parallel to the Pacific littoral, but between the latitudes of 34° and 41°, between San Buenaventura and Trinidad Bay, there runs to the west of the Sierra Nevada a small coastal range, the culmination point of which is the *Monte del Diablo* (3,448 ft). In the narrow valley between this coastal range and the great Sierra Nevada, the Rio de San Joaquin flows from the south, and the Rio del Sacramento from the north. Amidst the diluvial group along the latter lie the rich, currently active gold-panning camps.

Aside from the hypsometric leveling and barometric measurements (as mentioned above) of the area between the confluence of the Kansas River with the Missouri and the Pacific coast, over the tremendous expanse of 28° of longitude, a leveling begun by me in the equinoctial zone of Mexico has now also been successfully extended northward by Dr. Wislizenus to 35°38′, that is, to Santa Fe in Nuevo Mexico. It is an astonishing discovery that, despite what has been long believed, the high plain that forms the broad back of the Mexican Andes chain in no way declines to a low elevation. I present here for the first time, in accordance with the currently available measurements, the leveling from Mexico City to Santa Fe. This latter city lies barely four geographical miles from the Rio Grande del Norte.

| Mexico | 7,008 Par. ft | Ht. |
| Tula | 6,318 | Ht. |

San Juan del Rio	6,090	Ht.
Queretaro	5,970	Ht.
Celaya	5,646	Ht.
Salamanca	5,406	Ht.
Guanajuato	6,414	Ht.
Silao	5,546	Br.
Villa de Leon	5,755	Br.
Lagos	5,983	Br.
Aguas calientes	5,875	(San Luis Potosi 5,714 Par. ft) Br.
Zacatecas	7,544	Br.
Fresnillo	6,797	Br.
Durango	6,426	(Oteiza)
Parras	4,678	(Saltillo 4,917 Par. ft)Ws.
el Bolson de Mapimi	3,600–4,200	Ws.
Chihuahua	4,352	(Cusihuiriachi 5,886 Par. ft)Ws.
Passo del Norte	3577 Par. ft	Ws.
(on the Rio Grande del Norte)		
Santa Fe del Nuevo Mexico	6,612	Ws.

The accompanying letters Ht., Ws., and Br. indicate the barometric measurements that I made, those of Dr. Wislizenus, and those of Chief Mining Councillor Burkart. From Wislizenus we possess profile sketches accompanied by his comprehensive notes: from Santa Fe to Chihuahua over the Passo del Norte; from Chihuahua to Reynosa via Parras; from Fort Independence (somewhat east of the confluence of the Missouri and Kansas Rivers) to Santa Fe. The reckoning is based upon daily, corresponding barometric observations made by Engelmann in St. Louis and Lilly in New Orleans. When one considers that in a north–south direction the latitudinal difference between Santa Fe and Mexico City comprises over 16°, that the distance in a straight, meridional direction, ignoring the uneven surface of the path, covers 240 geographical miles, the question arises: is there, on the entire Earth, a similar ground formation of such immensity and elevation (between 5,000 and 7,000 feet above sea level)? Yet four-wheeled wagons do roll from Mexico City to Santa Fe. This highland, the leveling survey of which I introduce here, is formed by the broad, undulating, spreading back of the Mexican Andes chain itself; it is not the swell of a valley between two mountain chains, as is the case in the Northern Hemisphere with the Great Basin between the Rocky Mountains and the Sierra Nevadas of California, in the Southern Hemisphere with the high plateau of Lake Titicaca between the eastern and western ranges of Bolivia, or of the plain of Tibet between the Himalayas and the Kunlun.

5

Ideas for a Physiognomy of Plants

When a person possessed of an active mind explores Nature, or ponders in imagination the broad range of organic creation, no single one among the manifold impressions that occur to him has so deep and powerful an effect as that of the ubiquitous abundance of life. Everywhere, even near the ice-capped poles, the air rings with the songs of birds or the drone of buzzing insects. Not only the lower layers of the air, where the denser vapors hang, but the upper, ethereally pure layers are inhabited as well. For each time that some-one has ascended to the spine of the Peruvian cordilleras or to the summit of the White Mountain south of Lake Léman, animals were discovered even in these desolate spots. On Chimborazo,[1] nearly eight thousand feet higher than Aetna, we saw butterflies and other winged insects. Even if, driven by vertical air currents, they had merely gone astray to such places as the restive thirst for discovery had led the tentative footsteps of humans, their presence there still demonstrates that the more adaptable nature of animals endures where vegetation has long since met its limits. Higher than the cone of Tenerife, were it set atop the snow-covered spine of the Pyrenees, higher than all of the peaks of the Andes chain, there often soared above us the condor,[2] the giant among the vultures. Ravenous desire to prey upon the soft-wooled vicuñas that gather in herds like the chamois upon the snowy pasture lands lures the mighty birds to this region.

While even the naked eye sees the entire atmosphere to be inhabited, still greater wonders are revealed to the aided eye. Rotifers, brachioni, and a host of microscopic creatures are lifted by the winds from drying waters. Immobile and submerged in apparent death, they float through the air—until the thaw brings them back to nourishing earth, dissolves the hull that encases their transparent, gyrating bodies,[3] and (probably due to the life-giving material contained by all water) breathes new sensibility into their organs. The yel-lowish Atlantic "meteoric dusts" ("dust fogs") that from time to time press far eastward into North Africa, Italy, and Middle Europe from the sea around the

Cape Verde islands are, according to Ehrenberg's brilliant discovery, masses of microscopic organisms with siliceous shells. Many of them float, perhaps for years, in the highest layers of the atmosphere, and occasionally drift down, by way of the upper trade winds or by vertical air currents, engaged in organic self-division and capable of life.

Along with the developed creatures, the atmosphere also carries the germinal form of countless future living things: insect eggs and the eggs of plants that are sent, with crowns of hair or feather, upon long autumnal journeys. Even the life-giving dust, the pollen that among species with separate genders is scattered by the male blossom, is carried by winds and by winged insects[4] over land and sea to the lonely female. Wherever the glance of the researcher of Nature is directed, life or the germ of life has been disseminated.

While the moving sea of air, in which we are immersed and above whose surface we have not the capacity to lift ourselves, gives to many organisms their most vital nourishment, these creatures are also in need of the coarser fare offered only at the bottom of this gaseous ocean. This bottom is of two sorts: the lesser portion is formed by dry land in immediate contact with the air, while the greater portion is in the form of water—perhaps having once, millennia ago, coalesced from gaseous materials by means of electrical fire, and now relegated to unceasing activity within the workshop of the clouds, and within the pulsating vessels of animals and plants. Organic forms descend deep into the interior of the Earth—as deep as ever the atmospheric rainwater can penetrate into excavations or natural caverns. Early on, the realm of cryptogamic subterranean flora became a subject of my scientific works. At the highest temperatures, hot springs nourish small hydropores, confervae, and oscillatoriae. Close to the Arctic Circle, at Great Bear Lake in the New Continent, Richardson observed that the ground, which in summer was frozen to a depth of 20 inches, was adorned with blooming vegetation.

It is undecided where there is the greater variety of life: on the land or in the unsounded ocean. Through Ehrenberg's excellent work *Über das Verhalten des kleinsten Lebens*, in the tropical ocean and within the floating and stationary ice of the South Pole, the sphere of organic life, indeed the horizon of life itself, has broadened before our eyes. Siliceous-shelled polygastria, the coscinodiscuses with their green ovaries, have been found 12° from the pole, alive and encased in ice; in the same way, the Podurellae and the little black glacier flea, *Desoria glacialis*, inhabit narrow tubes in the ice of the Swiss glacier investigated by Agassiz. Ehrenberg has shown that upon several microscopic infusoria (*Synedra, Cocconeis*) still other creatures live in the manner of lice and that in the case of Gallionella, with their tremendous capacity for division and mass development, an invisible creature can produce, in four days, two cubic feet of Bilin polishing slate. In the ocean, gelatinlike sea microbes, now alive, now dying out, appear as glimmering stars.[5] Their phosphorescence transforms the greenish surface of the immense ocean to a sea of fire. I will never forget those tropical nights in the South Sea, when the constellations of the high-flying ship Argo and the setting Southern Cross poured out their mild planetary light from the blue gossamer of the heavens while the dolphins drew their glowing wakes in the foaming flood of the sea.

It is not only the ocean, however, but the waters of the swamps as well that hide countless tiny creatures of extraordinary form. Nearly unrecognizable to our eyes are the Cyclidia, the Euglenae, and the host of nymphs, which can be divided into separate branches, like the Lemna whose shade they seek. Surrounded by an air of a different admixture and ignorant of all light, the spotted ascaris breathe within the skin of the earthworm, the silvery leucophras in the body of the riverbank *Naidinae*, and a pentastoma in the wide-celled lungs of the tropical rattlesnake.[6] There are blood-dwelling creatures in frogs and salmon; indeed, according to Nordmann, there are creatures living within the fluids of fisheyes and in the gills of the bream. Thus are the most hidden realms of creation replete with life. But let us remain here a while on the subject of plant species, for upon their existence rests the existence of animal life. Plants ceaselessly organize the raw material of the Earth, preparing to mix together, through the force of life, that which after a thousand transformations is ennobled into animate nerve fibers. The same scrutiny that we devote to the spreading cover of plant life reveals for us the fullness of the animal life that is preserved and nourished by it.

This carpet spread by the blossom-rich flora over the naked body of the Earth is woven in varying ways: thicker where the sun climbs higher in the ever-cloudless sky, sparser toward the slow-moving poles, where the returning frost sometimes kills the grown bud or catches the ripening fruit. Yet everywhere, man joys in the nourishment from plants. If a volcano should part the boiling flood at the bottom of the sea and suddenly (as once occurred between the Greek isles) shove a slag-covered cliff into the air, or (to cite a more peaceful natural phenomenon) should the harmonious lithophytes[7] erect their alveolate dwellings upon the spine of an undersea mountain until, after thousands of years, looming upward over the surface of the water, they die and leave behind a flat coral island: the organic powers are immediately ready to bring the dead rock back to life. Whatever so suddenly brings the seeds, be it wandering birds, or winds, or the waves of the sea, is difficult to discern, given the great distances to the coasts. But as soon as the air first touches the naked stone in the northern countries, webs of silklike fibers form that look like colorful spots to the naked eye. Several of the spots are bordered by exquisite lines, sometimes single and sometimes double; some are cut through with small furrows dividing them into boxes. Their light color darkens with increased age. The vividly bright yellow turns brown, and the bluish-gray of the Lepraria transforms itself gradually to a dusty black. The borders of the aging covering flow into each other and on the dark ground form new, circular lichens of brilliant white. In this way, one organic web rests in layers upon another, and just as the colonizing group of humans must pass through certain stages of moral development, so too is the gradual spread of plants bound by physical laws. Where tall forest trees now lift their tops to lofty heights, there the soilless stone was once covered in such delicate lichens. Mosses, grasses, herbaceous plants, and shrubs fill the gulf of the long but unmeasured era in between. What lichens and moss accomplish in the North is done in the tropics by *Portulaca*, *Gomphrenae*, and other oily shore plants. Thus the history of the covering of plants and their gradual spread over the barren crust of the Earth has its epochs, as does the history of the migrating animal world.

But if the wealth of life is spread everywhere, organic life is also unceasingly occupied with connecting to new forms those elements liberated by death; this richness of life and its renewal differ, however, depending on differences in latitude. In the frigid zone, Nature periodically becomes frozen. However, fluidity being necessary to life, animals and plants (excepting mosses and other cryptogams) will lie sunk in hibernation here for many months. In a great portion of the Earth, then, only such organic beings have been able to develop as can withstand a considerable deprivation of heat and that are capable, without needles or leaves, of a long interruption of life functions. On the other hand, the nearer one comes to the tropics, the greater the variety of shapes, the gracefulness of forms and color combinations, and the perpetual youth and power of organic life.

This increase can easily be doubted by those who have never left our part of the world, or who have neglected the study of basic geography. When one passes out of our thickly leaved forests of oak and descends from the Alps into Welschland or the Pyrenees Range into Spain, indeed, when one merely directs his gaze to the African coast of the Mediterranean, one may easily but wrongly conclude that hot climates are treeless. But one forgets that Southern Europe had a different appearance when agrarian Pelasgic or Carthaginian peoples first gathered there; one forgets that early human civilization pushed back the forests, and that the drive of nations to re-create has gradually robbed the Earth of the sylvan adornment that so pleases those of us in the North and that is more indicative of the youthful period of our moral civilization than of any other part of history. The great catastrophe by which the Mediterranean took on its form when, as a swelling inland sea, it penetrated the strait of the Dardanelles and the Pillars of Hercules—this catastrophe seems to have robbed the bordering countries of much of their topsoil. What is mentioned by the Greek authors of the Samothracian legends[8] points to how recent this destructive natural change was. Also, in all countries that have been flooded by the Mediterranean and that are characterized by tertiary limestone and lower-stage chalk (nummulites and neocomian), a great portion of the ground surface is bare rock. The picturesque quality of regions of Italy rests primarily upon this delightful contrast between uninhabited, desolate stone and the luxuriant vegetation that springs up amidst it like islands. Where this stone is less fissured, keeping the water to the surface, and is covered with earth (as on the charming shores of Lake Albano), even Italy has its oak forests, as shady and green as the inhabitants of the North could wish.

The deserts beyond the Atlas and the immeasurable plains or steppes of South America may also be viewed as merely local phenomena. The latter may be found, at least during the rainy season, to be covered with grass and low, almost herbaceous mimosas; the former are oceans of sand within the Old Continent's interior, great spaces devoid of plants, surrounded by shores of perpetually green forest. Only solitary fan palms serve to remind the traveler that even these wastes are part of a living creation. In the deceptive play of the light brought about by the radiant heat, one might first see the foot of the palm hovering free in the air, and then see its reverse image reflected in the wavelike, quivering layers of air. To the west of the Peruvian Andes, too, on the coast of the Pacific, it took us weeks to traverse such waterless deserts.

The origin of these large expanses of earth devoid of plants in regions otherwise dominated by mighty vegetation is a little-acknowledged geognostic phenomenon that undoubtedly originated in ancient revolutions in Nature, such as floods or volcanic changes to the Earth's crust. If a region has lost all of its plants, if the sand is shifting and lacks all sources of water, then the hot, vertically rising air hinders the precipitation in clouds,[9] and millennia must pass before organic life presses into the interior of the waste from its green shores.

With this in mind, whoever is able to comprehend Nature with a single look and knows to abstract localized phenomena will see how, with the increase in invigorating heat from the poles to the equator, there is also a gradual increase in organic power and abundance of life. But with this increase, certain beautiful aspects are reserved to each different section of the Earth: to the tropics the diversity and immensity of plant forms; to the North the aspect of meadows and the periodic reawakening of Nature upon the first breaths of the spring airs. Besides its particular advantages, each zone has its own character. The old and profound power of organization, despite a certain liberty in the abnormal development of specific cases, binds all animal and vegetable life forms to firm, perpetually returning types. In the same way that one discerns a certain physiognomy in individual organic beings, just as descriptive botany and zoology, in the strict sense of the word, are the analysis of animal and plant forms, so too is there a physiognomy of Nature that applies, without exception, to each section of the Earth.

What the painter indicates with the expression "Swiss Nature" or "Italian sky" is the vague feeling of this localized character of Nature. The blue of the sky, illumination, fragrance borne from afar, the forms of the animals, the succulence of the plants, brilliance of the foliage, the outline of the mountains: all of these elements determine the overall impression of a region. Indeed, in all zones of the globe the same types of mountains—trachyte, basalt, porphyry shale, and dolomite—form cliff groups of a single physiognomy. The greenstone crags of South America and Mexico are identical to the German Fichtelgebirge, just as, in the case of animals, the Allco, or the original dog breed of the New Continent, is consistent in form with the breeds of Europe. For the inorganic crust of the Earth is everywhere independent of climatic influences, whether because the difference between climates that one encounters with changes of geographic latitude is a phenomenon newer than the stone, or the hardening, heat-conducting, and heat-discharging mass of the Earth provided its own temperature[10] without having captured it from elsewhere. All geologic formations thus belong to all regions and are manifested in the same way in all. Everywhere, the basalt builds twin peaks and truncated cones; everywhere trap-porphyry appears in grotesque rock masses, granite in gently rounded dome shapes. Similar plant forms too, firs and oaks, enwreathe the mountainsides in Sweden just as they do in the southernmost part of Mexico.[11] Yet amidst all this agreement in form, all this similarity in individual outlines, the groupings of these into a whole takes on the most diverse character.

In the same way that mineralogical knowledge of the types of stone is different from orology, so too is the general physiognomy of Nature different from individual natural descriptions. Georg Forster in his travels and in his shorter writings, Goethe in the natural descriptions found throughout his immortal works, Buffon, Bernardin de St. Pierre, and Chateaubriand have all portrayed with inimitable truth particular places on the planet. But such portrayals are not merely suited to provide enjoyment of the noblest sort; no, the knowledge of the natural character of different parts of the world is connected in the most intimate way to the history of humanity and to that of its culture. For even if the beginning of a culture is not determined by physical influences alone, still a culture's very direction, the character of a people, the bleak or cheerful attitude of humanity depend to a great degree on climatic conditions. How powerfully the sky over Greece affected its inhabitants! Where else but in the happy and beautiful region between the Euphrates, the Halys, and the Aegean Sea did the peoples who came to settle awaken so early to moral dignity and more tender sensibilities? And did not our forebears, when Europe was sinking into new barbarism, when religious zeal had suddenly opened the holy Orient, bring back once more from those gentle valleys gentler customs? The poetic works of the Greeks and the rougher songs of the old Nordic tribes owe much of their individual character to the forms of the plants and animals, to the mountains and valleys that surrounded the poets, and to the airs that swirled around them. Who does not feel a different mood in the dark shade of a beech tree, upon hills crowned with lonely firs, or in the middle of a grassy meadow, where the wind rustles in the trembling leaves of a birch? These native forms of plant life call forth within us images that are melancholy, solemnly uplifting, or merry. The influence of the physical world upon the moral, the mysterious interworking of the sensory and the extrasensory, bestows upon the study of Nature, when lifted to higher considerations, a charm that belongs to it alone, and that remains too little acknowledged.

But even if the character of different regions of the world depends on external phenomena—if the outline of the mountains, the physiognomy of the plants and animals, if the blue of the sky, the shape of the clouds, and the clarity of the atmosphere all exert their influence upon the overall impression—it still cannot be denied that the primary determining factor of this impression is the covering of vegetation. The animal organism lacks the mass: the mobility of individual creatures, and often their small size, keeps them from our view. The plant kingdom, however, impresses our imagination through a constant immensity. Its massive dimensions indicate its age, and in individual plants, age and the impression of a constantly renewing strength are paired with one another. The gigantic dragon tree[12] that I observed on the Canary Islands, which possesses a diameter of 16 feet, continues to bear blossoms and fruit as though in a state of perpetual youth. When French adventurers, the Bétencourts, conquered these happy islands at the beginning of the fifteenth century, the dragon tree of Orotava (sacred to the natives, as was the olive in the Acropolis of Athens or the elm in Ephesus) was of the same colossal size as today. In the tropics, a forest of Hymenaea and Caesalpinia stands as the monument to perhaps more than a millennium.

If one considers at a comprehensive glance all of the phanerogamic plant varieties that have so far[13] been incorporated into herbaria, the number of which is now estimated to be more than 80,000, one recognizes among this wondrous multitude certain primary forms to which many others may be traced back. For the determination of these types, upon whose individual beauty, distribution, and grouping the physiognomy of a country's vegetation depends, one need not concentrate (as happens for other reasons in botanical taxonomy) upon the tiny reproductive organs, the perianths, and the fruits, but rather upon the consideration of that which through its sheer mass individualizes the overall impression of a region. Among the primary forms of vegetation there are of course entire families of so-called natural taxonomies. Banana plants and palms, *Casuarinae* and conifers, are also individually itemized among these. But the botanical taxonomist separates a great number of plant groups that the physiognomist finds himself forced to bind together. Where plants appear in great number, the outlines and divisions of the leaves, the shapes of the trunks and twigs, run one into the other. In the background of a landscape, the painter (and it is exactly to the artist's refined feelings toward Nature that this statement applies) differentiates pines or palms from beech trees, but not beech forests from forests of other leafy hardwoods!

The physiognomy of Nature is determined primarily by sixteen plant forms. I am enumerating only those that I observed on my travels through both continents and over the course of years of attention to the vegetation of the various areas between the 60th degree of northern latitude and the 12th degree of southern. Certainly, the numbers of these forms will have noticeably increased once travelers have penetrated more deeply into the interiors of the continents and discovered new genera of plants. In Southeast Asia, in Inner Africa and New Holland (Australia), and in South America from the Amazon to the Chiquitos Province, the vegetation remains completely unknown to us. Imagine if someone were to discover a country where woody sponges, *Cenomyce rangiferina*, or mosses grew into tall trees. *Neckera dendroides*, a German moss, is in fact treelike, and the Bambuseae (treelike grasses), like the tropical treeferns, which are often taller than our lindens and alders, remain for Europeans as astonishing a sight as a forest of tall mosses would be to its first discoverer! The size and degree of development ultimately attained by organisms (plants and animals) belonging to a given family are governed by laws as yet unknown. In each of the great divisions of the animal kingdom— the insects, crustaceans, reptiles, birds, fish, or mammals—bodily dimensions oscillate between certain extreme limits. The measure of variability in size, as determined by observations made up to the present, can be corrected by new discoveries, by seeking out varieties of animals as yet unknown.

Temperature conditions, which are dependent upon latitude, seem above all else to have generated organic development among land animals. The small and slender form of our lizards stretches itself in the South to the colossal, ponderous armored body of the terrible crocodiles. In the immense cats of Africa and America, in the tiger, the lion, and the jaguar, the shape of our little house pet is repeated on a larger scale. If we but penetrate into the Earth's interior, if we stir up the grave sites of plants and animals, not only do the fossils reveal to us a category of forms that contradict the climates of today; they also show us gigantic forms that stand in no less contrast to those that currently surround us than do the dignified, simple, heroic natures of the Hellenic people contrast with that which we, in our time, describe with the term "greatness of character." If the temperature of the planet underwent significant, possibly periodically recurring changes, if the proportions of land to sea and even the height of the atmosphere and its pressure[14] have not always been the same, then the size and shape of the organism may likewise have been subject to manifold changes. Powerful pachyderms, elephantlike mastodons, Owen's *Mylodon robustus*, and the Colossochelys, a land tortoise six feet in height—these once inhabited the forests that then consisted of gigantic lepidodendrons, cactuslike stigmaria, and numerous varieties of cycad. As I am incapable of completely depicting this physiognomy of the aging planet according to its current traits, I will venture only to bring up those characteristics that best describe each plant group. Despite all the richness and flexibility of our mother tongue, it is still a difficult undertaking to represent in words that which better befits the painter's imitative art of depiction. Also, one should avoid the impression of tediousness that any enumeration of individual forms must invariably elicit.

We shall begin with the palms,[15] the tallest and noblest of all plants, for the peoples of the world (their earliest civilization having arisen in the Asiatic world of palms and in the area directly adjoining it) have always accorded the palm the prize for beauty. Tall, slender, ringed, and occasionally spiny shafts end in gleaming, outstretched foliage, sometimes fanning out, sometimes feathery. The leaves are often rippled like grass. The smooth trunk attains a height, as I carefully measured it, of 180 feet. The form of the palm diminishes in size and grandeur as one moves away from the equator toward the temperate zone. Among its native plants, Europe possesses but one representative of this form: the dwarf coastal palm, or chamaerops, which in Italy and Spain can be found as far north as the 44th parallel. The actual palm climate of the Earth has an average annual temperature of between 20½° and 22° Réaumur. But the date palm, which is exported to us from Africa and is considerably less attractive than other varieties in this group, will still vegetate in Southern Europe in regions with a mean temperature of 12° to 13½°. Palm trunks and elephant ribs lie buried in the ground in Northern Europe; the position in which they lie indicates that they probably did not wash up here in the North from the tropics but that during the great revolutions of our planet, the climate and the physiognomy of Nature that it determines were changed many times.

Together with the palms in all parts of the world occur varieties of the pisang or banana, the botanists' scitaminaea and musaceae, *Heliconia*, *Amomum*, *Strelitzia*—a low but succulent, nearly herbaceous stem, from the tip of which rise thin, loosely interwoven, delicately striped and silkily gleaming leaves. Pisang plants are the jewel of humid regions. The nourishment of nearly all of the inhabitants of the world's torrid zone relies upon their fruit. Like the farinaceous cereals or grains of the North, the pisang stalks have accompanied humanity since the earliest childhood of our culture.[16] Semitic legends place the point of this nourishing plant's origin on the Euphrates, others (with more probability) at the feet of the Himalayas in India. According to Greek legend, the fields of Enna were the fortunate fatherland of the cereal grains. While the Sicilian fruits of Ceres, spread by culture across the northern world, do little in their formation of uniform, widely spreading grasslands to beautify the countenance of Nature, the migrating inhabitant of the tropics, on the other hand, propagates in the cultivation of pisang plants one of most glorious and noble of plant forms.

The forms of the Malvaceae[17] and Bombaceae are represented by Ceiba, Cavanillesia, and the Mexican hand tree, *Cheirostemon*: colossally thick trunks with large, velvety, heart-shaped or crenelated leaves and gorgeous crimson blossoms. To this group of plants belongs the monkey bread tree, *Adansonia digitata*, which, while of moderate height, sometimes has a diameter of 30 feet and is probably the most ancient organic monument on our planet. In Italy it is the mallow forms that begin to give the vegetation a particularly southern character.

Our temperate zone in the Old Continent meanwhile is completely without the delicate pinnate foliage of the mimosa[18] forms dominated by *Acacia*, *Desmanthus*, *Gleditsia*, *Porlieria*, and *Tamarindus*. The United States of North America, in which the vegetation is more diverse and luxuriant than in the same latitudes in Europe, is not without this beautiful form. Typical for mimosas is an umbrellalike spreading of the branches, much like the Italian pines. The deep blue of the sky in the tropical climes shimmering through the delicate, feathery leaves creates an extremely picturesque effect.

One plant group whose species are found mostly in Africa comprises the varieties of heath plants.[19] To these belong, in terms of physiognomic character and general appearance, the Epicrideae and the Australian Acacia, the leaves of which are merely leafstalks (phyllodia): a group that has some similarity to the conifers, and which precisely because of this, creates an all the more charming contrast to the conifers thanks to its abundance of bell-shaped blossoms. The arborescent heaths, like some other African plants, reach as far as the northern shore of the Mediterranean. They adorn Welschland and the citrus groves of Southern Spain. I observed them growing at their most luxuriant in Tenerife, on the slopes of Teide Peak. In the Baltic countries and farther north, this plant form is feared as a harbinger of drought and infertility. Our heaths, *Erica (Calluna) vulgaris, E. tetralix, E. carnea*, and *E. cinerea*, are plants that grow socially, against whose progressive march agrarian peoples have battled for centuries with little success. How strange it is that the main representative of the family can be found on only one side of the planet! Of the 300 currently recognized species of *Erica*, only a single variety exists on the New Continent, from Pennsylvania and Labrador up to Nootka and Alaska.

On the other hand, to the New Continent alone belongs the cactus form:[20] sometimes spherical, sometimes with arms, sometimes in tall and crooked pillars standing upright like organ pipes. This group makes the most conspicuous contrast to the forms of the lily and banana plants. They belong among those plants that Bernardin de St. Pierre calls the "vegetable springs of the desert." In the waterless plains of South America, the animals, desperate with thirst, seek out the melon cactus: a ball-shaped plant that grows half-hidden in the sand and whose succulent interior is hidden behind fearsome needles. The columnar cactus stems achieve a height of 30 feet, and, divided like a candelabra and often covered in lichens, they are reminiscent in form of some African Euphorbia.

As these form green oases in deserts otherwise devoid of plants, so do the orchids[21] likewise enliven the tropical tree trunk blackened by the sunlight, or the cracks in the most desolate of rocks. The vanilla form distinguishes itself through light green succulent leaves and by multicolored blooms of remarkable construction. The orchid blooms sometimes resemble winged insects, sometimes the very birds that are attracted by the scent of the nectaries. The life span of a painter would be insufficient, even if concentrating upon only a narrow area, to depict the glorious orchids that adorn the deep-cut valleys of the Peruvian Andes chain.

Like nearly all cactus species, the Casuarineae[22] have a leafless form. They are a plant variety found only in the South Seas and the West Indies—trees with limbs similar to horsetail. But there are traces of this more unusual than attractive variety found in other areas of the world. Plumier's *Equisetum altissimum*, Forskål's *Ephedra aphylla* of North Africa, the Peruvian Colletia, and the Siberian *Calligonum pallasia* are closely related to the Casuarineae.

While the pisang plants accomplish the broadest expansion of the veins in the leaves, it is in the Casuarineae and the conifers[23] that they are most highly contracted. Firs, thujas, and cypresses belong to a northern form that is less common in the tropics and that in some species (*Dammara, Salisburia*) features broad, leaflike needles. Their eternally fresh green cheers the barren winter landscape. It also declares to the people of the polar region that when snow and ice cover the ground, the inner life of plants, like promethean fire, is never quenched on our planet.

Besides the orchids, varieties of Pothos,[24] too, spread parasitically over the aging trunks of the forest trees in the tropical world, much as mosses and lichens do in ours. Succulent herbaceous stalks lift up large leaves that, while they are sometimes arrowlike, sometimes palmate or oblong, are always thick-veined. The blossoms of the Aroideae, to increase their warmth, are wrapped in sheaths; without stems, they put forth roots in the air. Related forms are *Pothos, Dracontium, Caladium,* and *Arum,* the last progressing now as far as the coasts of the Mediterranean, and in Spain and Italy, along with succulent coltsfoot and with tall stands of thistle and *Acanthus,* indicating the luxuriance of southern plant life.

Growing socially with this Arum form are the tropical lianas,[25] displaying in the hot regions of South America the most exquisitely abundant vegetation: *Paullinia, Banisteria,* Bignonias, and Passiflora. Our tendrillar hops and grapevines are reminiscent of this plant form of the tropical world. By the Orinoco, the leafless branches of the Bauhinia often reach a length of 40 feet. They sometimes fall vertically from the tops of high Swieteniae, and sometimes they are suspended aslant like a mast line, and the tiger-cat has an admirable agility in climbing up and down them.

Contrasting with the supple and tendrillar lianas, with their fresh, light green, is the free-standing form of the bluish aloe plants:[26] the stems, if there are any, divide hardly at all and are marked by close-set rings winding around like a snake. At the tip are succulent, fleshy, long, and pointed leaves, bunched together and radiating outward. The tall-stemmed aloe plants do not form copses like other, socially growing plants; they stand alone in barren places, thus often lending to the tropical region its own melancholy (one might say "African") character. The following plants belong to this aloe form, due to the physiognomic similarity in their impressions upon the landscape:

from the Bromeliaceae, the Pitcairnias, which sprout from fissures in the rocks of the Andes chain, the great *Pournetia pyramidata* (Achupalla of the high plains of New Granada), the American aloe (agave), *Bromelia ananas* and *B. Karatas;*

from the Euphorbiaceae, the uncommon varieties with thick, short,
   candelabralike divided stems;
from the family of the Asphodeleae, the African aloe and the dragon tree
   Dracaena draco;
and finally, from the Liliaceae, the high-blooming yucca.

While the aloe form is characterized by a stolid quiet and firmness, the
grass form,[27] especially the physiognomy of the arborescent grasses, is defined by an expression of blithe lightness and mobile slenderness. Bamboo
groves form shady archways in both Indies. The smooth, often bowed and
swaying stem of the tropical grasses exceeds the height of our alders and
oaks. As far north as Italy this form begins to lift itself from the ground in the
Arundo donax, determining by its height and mass the natural character of
the land.

As in the case of the grasses, the form of the ferns[28] in the hot regions of
the Earth is also ennobled. Treelike ferns of up to 40 feet have an appearance
similar to palms, but their trunk is less slender, shorter, and more roughly
scaled than that of the palms. The foliage is more delicate, loosely intermingled, and translucent, with neatly serrated edges. These colossal ferns belong
almost exclusively to the tropics, but they prefer the tropical areas of more
moderate climate to the very hot ones. As the lessening of the heat is a function of increased altitude, it is accurate to describe mountain regions that rise
two to three thousand feet above sea level as the primary habitat of this form.
In South America, tall-stemmed ferns accompany the beneficial tree that provides the bark that is a remedy for fever. Both of these characterize that happy
region of the Earth in which prevails the eternal mildness of spring.

Yet to be named are the lily plants[29] (*Amaryllis, Ixia, Gladiolus, Pancratium*), with leaves like those of reed plants and glorious blooms, a form whose
primary place of origin is Southern Africa; also the willow form,[30] native to
all parts of the world, distinguishing itself on the high plains of Quito in *Schi-
nus molle* not by the shape of its leaves but by the manner of its branching;
the myrtle plants[31] (*Metrosideros, Eucalyptus, Escallonia mytilloides*); and the
Melastoma[32] and Laurel[33] forms.

It would be an undertaking worthy of a great artist to study the character
of all of these plant groups, not in greenhouses or in the descriptions of botanists but in the vast tropics of Nature itself. How interesting and instructive
to the landscape painter[34] would be a work that depicts for the eye each of the
sixteen enumerated primary forms, first individually, and then in contrast to
one another. What is more picturesque than a treelike fern spreading its ten-

derly woven leaves over the Mexican laurel oak? What is more charming than pisang plants shaded by tall Guadua and bamboo grasses? To the artist is left the task of separating the groups, and under his hand the great, magical image of Nature (if I may venture to use the expression) reveals itself, much like the written works of men, in a few simple strokes.

In the glowing sunshine of the tropical sky thrive the most splendid of plant forms. As the bark of trees in the North is covered with dry lichens and moss, just so in the tropics do Cymbidium and aromatic vanilla live upon the trunks of the Anacardia and the giant fig trees. The fresh green of the Pothos leaves and the Dracontia contrasts with the multicolored blooms of the orchids. Twining Bauhinia, Passiflora, and yellow-blossomed Banisteria wind about the trunks of the forest trees. Delicate blossoms unfold from the roots of the *Theobroma* and from the dense, rough barks of the Crescentia and *Gustavia*.[35] In this abundance of blossoms and leaves, this luxuriant growth with its confusion of tendrilous plants, it is often difficult for the natural scientist to recognize which blossoms and leaves belong to which stem. A single tree, adorned with Paullinia, Bignonia, and Dendrobium, comprises a group of plants that, if separated from one another, would cover a considerable space of earth.

The plants in the tropics are more extremely succulent, of a fresher green, and arrayed in larger, glossier leaves than are those in northern latitudes. Socially growing plants, which give to European vegetation such uniformity, are virtually absent at the equator. Trees, nearly twice as tall as our oaks, are resplendent with blossoms as grand and gorgeous as our lilies. On the shady banks of the Magdalena River in South America there grows a twining Aristolochia with flowers four feet in circumference, which the Indian boys at play take and pull over their heads.[36] In the southern portion of the Indian Archipelago, the bloom of the *Rafflesia* has a diameter of almost three feet and weighs over fourteen pounds.

Between the tropical lines, the extraordinary elevation to which rise not only single mountains but entire countries, along with the cold that is the result of this elevation, presents to the inhabitant of the tropics a curious perspective. Besides the palms and pisang trees, he is also surrounded by plant forms that would seem indigenous only to northern countries. Cypresses, firs, and oaks, Berberis shrubs and alders (closely related to ours) cover the high plateaus of Southern Mexico as well as the Andes chain below the equator. Nature has thus allowed the inhabitant of the torrid zone to see, without leaving his home, all of the plant forms of the Earth, just as the dome of the heavens[37] from pole to pole hides from him none of its luminous worlds.

The northern peoples are denied these and many other such enjoyments of Nature. Many celestial phenomena and many plant forms—and from these, indeed, the most beautiful (palms, long-stemmed ferns and pisang plants, treelike grasses and feathery mimosas)—remain forever unknown to them. The sickly plants within our greenhouses provide but a weak image of the majesty of tropical vegetation. But in the refinement of our language, in the incandescent imagination of the poet, in the depictive art of the painter there open rich wellsprings of compensation. From this, the power of our imagination creates a living picture of exotic Nature. In the cold of the North, in the starkness of the heath, the lone individual can acquire for himself that which is being explored in the most distant latitudes, and thus create within himself a world that is the work of, and is as free and immortal as, his own spirit.

Annotations and Additions

1. Chimborazo, nearly eight thousand feet higher than Aetna
During storms blowing off the land, small songbirds and even butterflies (as I myself observed several times in the Pacific) are often found out to sea, at great distances from the coasts. Equally involuntarily, insects will attain altitudes of 15,000 to 18,000 feet above the plains in the highest atmospheric levels. The warmed crust of the Earth gives off a vertical radiation by which lightweight bodies are driven upward. Mr. Boussingault, an excellent chemist who ascended the gneiss mountains of Caracas while still an instructor at the new school of mines in Santa Fé de Bogota, became, during his trip to the summit of Silla, an eyewitness to a phenomenon that, in a curious way, confirms these vertical currents of air. Along with his companion, don Mariano de Rivero, he watched at midday as small, whitish, bright bodies rose up from the valley of Caracas 5,400 feet to the summit of Silla and then descended in the direction of the nearby seacoast. This game lasted an hour without interruption, and what were first mistaken for a flock of small birds were soon observed to be little balls of accumulated blades of grass. Boussingault sent me some of these blades, which Professor Kunth immediately recognized as a variety of *Vilfa*, a grass genus that in the Caracas and Cumana Provinces often appears with *Agrostis*; it was the *Vilfa tenacissima* of our *Synopsis Plantarum aequinoctialium Orbis Novi*, vol. I, p. 205. Saussure found butterflies on Mont Blanc. Ramond noted them in the barren areas surrounding the peak of Mont Perdu. On June 23rd, 1802, when we—Bonpland, Carlos Montufar, and I—reached a point on the eastern slope of Chimborazo at an elevation of 3,016 toises (18,096 feet), a height at which the barometer sank to 13 inches, 11.2 lines, we saw winged insects flitting around us. We recognized them as flylike Diptera, but on a ridge (*cuchilla*) that was often only 10 inches wide between steeply sloping snowfields, it was impossible to catch these insects. The altitude at which we observed them was almost the same as that in which the bare trachyte rocks, jutting upward from the perpetual snow, presented to our eyes the last trace of

vegetation in *Lecidea geographica*. These little creatures buzzed about at an altitude of some 2,850 toises, 2,400 feet higher than the summit of Mont Blanc. Somewhat lower, at a height of around 2,600 toises, yet still above the snow line, Bonpland had seen yellowish butterflies flying close to the ground. Of the mammals, those that live closest to the line of perpetual snow in the Swiss Alps are hibernating marmots and a very small vole (*Hypudaeus navilis*), as described by Martins. On the Faulhorn, it stores caches of roots from phanerogamic mountain plants almost under the snow (*Actes de la Société Helvétique*, 1843, p. 324). That the beautiful rodent called the chinchilla, whose silky, gleaming coat is so desired, is likewise found at the greatest mountain heights of Chile is a widespread misconception in Europe. *Chinchilla lanigera* (gray) lives only in the milder lower zone and does not range southward beyond the 35th parallel (Claudio Gay, *Historia física y política de Chile, Zoologia*, 1844, p. 91).

In our European alpine mountains the Lecidea, Parmelia, and Umbilicaria colorfully but sparingly bedeck the stone not covered by snow, while in the Andes chain we found still beautifully blooming phanerogams (which we first described) at thirteen to fourteen thousand feet: the fuzzy frailejón varieties (*Culcitium nivale, C. rufescens,* and *C. reflexum, Espeletia grandiflora* and *E. argentea*), *Sida pichinchensis, Ranunculus nubigenus, R. Gusmanni* with red- or orange-colored blossoms, and the little mosslike umbel flora *Myrrhis andicola* and *Fragosa arctoides*. Described by Adolph Brongniart, the *Saxifraga boussingaulti* grows on loose boulders on the slopes of Chimborazo up to beyond the line of perpetual snow, 14,796 feet (2466 toises) high, as has been stated in two esteemed English journals. (Compare my *Asie centrale*, vol. III, p. 262, with Hooker, *Journal of Botany*, vol. I, 1834, p. 327, and *Edinburgh New Philosophical Journal*, vol. XVII, 1834, p. 380.) The Saxifraga, which was discovered by Boussingault, may for the present be considered the highest phanerogamic plant on Earth.

The vertical height of Chimborazo is, by my trigonometric measurement, 3,350 toises (*Recueil d'Observ. astron.*, vol. I, intro. p. LXXII). This result stands in the middle between those presented by the French and Spanish academics. The primary difference lies not in disparate accounts of light refraction but in the reduction of the measured baseline to the ocean horizon. In the Andes chain, this reduction occurred only by means of the barometer, and thus every trigonometric measurement is at the same time a barometric one, the result of which differs according to the stipulations of the formulas applied. With the enormous mass of the mountain chain, one will obtain very small elevation angles if one desires to determine the greater part of the mountain's complete elevation trigonometrically and makes the measurement from a low and distant point close to the level of the sea. On the other hand, not only is it difficult to find a comfortable line of bearing in the high mountains, but the portion that must be measured barometrically grows with each step closer to the mountain as well. Every traveler on the high plains that the Andes enclose, when he chooses the point from which he wishes to perform a geodetic operation, must contend with these obstacles. In the pumice-strewn plain of Tapia, west of the Rio Chambo, I measured Chimborazo from a barometrically determined elevation of 1,482 toises. The Llanos de Luisa and especially the high plain of Sisgun (which lies at 1,900 toises) would afford larger elevation angles. On the latter, I had already set up everything for making a measurement when the summit of Chimborazo became wrapped in thick clouds.

Perhaps it would not be unpleasant to the linguist at this point to examine some assumptions regarding the etymology of the celebrated name Chimborazo. "Chimbo" is the name of the *corregimiento* (district) where Chimborazo lies. La Condamine (*Voyage à l'Équateur*, 1751, p. 184) derives "Chimbo" from *chimpani*, "to cross a river." According to him, "Chimbo-raço" means "la neige de l'autre bord" (the snow of the other bank) because in the village of Chimbo one is in sight of the tremendous, snow-capped mountain, but across a stream. (In the Quechua language, *chimpa* means "the opposite bank; the other side"; *chimpani* "to cross—over a river, a bridge, etc.") Many natives of Quito Province have assured me that Chimborazo simply means "the snow of Chimbo." The same ending is found in "Cargai-razo." But *razo* seems to be a word of the province. I possess a copy of the Jesuit Holguin's excellent *Vocabulario de la Lengua general de todo el Peru llamada Lengua Qquichua ó del Inca* (printed in Peru, 1608), and he shows no knowledge of the word *razo* at all. The correct name of snow is *ritti*. On the other hand, my friend Professor Buschmann, who is learned in languages, notes that in the Chinchay Suyo dialect (from Cuzco northward, up to Quito and Pasto) the word *raju* (*j* most likely guttural) means "snow" (see Juan de Figueredo's Chinchay Suyo vocabulary list, appended to Diego de Torres Rubio, *Arte y Vocabulario de la Lengua Quichua*, repr. Lima, 1754, fol. 222, b). Since *chimpa* and *chimpani* do not seem to fit because of the *a*, we find for the first part of the mountain's name and for the village of Chimbo a specific interpretation in the Quechua word *chimpu*: an expression for a colorful thread or fringe (*señal de lana, hilo ó borlilla de colores*), for reddening of the sky (*arreboles*), and for the coronas around the sun and moon. One can attempt to derive the name of the mountain, without intermediate reference to the village or district, from this word. In any case, whatever the etymology of Chimborazo might be, one should write the word in the Peruvian manner "Chimporazo," since the Peruvians, as is well known, have no *b*.

But what if the name of this colossus among mountains has no connection whatever to the Inca language and instead has its origin in gray prehistoric times? Indeed, according to generally accepted tradition, the Inca or Quechua language was introduced not long before the arrival of the Spaniards into the kingdom of Quito, where until then the now completely extinct Puruay language was generally dominant. The names of other mountains too—Pichincha, Iliniza, Cotopaxi—have no meaning in the language of the Incas and are thus certainly older than the introduction of sun worship and the court language of the rulers from Cuzco. In all regions of the Earth, the names of mountains and rivers belong among the oldest and most certain of memorials of language, and my brother Wilhelm von Humboldt made astute use of these names in his investigations into the former distribution of Iberian tribes. Peculiar and unexpected is the recent assertion (Velasco, *Historia de Quito*, vol. I, p. 185) "that the Incas Tupac Yupanqui and Huayna Capac, upon their conquest of Quito, were astonished to discover a dialect of their Quechua language already being spoken among the natives there." Prescott, however, finds such an assertion to be highly suspect (*Hist. of the Conquest of Peru*, vol. I, p. 125).

If one were to place the Gotthard Pass, Athos, or Rigi upon the summit of Chimborazo, one would then reach the height that is presently ascribed to Dhaulagiri in the Himalayan Mountains. To the geognost, who aspires to broader views regarding the planet's interior, the relative altitudes of the ribs of rock that we call mountain chains (but not the directions in which they run) seems such a pitifully insignificant phenomenon that it would not astonish him if someone were to discover other peaks between the

Himalayas and the Altai Mountains that surpass Dhaulagiri and Djawahir by as much as these surpass Chimborazo. (See my *Vues des Cordillères et Monumens des peuples indigènes de l'Amérique*, vol. I, p. 116, and *Ueber zwei Versuche, den Chimborazo zu besteigen*, 1802 and 1831, in Schumacher's *Jahrbuch für 1837*, p. 176.) The great height to which the reflected heat from the mountain plateau of Inner Asia lifts the snow line on the north slope of the Himalayas in summer makes the mountain group there, despite a latitude of 29° to 30.5°, as accessible as the Peruvian Andes in the tropical region. On Tarke Kang, Captain Gerard was recently as high up, and perhaps (as is stated in the *Critical Researches on Philology and Geography*, 1824, p. 144) 110 feet higher, than I was on Chimborazo. Unfortunately, as I develop more completely in another work, these mountain expeditions beyond the line of everlasting snow (as much as they capture the imagination of the public) are of very little scientific value!

2. The condor, the giant among the vultures

The natural history of the condor (actually *cuntur* in the Incan tongue; *mañque* to the Araucans in Chile; *Sarcoramphus condor* to Duméril), which before my trip was largely imperfect, I have furnished elsewhere (see my *Recueil d'Observations de Zoologie et d'Anatomie comparée*, vol. I, pp. 26–45). I have made an actual-sized sketch of the head of the condor from life and had an engraving done. After the condor, our Swiss bearded vulture and the *Falco destructor* (*Daudin*; probably Linné's *Falco harpyia*) are the largest flying birds.

The region that may be considered the usual habitat of the condor begins at the elevation of Aetna. It comprises layers of the atmosphere that are between ten and eighteen thousand feet above sea level. And the hummingbirds, who make summer journeys as far as 61° latitude on the western coast of North America and as far back down as the archipelago of Tierra del Fuego, have been seen by Mr. von Tschudi (*Fauna Peruana, Ornithol.*, p. 12) flocking together in the punas at an altitude of up to 13,700 feet. It is enjoyable to compare the greatest and smallest denizens of the air. Of the condors, the largest individuals, which may be found in the Andes chain near Quito, have a wingspan of 14 feet, while the smaller measure 8 feet. Judging by this size, and by the angle at which the birds appeared often vertically above our heads, one can speculate upon the immense height to which the condor ascends in quiet skies. An angle of sight of just 4 minutes, for example, gives a vertical distance of 6,876 feet. The cave (*machay*) of Antisana, which stands opposite the mountain Chussulongo and over which we took measurements on the soaring bird in the Andes chain of Quito, is situated at 14,958 feet above the level of the Pacific. Accordingly, the absolute altitude reached by the condor was fully 21,834 feet, a height at which the barometer shows barely 12 inches but which still does not exceed the highest peaks of the Himalayas. It is a curious physiological phenomenon that the same bird that flies about in circles for hours at a time in a region of such rarefied air occasionally descends, on the western slope of the Pichincha volcano for example, all the way down to the seacoast and in the space of a few hours crosses through all types of climate. At heights of 22,000 feet, the membranous air sacs of the condor, having been filled in lower regions, must expand tremendously.

More than a hundred years ago Ulloa expressed his astonishment that the vulture of the Andes can soar at altitudes where the air pressure amounts to less than 14 inches

(*Voyage de l'Amérique méridionale* [*Viaje a la América Meridional*], vol. II, part 2, 1752; *Observations astronomiques et physiques*, p. 110). It was at that time believed, analogous to experiments employing an air pump, that no animal could live at such a low air pressure. I myself, as mentioned above, saw the barometer sink to 13 inches, 11.2 lines, on Chimborazo; my friend Mr. Gay-Lussac breathed for a quarter of an hour at an air pressure of 12 inches, 1.7 lines. Certainly a person finds himself, when he is already fatigued from muscular exertion, to be in a frightening, asthenic condition at such altitudes. The condor, on the other hand, seems to complete the function of breathing with equal ease whether the air pressure is 12 inches or 28! Of all living creatures, he is probably the one that voluntarily puts the greatest distance between himself and our planet. I say "voluntarily," for small insects and siliceous shelled microbes, as I have mentioned numerous times, are driven even higher up by the rising air currents (*courant ascendant*). It is probable that the condor flies even higher than we found through our calculations. I can recall, while on Cotopaxi in the pumice plateau of Suniguaicu, at 13,578 feet above sea level, having seen the soaring bird at such an altitude that he appeared as a small black dot. But what is the smallest angle at which one may recognize weakly lighted objects? Their form (extension in length) has a great influence upon the minimum size of this angle. The clarity of the air at the equator is, by the way, so great that in the Province of Quito (as I have demonstrated elsewhere) one could differentiate with the naked eye the white coat (*poncho*) of a rider at a horizontal distance of 84,132 feet, that is, under an angle of 13 seconds. It was my friend Bonpland that we could see from the charming country estate of the Marques de Selvalegre, moving along a black rock wall of the volcano Pichincha. Lightning rods, being thin, elongated objects, are often visible at the greatest of distances, and under the smallest angles, as Arago has already observed.

What I described in my monograph on the condor (pp. 26–45) in regards to the behavior of the great bird in the mountain regions of Quito and Peru is corroborated by a more recent traveler, Mr. Gay, who explored the entirety of Chile and describes it in his excellent *Historia Fisica y Politica de Chile*. The bird, which, oddly enough, like the camelids (llamas, vicuñas, alpacas, and guanacos) does not range above the equator to New Granada, presses southward as far as the Strait of Magellan. In Chile as in the high plains of Quito, the condors, who normally appear in pairs or as single individuals, assemble in bunches to attack lambs or calves, or to prey upon young guanacos (*guanacillos*). The damage done annually by the condors to the herds of sheep, goats, and cattle, as well as to the wild vicuñas, alpacas, and guanacos of the Andes chain, is quite substantial. The inhabitants of Chile assert that in captivity, the bird can survive without food for 40 days; in free conditions, however, his voraciousness is monstrous, with a preference in the way of vultures for dead flesh.

As in Peru, the corralling method of trapping that I described is also successful in Chile, for to take flight, the birds, now grown heavier from being sated with meat, must first run for a stretch with half-spread wings. A portion of slaughtered beef, already beginning to grow rotten, is encircled by a secure fence; the condors gather together within the narrow space, and since, as stated, they cannot take off because the fence limits their room to run and because of the overlarge quantity of food they have consumed, they are either clubbed to death by the enclosing natives or taken alive by means of thrown nooses (*lazos*). Upon the coinage of Chile, the condor appeared as a symbol of strength immediately after the first declaration of the country's independence. (Claudio Gay, *Historia Fisica y Politica de Chile, publicada bajo los auspicious del Supremo Gobierno: Zoologia*, p. 194–198.)

In Nature's great household, the varieties of *gallinazos* (vultures), of which individuals are quite numerous, are far more useful than the condors for the disintegration and removal of decomposing animal substances, and thus also for the purification of the air in the vicinity of human habitation. In tropical America, I have seen 70 to 80 gathered simultaneously around a dead cow; I can also corroborate as an eyewitness the fact, recently and unfairly doubted by ornithologists, that the appearance of a single king vulture, though it was no larger than the *gallinazo*, drove the entire assemblage to take flight. A battle never develops, but the *gallinazos*, whose two species (*Cathartes urubu* and *C. aura*) are often confounded thanks to an unfortunately shaky nomenclature, are frightened by the sudden appearance and bolder demeanor of the colorful *Sarcoramphus papa*. Much as the ancient Egyptians protected the Percuoptera that cleaned the air, so too in Peru does the heinous killing of the *gallinazo* meet with a punishment (*multa*) which in certain cities can be, according to Gay, as high as 300 piasters per bird. It is also peculiar that this variety of vulture, as has been witnessed by don Felix de Azara, will, when raised from a young age, become attached to the person who feeds it to such a degree that it will follow him on journeys of many miles, flying behind his wagon as it rolls across the grassy steppe (*pampa*).

3. Encases their transparent, gyrating bodies

In his outstanding work on viper's poison (vol. I, p. 62), Fontana relates that he once had the good fortune to reinvigorate with a drop of water and in the space of two hours a rotifer that for 2½ years had lain dried and immobile. For more on the effects of water, see my *Versuche über die gereizte Muskel- und Nervenfaser*, vol. II, p. 250.

Since more precise observations are now being performed, and that which is observed is viewed in a manner which is more rigorously critical, the so-called reanimation of rotifers has in recent times become an object of lively discussion. Baker has claimed that in the year 1771 he reawakened paste eelworms that Needham gave to him in 1744! Franz Bauer saw his *Vibrio tritici*, which had lain dry for 4 years, move again upon being moistened. In his *Mémoire sur les Tardigrades et sur leur propriété de revenir à la vie* (1842), Doyère, a thoroughly meticulous and experienced observer, derives from his fine experiments the following results: rotifers will revive, i.e., can transition from a motionless state back to a moving state, even after they have first been cooled to a temperature of 19.2° Réaum. below freezing or heated to as much as 36°. They seem to retain the capacity to be revived in dry sand up to 56.4°; they lose this characteristic, however, and remain immobile when they are heated in moist sand up to only 44° (Doyère, p. 119). A 28-day drying period in an airless barometric chamber, even with the introduction of chloride of lime or sulfuric acid (pp. 130–133) does not hinder the possibility of the so-called reanimation.

Even when they are dried without sand (*desséchés à nu*), Doyère has seen rotifers slowly revive, which Spallanzani denied (pp. 117 and 129). "Any desiccation done at the usual temperature would be able to withstand objections to which the use of the dry vacuum might not have fully responded: but seeing the Tardigrades die irrevocably at a temperature of 44° if their tissues are soaked with water, whereas dried out they can tolerate, without dying, a heat that one can estimate at 96° Reaumur, one must be willing to admit that revitalization has no other condition in the animal than the integrity of composition and organic structure." The *sporulae* as well, the gametes or germ cells of cryptogamic plants, which Kunth compares to the reproduction of certain phanerogamic plants by means of bulbs (*bulbillae*), retain their capacity to germinate in the highest temperatures. According to the latest experiments by Payen, the spores (*sporulae*) of a small fungus (*Oidium aurantiacum*), which covers the breadcrumbs with a reddish, feathery coating, do not lose their generative power even if, before being scattered, they have been placed on unspoiled bread dough in closed tubes at 67° to 78° (84–97.5° C.) for one half-hour. Might not the newly discovered "wonder monad" (*Monas prodigiosa*), which creates bloodlike flecks in farinaceous substances, have been mixed in beneath these fungi?

In his great work on infusoria (pp. 492–496), Ehrenberg presents the entire history of the research on the so-called reanimation of rotifers. He believes that in spite of all desiccating agents one might employ, organic fluid remains in the seemingly dead little creatures. He disputes the hypothesis of "latent life"; death is not "life held bound, but absence of life."

Hibernation in both the warm and cold-blooded classes of animals gives witness to the reduction, if not the complete suspension, of organic functions: in the cases of dormice, marmots, bank swallows (*Hirundo riparia* as reported by Cuvier, *Règne animal*, 1829, vol. I, p. 396), frogs, and toads. The frogs awakened from their winter sleep by the warmth can spend an eightfold longer time underwater without drowning than frogs during mating season. The returning respiratory function of the lungs seems after such long periods of dormant vitality to require a reduction in activity for some time. The seemingly indubitable winter descent of the bank swallow into the marsh is a still more astonishing phenomenon considering how the function of respiration in the bird class is so completely energetic, as illustrated by Lavoisier's experiments wherein two small sparrows in normal living conditions break down as much atmospheric air as a guinea pig (Lavoisier, *Mémoires de Chimie*, vol. I, p. 119). And the winter sleep of the bank swallow is supposed to have been observed not in the entire species but only in single individuals (Milne-Edwards, *Élémens de Zoologie*, 1834, p. 543).

Much as the deprivation of warmth in the frigid zone spurs the hibernation of some animals, the hot tropical lands afford an analogous and insufficiently examined phenomenon that I have documented with the name "summer sleep" (*Relation historique*, vol. II, pp. 192 and 626). Drought and persistent high temperatures function much like the winter cold toward inhibition of mobility. Except for a small portion of its southern tip, Madagascar lies entirely in the tropical zone, and as Bruguière has already noted, the hedgehoglike tenrecs (*Centenes illiger*), one species of which (*C. ecaudatus*) has been introduced to Ile de France, fall asleep during periods of high heat. Desjardin's objection that the period of their slumber is winter in the Southern Hemisphere (in a land where the mean temperature of the coldest month exceeds by 3° the mean temperature of the hottest month in Paris) probably does not qualify the three-month summer sleep of the tenrec on Madagascar and in Port Louis on Ile de France as an actual hibernation.

The crocodile in the llanos of Venezuela, the land and water turtles of the Orinoco, the enormous boa, and several smaller varieties of snakes lie in a similar way, moribund and motionless, under the hardened earth during the hot and dry season. The missionary Gilij relates that the natives, when they search for the terekai (land tortoises that lie in dried mud at a depth of 15 to 16 inches), are bitten by suddenly awakening snakes that have buried themselves together with the tortoises. An excellent observer, Dr. Peters, who has just returned from the East African coast, writes me the following: "I was able to draw no certain conclusions about the tenrec during my short stay on Madagascar; however, I have come to know well that in the part of East Africa where I lived for several years, various species of turtles (*Pentonyx* and Trionychidae) lie for months without sustenance, encased in the dry, hard earth during the arid season of this tropical country. The *Lepidosiren* too, in the places where the swamp dries out, spends the time from May to December rolled up and motionless in stone-hard earth."

Thus do we find the weakening of certain life functions in many and quite disparate classes of animals, and what is particularly conspicuous is that closely related organisms belonging to the very same family do not display these same aspects. Unlike his relative the badger (*Meles*), the wolverine (*Gulo*) of the North does not hibernate, yet according to Cuvier's assertion, "A *Myoxis* (dormouse of Senegal, *Myoxis coupeii*), which in his tropical homeland probably never entered into a winter sleep, in its first year in Europe fell immediately into hibernation at the advent of winter." The weakening of life functions and vital activity crosses through graduated levels, each according to the extent to which the processes of nourishment, respiration, and muscle activity are affected, or to which brain and nerve systems are depressed. The winter sleep of the hermitic bear and of the badger is accompanied by no paralysis; it is for this reason that these animals are so easily awakened and, as I was often told in Siberia, present such a danger to the hunter and woodsman. The recognition of the graduated sequence and interconnectedness of these phenomena leads all the way to the so-named *vita minima* of the microscopic organisms that fall from the Atlantic meteoric dust, some having green ovaries and engaged in self-division. The apparent reanimation of the rotifers, as with the siliceous shelled infusoria, is merely the renewal of long-weakened life functions, the state of a life whose flame was never completely extinguished but instead, through stimulation, has been fanned up anew. Physiological phenomena can be comprehended only when one traces them through the entire graduated process of analogous modifications.

4. Winged insects

Previously, the pollination of blossoms of separate sexes was ascribed primarily to the wind. Kölreuter and, with great acuity, Sprengel have shown that bees, wasps, and a great number of small winged insects play the leading role in this. I say the "leading" role, for the opinion that no pollination of the stigmata is possible without the intervention of these little animals does not seem consistent with Nature, as Willdenow has also convincingly demonstrated (*Grundriß der Kräuterkunde*, 4th ed., Berlin, 1805, pp. 405–412). But dichogamy, nectar guides (*maculae indicantes*), colored spots indicative of nectaries, and pollination through insects do seem to be inextricably connected (compare Auguste de St. Hilaire, *Leçons de Botanique*, 1840, pp. 565–571).

The contention, often repeated since Spallanzani, that the dioecious common hemp (*Cannabis sativa*) introduced to Europe from Persia bears ripe seeds outside the proximity of stamens has been sufficiently refuted by recent experiments. In cases where seeds were yielded, stamens of a rudimentary form capable of producing a few grains of fructifying pollen were discovered next to the ovary. Such hermaphroditism is common throughout the family of Urticaceae, but in the greenhouses in Kew a unique and as yet unexplained phenomenon occurs in a small Australian shrub, Smith's *Coelebogyne*. In England, this phanerogamic plant produces mature seeds without a trace of male organs and without the bastard introduction of foreign pollen. "A type of Euphorbiaceae (?) rather recently described, but cultivated for several years in greenhouses in England, the *Coelebogyne*, has yielded fruit there several times, and its seeds were obviously perfect, since not only has one observed a well-formed embryo there, but when it is planted, that embryo has developed into a similar plant. Now the flowers are dioecious, but one does not know of or possess (in England) any male plants, and the most meticulous research, done by the best observers, has thus far failed to discover the slightest trace of anthers or even of pollen. Therefore, the embryo had not come from such pollen, which is entirely lacking: it must have formed itself from scratch in the ovule." So writes a gifted botanist, Adrien de Jussieu, in his *Cours élémentaire de Botanique*, 1840, p. 463.

To obtain a newer corroborative elucidation of this very important and isolated physiological phenomenon, I recently turned to my young friend Mr. Joseph Hooker, who, after the Antarctic journey with Sir James Ross, has now attached himself to the great Tibetan Himalayan expedition. Mr. Hooker writes to me upon his arrival in Alexandria at the end of December 1847, prior to his embarking at Suez: "Our *Coelebogyne* continues to bloom at my father's in Kew, just as it did in the garden of the Horticultural Society. It regularly produces mature seeds. I have repeatedly and painstakingly examined it, and have been able to find neither an intrusion of pollen tubes into the stigmata, nor any trace of the presence of these tubes in the pistil and micropyle. In my herbarium the male blossoms are situated in small catkins."

5. As glimmering stars

The luminosity of the ocean counts as one of those gorgeous occurrences in Nature that give rise to wonderment, even when one has watched it return nightly for months. In all climates the sea phosphoresces, but one who has not observed the phenomenon in the tropical latitudes (especially in the Pacific) has only an incomplete notion of the majesty of this grand spectacle. When a warship under a freshening wind cuts across the foaming waters, the observer standing at the rail can never tire of the sight that the nearby waves afford. Whenever the exposed side of the ship lays itself over, it appears as if bluish and reddish flames like lightning are shooting from the keel upward. Likewise indescribably glorious is the spectacle that a pod of frolicking dolphins create in the tropical seas by the dark of night. Wherever their long lines plow circles in the foaming flood, one can see their paths traced out in sparks and intense light. In the Gulf of Cariaco between Cumana and the Maniquarez peninsula I have enjoyed this sight for hours at a time.

Le Gentil and the elder Forster explained these flames through electrical friction of the water on moving objects as they glided along: an explanation that by the current state of our understanding of physics must be considered insupportable (Johann Reinhard Forster's *Bemerkungen auf seiner Reise um die Welt*, 1783, p. 57; Le Gentil, *Voyage dans les mers de l'Inde*, 1779, vol. I, pp. 685–698).

There are perhaps few objects of observation in Nature that have been debated so much and for so long as the luminosity of the ocean. That which is currently known with certainty may be reduced to the facts that follow. There are several luminous mollusks that, while alive, can give off a weak phosphorescent light at will—a light that falls primarily in the blue range, as is the case with *Nereis noctiluca, Medusa pelagica var. β* (Forskål, *Fauna aegyptiaco-arabica, s. Descriptiones animalium quae in itinere orientali observavit,* 1775, p. 109), as well as in the case of the one discovered by the Baudin expedition, the tubelike *Monophora noctiluca* (Bory de St. Vincent, *Voyage dans les Iles des Mers d'Afrique,* 1804, vol. I, p. 107, pl. VI). The luminance of seawater is created partly by living light sources and partly by organic fibers and membranes that owe their origin to the destruction of these living light sources. The first of these causes of the ocean's luminance is indisputably the most common and widespread. The more active and practiced the traveling researcher of Nature has become in the implementation of high-quality microscopes, the more numerous in our zoological systems have grown the members of the groups of mollusks and infusoria in whom the capacity to produce light, whether by mere force of will or by external excitation, has been recognized.

Regarding the luminosity of the sea insofar as it is caused by living organisms, these are among the primary contributors: in the zoophyte class, the acalephs (family of the *Medusozoa* and *Cyaneae*), some mollusks, and an innumerable host of infusoria. Among the small acalephs (jellyfish), *Mammaria scintillans* presents in the ocean's surface a reflection, as it were, of the glorious spectacle of the star-filled firmament. Fully grown, the creature barely reaches the size of the head of a stickpin. The existence of siliceous shelled luminescent infusoria was first proved by Michaelis in Kiel; he observed the flashing light of the *Peridinium,* a ciliate, of the armored monad *Prorocentrum micans,* and of a rotifer that he named *Synchata baltica* (Michaelis, *Über das Leuchten der Ostsee bei Kiel,* 1830, p. 17). The same *Synchata baltica* was later found again by Focke in the lagoons of Venice. Ehrenberg, my renowned friend and companion in Siberia, succeeded in keeping luminescent infusoria from the Baltic Sea alive in Berlin for almost two months. I saw them there in 1832, emitting light in a drop of seawater under a microscope in a darkened room. When the luminous infusoria (the largest of which is ⅛ of a Parisian line long, while the smallest are ⅟₄₈ to ⅟₉₆), having become exhausted, ceased to produce flashing sparks, they would start again if stimulated by pouring acids or mixing alcohol into the seawater.

Through repeated filtration of freshly poured seawater, Ehrenberg succeeded in producing a liquid that contained a higher concentration of luminous microbes (*Abhandlungen der Akad. der Wiss. zu Berlin,* 1833, p. 307; 1834, pp. 537–575; 1838, pp. 45 and 258). Within the luminescing organs of the *Photocharis,* whether flashing at will or through stimulation, the acute observer found a large-celled structure with jellylike properties that shows some resemblance to the electrical organs of gymnotes and electric rays. "When the *Photocaris* is stimulated, there appears on each cirrus a flicker and glow of individual sparks, which gradually increase in strength and illuminate the entire cirrus; finally, the living fire also runs over the back of the nymphlike microbe, so it looks under the microscope like a burning sulfur match under green-yellow light. In *Oceania (Thaumanthias) hemisphaerica,* the number and the position of the sparks on the thickened base correspond exactly—a fact truly worthy of note—with the larger cirri or organs that alternate with them. The appearance of this wreath of fire is an act of life, the entire evolution of light an organic life process that among the infusorial creatures appears as a single, momentary spark of light, but which, after a brief space of quiet, returns." (Ehrenberg, *Über das Leuchten des Meers,* 1836, pp. 110, 158, 160, and 163.)

The luminescent creatures of the ocean would reveal, by these assumptions, the existence of an electromagnetic, light-producing life process in animal classes apart from fishes, insects, mollusks, and acalephs. Is the secretion of the glowing fluid, which in the case of some luminous creatures flows forth and, without further influence from the organism, continues to glow for a long period (as with, for example, Lampyridae and Elateridae, the German and Italian glowworm, and the South American cucuyo of the sugar cane), merely the result of the first electrical discharge, or does it depend only upon the chemical mixture? The luminosity of the insects surrounded by air certainly has different physiological origins from that of the luminous water creatures, the fishes, Medusae, and infusoria. Surrounded with a coating of seawater, a highly conductive liquid, the tiny infusoria of the sea must be capable of a tremendous electrical tension of their light organs to be able, as water creatures, to glow so powerfully. Like the torpedo rays, the gymnotes, and the Nilotic electric catfish, they penetrate the layer of water; while electric fish, on the other hand, though they churn up the water and give off steel needles of electric power through galvanic chain connections, do not produce a flame that will span even the smallest insulating layer, as I showed a half-century ago (*Versuche über die gereizte Muskel- und Nervenfaser*, vol. I, pp. 438–441; comp. *Obs. de Zoologie et d'Anatomiecomparée*, vol. I, p.84), and as John Davy (*Philosophical Transactions for the Year 1834 Part II*, pp. 515–517) has corroborated.

The considerations developed here indicate that it is probable that in the smallest living organisms, which escape the sight of the naked eye, in the struggles of snakelike gymnotes, in the flashing luminescent infusoria that make glorious the phosphorescence of the sea—just as in the thundering cloud or in the polar lights (the silent magnetic sheet-lightning), which, as a result of increased tension in the Earth's interior, predict hours ahead of time the suddenly changed bearing of the compass needle—all proceed from one and the same process. (See my letter to the publisher of the *Annalen der Physik und Chemie*, vol. XXXVII, 1836, pp. 242–244.)

Occasionally, one might even find under magnification no creatures at all in glowing water, and yet wherever the waves strike against a hard body and fly into foam, wherever the water is violently disturbed, a lightninglike flash gleams forth. The cause of this phenomenon probably lies with decomposing fibers of dead mollusks, which are scattered in countless numbers in the water. If glowing water is filtered through tightly woven towels, these fibers and membranes are separated out as glowing specks. When we bathed at Cumana on the Gulf of Cariaco and walked about naked in the pleasant night air along the lonely shore, various places on our bodies continued to glow. The glowing fibers and organic membranes had stuck to our skin, and the light died after a few minutes. Perhaps one should not be amazed, considering the enormous number of mollusks that inhabit all tropical seas, that the seawater itself glows there, where by sight one cannot differentiate individual fibers. In the never-ending disintegration of masses of deceased Dagysae and Medusae, the sea might be considered a gelatin-filled fluid that, as such (and glowing), would be repellent and undrinkable to humans but nutritious for many fish. When one strokes a board with part of a *Medusa hyosella*, the place that was stroked will light up again as soon as it is rubbed with a dry finger. During my voyage across to South America, I occasionally placed a Medusa on a tin plate. If I struck the plate with another metal object, the smallest vibrations of the tin were sufficient to cause the animal to luminesce. How are this collision and vibration effective?

Has one created a momentary increase in temperature? Or created new surfaces? Or has one, through the collision, expressed some fluid like phosphorized hydrogen gas, so that in coming into contact with the oxygen in the atmosphere or with the air from the mollusk's respiration in the seawater, it burns? This light-producing action of the collision is most apparent in a choppy sea (*mer clapoteuse*), when waves moving in opposing directions cross one another.

I have seen the sea between the tropical lines glowing during times of greatly differing weather, most strongly at times of approaching storms or under muggy, hazy, heavily clouded skies. Heat and cold seem to have little influence on the phenomenon, for on the banks off Newfoundland the phosphorescence is often very strong in coldest winter. Occasionally, under seemingly identical outward conditions the sea will glow very strongly one night, and not at all the next. Does the atmosphere promote the manifestation of this light, or are all of these variables connected only by the coincidence of someone sailing through a sea which is impregnated to a greater or lesser degree with mollusk jelly? Perhaps the socially luminous microbes come together only under a particular condition of the air currents on the ocean's surface. The question has been raised: why do we never see our fresh, polyp-filled swamp water glowing? It seems in terms of animals and plants to possess a mixture of its own that would be conducive to the generation of light. Willows, after all, are seen glowing more often than oaks! There has been success in England in making saltwater glow by the addition of herring brine. Through galvanic experiments one can easily see, by the way, that the luminescence of living creatures depends upon nerve stimulation. I have seen a dying *Elater noctilucus* glow strongly when I touched the ganglion of his foreleg with zinc and silver. Medusae too sometimes give off a stronger light at the moment when the galvanic chain is closed (Humboldt, *Relat. hist.*, vol. I, pp. 79 and 533).

For information on the wondrous mass-development and propagative powers of infusorial creatures as mentioned in this text, see Ehrenberg, *Infus.*, pp. XIII, 291, and 512. "The Milky Way of the smallest organisms," as he puts it, "comprises the species *Monas* (often only $\frac{1}{3,000}$ of a line), *Vibrio*, and *Bacterium*" (pp. XIX and 244).

6. Which inhabits the lungs of the tropical rattlesnake

The animal that at that time I named *Echinorhynchus* or even *Porocephalus* seems upon closer examination (according to Rudolphi's more informed judgment) to belong to the category of Pentastoma (Rudolphi, *Entozoorum Synopsis*, pp. 124 and 434). It inhabits the abdominal cavity and the wide-celled lungs of a *Crotalus* form that lives in Cumana, sometimes even inside the houses, and hunts mice. *Ascaris lumbrici* (*Gözen's Eingeweidewürmer*, tbl. IV, fig. 10) lives under the skin of the common earthworm and is the smallest of all of the *Ascaris* forms. *Leucophra nodulata*, Gleichen's pearl microbe, has been observed by Otto Friedrich Müller in the interior of the reddish *Nais littoralis* (Müller, *Zoologia danica*, fasc. II, tbl. LXXX, a–e). These microscopic animals are probably inhabited by still others. All are surrounded by air pockets that are poor in oxygen, and heavily mixed with hydrogen and carbon dioxide. It is very doubtful that any animal lives in pure nitrogen. It was previously thought plausible of Fischer's *Cistidicola farionis*, for according to Fourcroy's experiments, the swim bladder of these fish seemed to contain an air devoid of any oxygen. Erman's experiments and my own indicate, however, that freshwater fish never have pure nitrogen in their swim bladders

(*Humboldt et Provençal sur la respiration des Poissons*, in *Recueil d'Observ de Zoologie*, vol. II, pp. 194–216). In ocean fish there is 0.80 oxygen, and according to Boit, the purity of the air seems to depend upon the depth in which the fish live (*Mémoires de physique et de chimie de la Société d'Arcueil*, vol. I, 1807, pp. 252–281).

7. The harmonious lithophytes

According to Linné and Ellis, the calciferous zoophytes, among which especially the Madrepores, Maeandrines, Astraea, and Pocillopores generate wall-like coral reefs, are surrounded and inhabited by tiny creatures that were long believed to be related to the Nereids that are classified among Cuvier's annelids (ringed worms). Through the perceptive, comprehensive work of Cavolini, Savigny, and Ehrenberg, the anatomy of these small, gelatinlike animals has been elucidated. It has been learned that in order to understand the entire organism of the so-called rock-building corals, one must not view these monuments that survive after their deaths, these calcified lamellae that are exuded in the course of life functions in the form of delicate leaves, as something foreign to the soft membranes of the nutrient-absorbing animal itself.

Along with the increased understanding of the wondrous creation of inhabited coral formations, a more accurate view has gradually developed regarding the tremendous influence that the world of corals has exerted upon the emergence of small island groups from the surface of the ocean, upon the migration of land plants and the successive expansion of the realm of flora, and indeed, in certain parts of the ocean basin, upon the spread of human races and languages. Though small, socially living organisms, the corals play an important role in the general economy of Nature: even if they do not, as some have falsely suspected since the time of the voyages of Captain Cook, build islands or enlarge continents from the unfathomable depths of the ocean, still they excite the most lively interest, whether as the object of physiological study and the teaching of the successive stages of animal forms or in the context of the geography of plants and the geognostic conditions of the Earth's crust. The entire Jura formation originates, in the sweeping view of Leopold von Buch, "in great, up-thrust coral banks of the prehistoric world that surround at a certain distance the old mountain chains."

In Ehrenberg's classification (*Abhandlungen der Akad. der Wissenschaft zu Berlin*, 1832, pp. 393–432) of the corals (in English works often inaccurately called "coral insects") the monostomous Anthozoa appear: either as those that are free or possessing the capacity to detach themselves, the "animal corals," or as those that are bound in plantlike fashion, the "phytocorals." To the first order (*Zoocorallia*) belong the hydras or arm-polyps of Trembly, the Actinia, resplendent with the most exquisite colors, and the mushroom corals; to the second order belong the Madrepores, Astraea, and Ocellinae. The polyps of this second order are the primary focus of this annotation due to their cellular, wave-breaking walls. The wall is the aggregate of coral formations in which the collective life has died out, but not all at once, as with a dead tree of the forest.

Every coral formation is a whole that has grown through the formation of buds according to distinct laws—a whole whose parts are made up of a multitude of organically self-contained individual animals. Those in the group of plant corals cannot freely separate from one another but remain bound to one another by lamellae of calcium carbonate. Each coral formation thus in no way has a central point of common life (Ehrenberg, op.cit., p. 419). Reproduction of corals occurs, depending on variety, through eggs, spontaneous division, or gemmation. This last method is the most common in the development of new individuals.

The coral reefs (in Dioscorides' terms, "ocean growths," a "forest of stony trees," "Lithodendra") are of three sorts. First are the coastal reefs (shore reefs, fringing reefs), which are directly connected to the shore of the continent or island, as are found on Australia's northeastern coast between Sandy Cape and the formidable Torres Strait, and like nearly all of the coral banks of the Red Sea, which Ehrenberg and Hemprich studied for eighteen months. Next are the reefs that surround an island (barrier reefs, encircling reefs), as, for example, Vanikoro in the small archipelago of Santa Cruz north of the New Hebrides, or Puynipete, one of the Carolines. Finally, there are the coral banks that enclose lagoons, called lagoon islands or atolls. This division and nomenclature, reflecting Nature, has been introduced by Charles Darwin, and coincides most closely with the astute explanation that this gifted natural philosopher has provided regarding the gradual evolution of such wondrous species. Just as Cavolini, Ehrenberg, and Savigny have greatly improved the scientific and anatomical knowledge of the organization of the coral animals, so too have the geographical and geological conditions of the coral islands been considered, first by Reinhold and Georg Forster on the second voyage of Cook, and then, after a long interval, by Chamisso, Péron, Quoy and Gaimard, Flinders, Lütke, Beechey, Darwin, d'Urville, and Lottin.

The coral animals with their stony cellular scaffolding are mainly at home in the warm seas of the tropics; indeed, reefs appear in greater numbers in the Southern Hemisphere. Atolls, or lagoon islands, appear in close groups: in the so-called Coral Sea between the northeast coast of Australia, New Caledonia, the Solomon Islands, and the Louisiade Archipelago; in the group known as the Low Archipelago, eighty in number; in the Fiji, Ellice, and Gilbert Islands; and in the Indian Ocean northeast of Madagascar under the name of the Saya de Malha atoll group.

The Great Chagos Bank, whose structure and deceased coral formations have been exhaustively investigated by Captains Moresby and Powell, is all the more worthy of interest in that one may consider it a continuation of the more northerly Laccadives and Maldives. In a previous work (*Asie centrale*, vol. I, p. 218) I have already pointed out how significant the succession of the atolls, in the exact direction of the meridian down to 7° south latitude, is to the general mountain system and topography of Inner Asia. In Trans-Ganges India on the opposite side, the meridional chains that are characterized by their crossing of several east-west mountain systems at the great bend of the Tibetan Tsangpo River correspond to the meridional mountain walls of the Ghats and the northern Bolor. Here lie the parallel ranges of Cochinchina, Siam, and Malacca, of Ava and Arracan, which together, with their various lengths, end in the bays of Siam, Martaban, and Bengal. The Bay of Bengal appears to be a thwarted attempt by Nature at creating an inland sea. A deep incursion between the simple western system of the Ghats and the highly concentrated eastern Trans-Ganges systems swallowed a great portion of the lowlands in the east but found obstacles less easy to overcome in the form of the long-existing and extensive highland of Mysore.

A similar oceanic incursion gave rise to two pyramidal peninsulas of very different lengths and widths, and the continuation of two meridional systems standing opposite one another (the mountain systems of Malacca in the east and of the Ghats of Malabar in the west) reveals itself in the ocean in symmetrical rows of islands—on one side under the name of the coral-poor Andaman and Nicobar Islands, and on the other side in three far-reaching archipelagos of atolls: the Laccadives, Maldives, and Chagos. These last, called the Chagos Bank by mariners, form a lagoon surrounded by the narrow reef with its frequent gaps. The average length and width of these atolls comes to 22 and 18 geographical miles respectively. While the enclosed lagoon has a depth of only 17 to 40 fathoms, at a short distance from the outer edge of the seemingly sinking coral wall one can hardly find bottom at 210 fathoms (Darwin, *Structure of Coral Reefs*, pp. 39, 111, and 183). Near the coral lagoon of Keeling Atoll south of Sumatra, according to Captain FitzRoy, at only 2,000 yards' distance from the reef, the sounding line had still found no bottom, even at a depth of 7,200 feet.

"The coral forms that build thick, wall-like masses in the Red Sea are Meandrina, Astraidae, Favia, Madrepores (Porites), *Pocillopora hemprichii*, Millepora, and Heteropora. The last are among the most massive, although they are branched. The deepest coral formations here, which when enlarged by refraction appear to the eye like the dome of a cathedral, are, as best one can judge, Meandrina and Astraea" (Ehrenberg; handwritten notes). One must differentiate between the single and to some extent free polyp formations and those that, building together, create wall-like mountainous structures.

If the plenitude of aggregate polyp constructions is so conspicuous in some regions, no less astonishing is the complete absence of these structures in other regions that very often lie quite nearby. It must be the case that plenitude or absence is determined by specific conditions of current, local water temperature, and available nourishment which as yet have not been discerned. That certain thin-branched coral forms, having less accumulation of calcareous minerals on their reverse side (that is, not the side with the mouth opening), prefer the stillness of the inner lagoons cannot well be denied. But this inclination to still waters should not, as happens only too often (*Annales des Sciences naturelles*, vol. VI, 1825, p. 277) be construed as a characteristic of the entire class of animals. According to Ehrenberg's and Chamisso's experiences in the Red Sea and in the myriad atolls of the Marshall Islands east of the Carolines, as well as Capt. Bird Allen's and Moresby's observations in the West Indies and the Maldives, living Madrepores, Millepores, Astraea, and Meandrina can withstand "a tremendous surf" (Darwin, *Coral Reefs*, pp. 63–65); indeed, they seem to prefer such stormy exposure. The life forces of the organism, being of a cellular structure that transforms to stonelike hardness, withstand the mechanical forces in the pounding of the moving waters with marvelous success.

In the Pacific, the Mendaña or Marquesas Archipelago, the Galapagos Islands, and the entire western coast of the New Continent, despite the proximity of so many atolls among the lower islands, are quite without coral reefs. Certainly, the Pacific current, which washes along the coasts of Chile and Peru and whose low temperature I discovered in 1802, averages only 12.5° Réaumur, while the quiet waters outside of the cold current, which turns westward near the Punta Barima, have a temperature of 22° to 23°. Near the Galapagos as well, the small currents between the islands have a temperature of 11.7° Réaum. But this low temperature is not prevalent farther north on the Pacific coasts of Guayaquil and on to Guatemala and Mexico; it is also not prevalent in the Cape Verde Islands nor along the entire western coast of Africa around the small islands of St. Paul, St. Helena, Ascencion, and Fernando de Noronha, all of which likewise are without coral reefs.

Though this absence of reefs is characteristic of the western coasts of America, Africa, and Australia, reefs are plentiful on the eastern coast of tropical America, the African coast of Zanzibar, and the Australian coast of New South Wales. I had the most opportunity for exploring coral banks in the Gulf of Mexico and south of the island of Cuba in the so-called Gardens of the King and Queen (*Jardines y Jardinillos del Rey y de la Reina*). Christopher Columbus himself gave the name to this small island group on his second voyage, in May of 1494, for with their graceful mingling of the silver-leaved, treelike *Tournefortia gnapholoides*, of blossoming varieties of Dolichos, of *Avicennia nitida* and mangrove hedges (*Rhizophora*), the coral isles create a sort of archipelago of floating gardens. "Son Cayos verdes y graciosos, llenos de arboledas," says the Admiral. I spent several days in these gardens east of the large and mahogany-rich isle of pines, the *Isla de Pinos* (while on the voyage from Batabano to Trinidad de Cuba), determining the length of the individual islands.

The Cayos—Flamenco, Bonito, de Diego Perez, and de Piedras—are coral islands that rise a mere 8 to 14 inches above the surface of the sea. The uppermost edge of the reef does not consist only of dead polyp formations; it is, more accurately, formed of a true conglomerate into which are baked jagged pieces of coral, cemented together in various directions by grains of quartz. In the Cayo de Piedras, some of these conglomerated coral pieces that I saw measured up to three cubic feet. Several of the smaller coral islands of the West Indies have freshwater, a phenomenon that, wherever it occurs, e.g., around the Radak Islands in the South Pacific (Chamisso, in Kotzebue's *Voyage of Discovery*, vol. III, p. 108), is worthy of more exhaustive investigation, for it is sometimes attributed to a hydrostatic pressure exerted by a remote coast (as in Venice and in the Bay of Xagua, east of Batabano) and sometimes to the filtration of rainwater (see my *Essai politique sur l'Île de Cuba*, vol. II, p. 137).

The living, gelatinous coating of the calciferous scaffolds of the coral formations attracts fish and even sea turtles that seek nourishment. In Columbus's time, this now so lonely region of the Gardens of the King teemed with a peculiar sort of industry on the part of the coastal people. They employed, as it were, a "fishing fish," as a means of capturing sea turtles: the remora, the so-called *Schiffhalter* (ship grasper), probably *Echeneis naucrates*. To the tail of this fish a long cord of palm fiber was attached. The remora (in Spanish *reves*, the "reversed one," for at first glance, one mistakes the back for the abdomen) attaches itself firmly through suction to the turtle by means of the serrated and movable plate of cartilage on its upper skull. It would sooner be ripped to pieces, says Columbus, than release its prey. The small fish and the turtle were then pulled out together. "Our countrymen call the fish 'Turned Around' because it hunts turned in a specific direction. In the same way as we pursue hares with hunting dogs across the level expanse of a field, they [the inhabitants of the island of Cuba] would capture other fish by means of a hunting fish." (Petr. Martyr, *Oceanica*, 1532, Dec. I, p. 9; Gomara, *Hist. de las Indias*, 1553, fol. XIV.) From Dampier and Commerson we learn that this method of hunting, the use of a fishing suckerfish, is also quite usual on the eastern coast of Africa at Cape Natal and in Mozambique, as well as on the island of Madagascar (Lacépède, *Hist. nat. des Poissons*, vol. I, p. 55). Among races of people who have nothing in common, similar needs and familiarity with the behavior of animals give rise to identical methods of hunting.

While it is true, as we mention above, that the actual habitat of wall-building lithophytes is the zone between 22° and 24° north and south of the equator, there are also coral reefs (encouraged, it is believed, by the warm gulf stream) to be found around the Bermudas (lat. 32°23′), which Lieutenant Nelson has admirably described (*Transactions of the Geological Society*, 2nd Series, vol. V, part 1, 1837, p. 103). In the Southern Hemisphere, single corals (Millepores and Cellepores) have been found as far down as Chiloé Island and the Chonos Archipelago, and in Tierra del Fuego as far down as 53°; indeed, Retepores have been found as far down as 72½°.

Since the second voyage of Captain Cook, the hypothesis that he proposed, together with Reinhold and Georg Forster, according to which the flat-topped coral islands of the Pacific were built up by living beings from the depths of the sea floor, has won many adherents. The excellent natural explorers Quoy and Gaimard, who accompanied Captain Freycinet on his circumnavigation of the globe in the frigate *Uranie*, first expressed in 1823, and with great candor, their opposition to the views of the two Forsters, father and son, and of Flinders and Péron (*Annales de Sciences naturelles*, vol. VI, 1825, p. 273). "By drawing the attention of the naturalists to the animalcules of the corals, we hope to show that everything that one has said or believed to have observed until now, relative to the immense work that they are likely to do, is more often than not inaccurate and always excessively exaggerated. We think that the corals, far from lifting, from the depths of the ocean, perpendicular walls, only form layers or fossilizations a few toises thick." Quoy and Gaimard also expressed the assumption (p. 289) that the atolls (coral walls that enclose a lagoon) owe their origin to undersea volcanic craters. Their estimate of the depth in which the reef-building corals (the Astraea, for example) can live is certainly too low, for they assert that it is at most 25 to 30 feet below the surface. A natural explorer who was able to increase his own observations through comparison with those gathered by others in many regions of the world, Charles Darwin sets the region of living coral, and with greater certainty, to be 20 to 30 fathoms (Darwin, *Journal 1845*, p. 467; *Structure of Coral Reefs*, pp. 84–87; Sir Robert Schomburgk, *Hist. of Barbados*, 1848, p. 636). This is also the depth at which Professor Edward Forbes found the most corals in the Greek Sea. It is his "4th region of sea creatures" in his insightful work on the *Provinces of Depth* and the geographical distribution of mollusks over vertical distances from the water's surface (*Report on Aegean Invertebrata* in the *Report of the 13th Meeting of the British Association, held at Cork in 1843*, pp. 151 and 161). It appears, however, that, depending upon the differences in coral species, especially in the case of the more delicate ones that build less solid formations, the depths to which they can survive differ greatly.

On his expedition to the South Pole, Sir James Ross pulled up corals from great depths by use of a sounding bob, entrusting the corals to Mr. Stokes and Professor Forbes for closer examination. *Reptora cellulosa*, a *Hornera*, and *Prymnoa Rossii* (the last being quite analogous to a variety found on the Norwegian coast) were found alive and in perfectly fresh condition west of Victoria Land near Coulman Island, at 72°31′ south latitude and at a depth of 270 fathoms. (Cf. Ross, *Voyage of Discovery in the Southern and Antarctic Regions*, vol. I, pp. 334 and 337.) And in the high north, the Greenland sea pen (*Umbellaria groenlandica*) has been brought up alive by whalers from a depth of 236 fathoms (Ehrenberg in the *Abhandlungen der Berliner Akademie aus dem Jahr 1832*, p. 430). We find this same relationship of species and habitat among the sponges, which admittedly are now more often counted among the plants than the zoophytes. On the coast of Asia Minor, the common sea sponge is fished from depths of 5 to 30 fathoms, while a very small species of the same family is not found at less than 180 fathoms (Forbes and Sprutt, *Travels in Lycia*, 1847, vol. II, p. 124). It is difficult to guess what it is that hinders Astraea, Madrepores, Maeandrines and the entire group of tropical plant-like corals that are capable of building calciferous walls from living in very deep waters. The decrease in temperature occurs but slowly, the lack of light nearly the same; and the presence of bountiful living infusoria at great ocean depths indicates that even there the polyp formations would not want for nourishment.

Contrary to the previously widespread assumption of the complete absence of all organisms and living creatures in the Dead Sea, it is worth mentioning here that my friend and collaborator Mr. Valenciennes has received, through the Marquis Charles de l'Escalopier and through the French consul Botta, excellent specimens of *Porites elongata* taken from the Dead Sea. This fact is of even greater interest considering that this species does not occur in the Mediterranean but is found in the Red Sea, which, according to Valenciennes, has few organisms in common with the Mediterranean. Much like a species of *Pleuronectes*, an ocean fish that has penetrated deep into the interior of France and has adapted to gill respiration in freshwater, we find with the above-mentioned coral animals (*Porites elongata*, Lamarck) a similarly peculiar flexibility of organization, for this same variety lives simultaneously in the intensely saline waters of the Dead Sea and in the open ocean near the Seychelles (see my *Asie centrale*, vol. II, p. 517).

According to the latest chemical analyses of the younger Silliman, the genus *Porites*, like many other cellular coral formations (Madrepores, Astraea, and Maeandrines from Ceylon and the Bermudas), contains (along with 92–95 percent calcium carbonate) some fluoric and phosphoric acids (cf. James Dana, geologist for the United States Exploring Expedition under the command of Capt. Wilkes, *Structure and Classification of Zoophytes*, 1846, pp. 124–131). The presence of fluorine in the scaffolding of the polyps is reminiscent of the calcium fluoride in fish bones, as found in Morechini's and Gay-Lussac's experiments in Rome. In the coral formations, silica is only found mixed into the very small amounts of fluoric and phosphoric lime, but one coral animal related to the horn corals, Gray's *Hyalonema* (the glass thread), has an axis of pure silica fibers, rather like a hanging pigtail. Professor Forchhammer, who has recently conducted exhaustive analyses of seawater from the various regions of the world, finds the calcium content of the Antillean Sea to be strangely low. Lime amounts there to but 247/10,000, while in the Kattegat it rises to 371/10,000. He is inclined to ascribe this difference to the many coral banks of the West Indian islands, which absorb the lime and exhaust it from the seawater (*Report of the 16th Meeting of the British Association for the Advancement of Science, held in 1846*, p. 91).

Charles Darwin has illustrated in an astute manner the probability of the genetic connection between coastal reefs, reefs surrounding islands, and lagoon islands, that is, narrow ring-shaped coral banks that surround a lagoon. He asserts that these three types of construction are dependent upon the oscillations of the sea floor in its periodic heaving and sinking. The oft-repeated hypothesis according to which the closed circular coral reefs of the lagoon islands or atolls supposedly indicates the form of an undersea crater or the rim of a submerged volcanic cone is refuted by their great diameters of 8, 10, or even 15 geographical miles. Our fire-spewing mountains do not possess such craters; should one wish to compare the lagoon with the sunken plain and the narrow reef with the ring-mountains of the Earth's moon, one must not forget that these mountainous lunar rings are not volcanoes but landscapes encircled by a wall. According to Darwin, this is the process by which these constructions formed: an island mountain closely encircled by a coral reef sinks, and its "fringing reef" likewise sinks at the same rate with it. But the reef is also, at the same time, growing vertically by means of the efforts of the coral creatures striving upward toward the surface. This island mountain, then, first becomes an island surrounded at a distance by a reef and later, as the sinking progresses and the island eventually disappears, it becomes an atoll. In this view, where islands are seen as the extreme heights of a land beneath the sea (the culmination points), the relative positions of the coral islands would reveal to us what we could scarcely hope to ascertain with the sounding bob: the former topography and divisions of the land. This attractive prospect and, as mentioned at the beginning of this note, its relationship to the migration of plants and the spread of the races of Man will first become truly clear when we succeed in gathering more knowledge regarding the depth of the substratum and the nature of the mountain masses that serve as the foundations for the bottommost, long-dead layers of the polyp formations.

8. Of the Samothracian legends

Diodorus preserved for us these strange legends, the veracity of which has nearly become historical certainty to geognosts. The island of Samothrace, once also called Ethiopia, Dardania, Leucania, or Leucosia in the scholium of Appollonius of Rhodes, seat of the mysteries of the Cabeiri, was once inhabited by the remnants of an ancient race, from whose singular language many terms were afterward preserved in sacrificial ceremonies. The island's position, across from the Thracian Hebrus and near the Dardanelles, explains why it is here that there remained among the people a more formal tradition regarding the great catastrophe of a breach of the inland waters of the Black Sea. Upon certain altars bordering the flood, holy customs were instituted, and in Samothrace as among the Boeotians, the belief in a periodic destruction of the human race (a belief that also exists among the Mexicans as a myth of four periods of global destruction) was attached to the historical remembrance of particular floods (Otfried Müller, *Geschichten Hellenischer Stämme und Städte*, vol. I, pp. 65 and 119).

The Samothracians related, according to Diodorus, that the Black Sea was an inland sea that, swelling from the rivers that flowed into it (long before the deluges that befell other peoples), broke first through the Straits of Bosporus and afterward through those of the Hellespont (Diodorus Siculus, lib. V, cap. 47, p. 369, Wesseling). The details of these ancient revolutions of Nature, which Dureau de la Malle discusses in a separate work, are collected in Karl von Hoff's important work *Geschichte der natürlichen Veränderungen der Erdoberfläche*, part I, 1822, pp. 105–162, and in Creuzer's *Symbolik*, 2nd ed., part II, pp. 285, 318, and 361. The Samothracean legends are reflected as it were in the sluice theory of Strato of Lampsacus, which maintains that the rising of the waters of the Black Sea first caused the penetration of the Dardanelles and then the opening of the Pillars of Hercules. In the first book of his geography, Strabo has preserved for us, amidst the critical excerpts of Eratosthenes, a peculiar fragment of the lost writings of Strato. It presents views that touch upon nearly the entire perimeter of the Mediterranean.

"Strato of Lampsacus," it says in Strabo (lib. I, pp. 49 and 50, Casaubon), "goes further even than Xanthus of Lydia (who describes impressions of shells far from the ocean) in the analysis of the causality of these phenomena. He maintains that the Euxinus [the Black Sea] once had no mouth at Byzantium, but that the rivers that flowed into this sea, through the pressure of the swelling water mass, caused it to open, whereupon the water flowed off into the Propontis [the Sea of Marmara] and the Hellespont. The same thing occurred to our sea [the Mediterranean], for here too the isthmus by the Pillars was penetrated when the sea was filled up by the streams that, in their runoff, uncovered [dried up] the former coastal marshlands. As evidence Strato submits, first, that the outer and inner seabeds are different, such that even now a submarine earth bank extends from Europe to Libya as though at that time the inner sea and the outer were not one. Also, the Pontus is the shallowest, while the Cretan, Sicilian, and Arabian Seas are, on the other hand, very deep. This is due to the fact that the Pontus is filled with silt by the many great rivers flowing into it from the north, while the other seas remain deep. For this reason, the water of the Pontus has the lowest salinity too, and the outflows occur in the direction of regions where the bottom slopes downward. Also, it seems that the entire Pontus, should these influxes continue, will one day become completely filled with mud. For even now the left side of the Pontus near Salmydessus [the Thracian Apollonia], which the Mariners call the 'breasts' near the mouth of the Ister [Danube] and the desert of Scythia, is becoming a swamp. Perhaps even the [Libyan] Temple of Ammon once stood by the sea, though now, after the runoff of the waters, it is found deep in the land's interior. Strato also assumes that the reason the oracle [of Ammon] became so clearly distinguished and famous was because it stood at the sea; a great distance from the coast would make its current distinction and fame unexplainable. Egypt too, ages ago, was covered by the sea as far as the swamps of Pelusium, Mount Casius, and Lake Serbonis, for even now, when one strikes saltwater when digging, the sides of the excavation are layered with sea sand and shell creatures, as though the land had been flooded and the entire region around Mount Casius and the so-called *Gerrha* had been an ocean marsh that stretched to the gulf of the Red Sea; when the sea [the Mediterranean] receded, however, the land was exposed, but Lake Serbonis remained. Later, this lake also broke through and diminished into a swamp. Thus are the banks of Lake Moeris more similar to seashores than to riverbanks." A falsely corrected version by Großkurd (based on Strabo lib. XVII, p. 809, Casaubon), instead of "Moeris," has "Lake Halmyris." This lake, however, lay not far from the southern mouth of the Danube.

Eratosthenes of Cyrene, the most famous of the succession of chief librarians of Alexandria, though not as successful as Archimedes in writing on submerged bodies, was directed by Strato's sluice theory in his investigations of the problem of the equality of the level of all outer oceans flowing around the continents (Strabo lib. I, pp. 51–56, lib. II, p. 104 Casaubon). The articulation of the northern coasts of the Mediterranean, like the shapes of the peninsulas and islands, had given rise to the geognostic myth of the ancient land of Lyctonia. The origin of Syrtis Minor and Lake Tritonis (Diodorus III, 53–55) and of the entire western Atlas (Maximus Tyrius VIII, 7) were tied to a fantasy involving eruptions of fire and earthquakes (cf. my *Examen crit. de l'hist. de la Géographie*, vol. I, p. 179, vol. III, p. 136). I have quite recently commented in greater detail on this subject that so closely touches upon the ancestral seat of our civilization (*Kosmos*, vol. II, p. 153) and so take the liberty as I complete this note of inserting, if fragmentarily, the following.

The northern shore of the Inner or Mediterranean Sea has the virtue, already mentioned by Eratosthenes, of being more richly formed, more multifarious, more articulated than the Libyan coast to the south. Here three peninsulas stand out, the Iberian, the Italian, and the Hellenic, which, abundantly indented with bays, form straits and isthmuses with the nearby islands and opposite coastlines. Such formation of the Continent and of the islands, some torn from the mainland and some volcanic in nature, lying in rows as though thrown up by long, far-reaching fissures, early on gave rise to geognostic views regarding breaches, revolutions of the Earth, and the pouring forth of the swollen high seas into the lower-lying waters. The Pontus, the Dardanelles, the Strait of Gades (Gibraltar), and the island-rich Mediterranean were especially well suited to bring forth views of such a sluice system. The Orphic Argonaut, probably of Christian times, has woven in old myths; he sings of the fragmentation of the ancient Lyktonia into individual islands: "Poseidon with the dark curls, scorning Father Kronion, smote Lyktonia with the golden trident." Similar fantasies, which admittedly could often have arisen from an imperfect knowledge of spatial relationships, were spun in the richly erudite Alexandrine School, which turned its attentions to all ancient topics. Whether the myth of the destruction of Atlantis is a distant Western reflection of the myth of Lyktonia, as I believe I have shown in another work to be probable, or whether, as asserted by Otfried Müller, "the fall of Lyktonia (Leukonia) points to the Samothracian legend of a great flood that transformed that region" is a question that need not be decided here.

9. Precipitation in clouds

The current of vertically rising air is a primary cause of the most significant meteorological phenomena. If a desert, a sandy plain devoid of plants, is bordered by a tall mountain chain, one may observe how the sea wind drives thick clouds over the desert, without the rain falling until they reach the mountains. This phenomenon was in former times explained most unfittingly as the result of an attraction exerted by the mountains upon the clouds. The true cause seems to lie in the column of warm air that rises vertically from the plain of sand and hinders the collected vapors from condensing. The more the surface is empty of vegetation and the more the sand heats up, the higher up the clouds rise and the less the precipitation can take place. Above the slopes of the mountains, these causes cease to be. The effect of the vertical current of air is weaker there, the clouds descend, and condensation occurs in the cooler layer of air. Thus do the lack of rain and the absence of plants share a mutual causality. It does not rain, because the naked and barren plain of sand grows hotter and radiates more heat. The desert does not become a steppe or a grassland, because no organic development is possible without water.

10. The hardening, heat-discharging mass of the Earth

If, according to the long-since antiquated hypothesis of the Neptunists, the so-called primordial rock forms also precipitated from a liquid, then there must have been a tremendous amount of heat released during the transition of the Earth's crust from a liquid to a solid state, which would in turn have been the cause of new evaporation and new precipitations. The later these formed, the more quickly and tumultuously they would have occurred and the less crystalline they would have been. Such a sudden discharge of heat from the Earth's hardening crust, independent of the pole height of the place and independent of the position of the Earth's axis, could accordingly have brought about localized temperature increases in the atmosphere which would have acted upon the distribution of plant life. It could also have caused a sort of porosity, as is indicated by many puzzling geognostic phenomena in stratified rock formations. I have developed these assumptions in detail in a small treatment titled *Über ursprüngliche Porosität* (see my work *Versuche über die chemische Zersetzung des Luftkreises*, 1799, p. 177, and Moll's *Jahrbücher der Berg und Hüttenkunde*, 1797, p. 234). In the primitive age, according to my more recent views, the oft-shaken, many-fissured Earth with its molten interior can long have given its oxidized surface a high temperature, regardless of its latitude or position relative to the sun. What influence upon the climate of Germany could not be exerted today and for centuries by an open fissure a thousand fathoms deep stretching from the Adriatic Gulf to the northern coast? While it is true that with the Earth in its present state (in the stability relationship first calculated by Fourier in his *Théorie analytique de la chaleur*, a state achieved almost solely by means of a long period of heat radiation), the Earth's outer atmosphere now comes into direct contact with its molten interior only by way of the insignificant openings of a few volcanoes, in the primitive age this interior poured streams of hot air into the atmosphere through myriad crevasses and fissures formed by the often self-renewing foldings of the mountain strata. These emissions of hot air were independent of distance from the equator. Each newly formed planet must likewise, in its earliest phases, have created for itself a temperature that only later would be determined by its position relative to the central celestial body, the sun. The moon too shows traces of this reaction of its interior upon its crust.

11. Mountainsides in southernmost Mexico

Similar to greenstone, the orbicular rock of the mountain district of Guanaxuato is quite the same as the orbicular rock of the Franconian Fichtelgebirge. Both form grotesque dome shapes that burst through the upper layer of clay and stand above it. Likewise, perlite, porphyritic schist, trachyte, and pitchstone porphyry compose rocks of the same form in the Mexican mountains of Zinepecuaro and Moran, in Hungary, in Bohemia, and in Northern Asia.

12. The dragon tree of Orotava

The colossal dragon tree, *Dracaena draco*, stands in the garden of Mr. Franqui in the small city of Orotava (formerly Taoro), one of the most charming places on Earth. In June of 1799, when we also climbed the peak of Tenerife, we measured the circumference of the dragon tree to be 45 Parisian feet. Our measurement was taken several feet above the roots. Lower down and closer to the ground, the figure given by Le Dru for the circumference of the enormous tree is 74 feet. According to George Staunton, when measured at a height of 10 feet, the trunk has a diameter of 12 feet. The tree's height is little more than 65 feet. The legend states that this dragon tree was worshiped by the Guanches (as was the ash tree at Ephesus by the Greeks, the sycamore ornamented by Xerxes in Lydia, or the holy banyan fig on Ceylon) and that in 1402, during the Bétencourts' first expedition, it was already as thick and as hollow as it is today. Considering that the Dracaena grows exceedingly slowly, one can estimate the great age of the tree at Orotava. In his description of Tenerife, Berthelot says, "While comparing the young Dragon Trees, neighbors of the gigantic tree, the calculations that one makes on the age of the latter stagger the imagination" (*Nova acta Acad. Leop. Carol. Naturae Curiosorum*, vol. XIII, 1827, p. 781). The dragon tree has been cultivated on the Canary Islands, on Madera, and on Porto Santo since the most ancient times, and a keen observer, Leopold von Buch, even found it growing wild on Tenerife near Igueste. Its original fatherland is thus not the East Indies, as has been long believed, and its presence does not contradict the contention of those who consider the Guanches to be a completely isolated aboriginal Atlantic people who had no traffic with the nations of Africa and Asia. The form of the Dracaenas occurs again at the southern tip of Africa, on Île Bourbon, in China, and in New Zealand. In these disparate parts of the world one finds species of this same family, but none in the New Continent, where their form is replaced by the yucca. Aiton's *Dracaena borealis* is a true Convallaria, both of them sharing the same physical characteristics (Humboldt, *Relation historique*, vol. I, pp. 118 and 639). On the last plate of the pictorial atlas of my American journey (*Vues de Cordillères et Monumens des peuples indigenes de l'Amérique*, pl. LXIX) I included a rendering of the dragon tree of Orotava after a sketch done in 1776 by F. d'Ozonne. I found this drawing among the posthumous papers of the famous Borda, in the as yet unpublished travel journal that was entrusted to me by the *Dépôt de la Marine* and from which I borrowed important astronomical and geographical as well as barometric and trigonometric notes (*Relation historique*, vol. I, p. 282). The measurement of the Dracaena in the Villa Franqui was taken during Borda's first journey (1771), with Pingré, not on the second (1776), which was with Varela. It is believed that in the 15th century, in the earliest days of the Norman and Spanish *conquista*, that a mass was read within the hollow tree trunk upon a small altar that had been erected there. Sadly, the Dracaena of Orotava suffered the loss of one side of its crown (the treetop) in the storm of July 21, 1819. There is a large and attractive English copper etching that presents the current condition of the tree in a manner that is quite true to life.

The monumental aspect of these colossal life forms and the impression of dignity that they evoke in all peoples have given rise in recent times to more care being taken in the numerical determination of their ages and the size of their trunks. The results of these investigations have demonstrated to the author of the important treatise *De la longévité des arbres*, the elder Decandolle, and to Endlicher, Unger, and other gifted botanists, that it is not improbable that the age of several still-living individuals goes back to the earliest historical times—if not as far back as the civilization of the Nile country, still as far back as that of Greece or Italy. "Several examples," so it says in the *Bibliothèque universelle de Genève*, vol. XLVII, 1831, p. 50, "seem to confirm the idea that there are still trees on the globe of a tremendous antiquity and perhaps witnesses to its most recent physical revolutions. If one considers a tree as an aggregate of so many individuals fused together, having developed from buds on its surface, one cannot be surprised if, from new buds constantly adding themselves to the older ones, the resulting aggregate has no necessary end to its existence." In much the same way, Agardh says, "If within the plant, with each passing year, new parts grow, and the older, hardened parts are replaced by new ones capable of moving the sap, then an image arises of plant growth that is limited only by external causes." The short lifespan of herbaceous plants he ascribes to "the disproportionate amount of blossoming and the creation of fruit over the production of leaves." For a plant, infertility extends longevity. Endlicher offers the example of a specimen of *Medicago sativa*, variation β versicolor, that lived for 80 years because it bore no fruit (*Grundzüge zur Botanik*, 1843, § 1,003).

Along with the dragon trees (which despite the gigantic development of their closed vascular bundles must be classified, due to the nature of their blossom structures, within the very same natural family as asparagus and garden onions), the *Adansonia* (monkey bread tree; baobab) likewise belongs among the largest and oldest inhabitants of our planet. As early as the first exploratory journeys of the Catalans and the Portuguese, mariners had adopted the custom of carving their names into these trees—not always merely as a notable remembrance but as *marcos*, that is, as signs of possession, of the right that a nation assumes for itself by virtue of earliest discovery. The Portuguese mariners often preferred as a *marco*, or sign of possession, to carve that popular French motto that the Infante don Henrique the Navigator was fond of using: *talent de bien faire*. As expressly stated by Manuel de Faria y Sousa in his *Asia Portuguesa* (vol. I, cap. 2, pp. 14 and 18): "It was the practice of the first navigators to leave the motto of the Infante, *talent de bien faire*, inscribed in the bark of the trees." (Cf. Barros, *Asia*, Dec. I, liv. II, cap. 2, vol. I, Lisbon 1778, p. 148.)

The aforementioned motto, carved into two trees by Portuguese mariners in the year 1435, that is, 28 years before the death of the Infante don Henrique, Duke of Viseu, is connected in the history of discovery in a curious way with the debate that arose from the comparison of Vespucci's fourth voyage with Gonzalo Coelho's voyage of 1503. Vespucci relates that Coelho's flagship was wrecked on an island that was variously thought to be San Fernando Noronha, Peñedo de San Pedro, or the problematic Isle of St. Matthias. This last island was discovered by Garcia Jofre de Loaisa on the 15th of October, 1525, at 2.5° south latitude on the meridian of Cape Palmas, almost in the Gulf of Guinea. He sat at anchor there for 18 days, found crosses, orange trees that had grown wild, and two tree trunks with inscriptions that were already 90 years old (Navarrete, vol. V, pp. 8, 247, and 401). I have illuminated this problem in more detail elsewhere (*Examen critique de l'hist. de la Géographie*, vol. V, pp. 129–132) within the investigation of the credibility of Amerigo Vespucci.

The oldest description of the baobab tree (*Adansonia digitata*) is that which the Venetian Aloysius Cadamosto (his actual name was Alvise da Ca da Mosto) gave in 1454. At the mouth of the Senegal River, where he joined with Antoniotto Usodimare, he found trunks, the circumference of which he estimated to be 17 fathoms, that is, around 102 feet (Ramusio, vol. I, p. 109). It is possible he could have compared them to the dragon trees, which had been observed earlier. Perrottet states in his *Flore de Sénégambie* (p. 76) that he had seen monkey bread trees that, standing but 70 to 80 feet tall, had a diameter of 30 feet. The same dimensions were presented by Adanson during his journey of 1748. The largest trunks of the monkey bread tree that he himself saw (1749), some on one of the small Magdalen Islands close to the green foothills and some at the mouth of the Senegal, had diameters of 25 to 27 feet with a height of 70 feet, while the spread of the treetop was 170 feet wide. Adanson, however, also notes that other travelers have found trunks of 30 feet in diameter. Dutch and French mariners had carved their names in letters six inches tall into the trees. One of these inscriptions dates to the 15th century (in Adanson's *Familles des Plantes* of 1763, part I, pp. CCXV–CCXVIII, it states—probably erroneously—that it dates to the 14th), while the rest are from the 16th. Judging by the depth of the inscriptions, which are overlaid with new wood (Adrien de Jussieu, *Cours de Botanique*, p. 62), and by comparing the thickness of such trunks whose various ages were known, Adanson calculated the age and found that a tree with a trunk 30 feet in diameter had been alive for 5,150 years (*Voyage au Sénégal*, 1757, p. 66). He cautiously adds (and I will not change his bizarre orthography): "Le calcul de l'age de chake couche n'a pas d'exactitude géométrike." In the village of Grand Galarques, also in Senegambia, the Negroes have decorated the entrance of a hollow baobab with sculptures cut into the still-living wood. The inner space is used for community gatherings, where people conduct disputes over their various interests. This hall is reminiscent of the "cavern" (*specus*) in a sycamore in Lycia in which the former consul Licinius Mucianus dined with 21 visitors. Pliny (XII, 3) provides the somewhat generous figure of 80 Roman feet for one such hollow in a tree. René Caillié has found the baobab in the Niger valley near Jenne, Cailliaud has found it in Nubia, and Wilhelm Peters has found it all along the eastern African coast, where it is called *Mulapa*, that is, nlapa-tree (actually *muti-nlapa*), and can be found as far down as Lourenzo-Marques, almost to 26° south latitude. The oldest and thickest trees seen by Peters "were 60–70 feet in circumference." Though Cadamosto stated in the 15th century "eminentia non quadrat magnitudini," and even though Golberry (*Fragmens d'un Voyage en Afrique*, vol. II, p. 92) found trunks in the *Vallée des deux Gagnacks* with a diameter of 34 feet at the root, but which stood only 60 feet tall, this disproportion of thickness and height must not be taken as a general rule. "Very old trees," says the learned traveler Peters, "through a gradual dying process, lose their crowns and continue to grow in circumference. In the littoral region of East Africa, one often sees trunks 10 feet thick attain a height of 65 feet."

While the bold estimates of Adanson and Perrottet thus ascribe an age of 5,150 to 6,000 years to the *Adansonia* that they measured, which indeed places them back in the era of the building of the pyramids or even in the time of Menes, i.e., into an epoch when the Southern Cross was still visible in Northern Germany (*Kosmos*, vol. II, pp. 402 and 487), the more certain estimates made in our northern temperate zone, on the other hand, which are based on annual rings and on the relationship that has been discovered between the thickness of the wood layers and the duration of growth, present us with shorter periods. Decandolle finds that among all European tree species, the English yew tree attains the greatest age. Examination of the bole of a *Taxus baccata* yielded results of 30 centuries in Brabourn, County Kent; of 25 to 26 for one from Fotheringall in Scotland; of 14½ for one from Crowhurst in Surrey and 12 for one from Ripon in Yorkshire (Decandolle, *De la longévité des arbres*, p. 65). Endlicher recalls "that in a churchyard in Grasford in North Wales there is another yew tree that below the boughs measured 49 feet in circumference, and which is 1,400 years old, and that one in Derbyshire is estimated to be 2,096 years old. In Lithuania, linden trees have been felled which had a circumference of 82 feet and in which 815 annual rings were counted" (Endlicher, *Grundzüge der Botanik*, p. 399). In the temperate zone of the Southern Hemisphere the eucalyptus species grow to a monstrous circumference; as they also rise to heights over 230 Parisian feet, they contrast extraordinarily with our oaks (*Taxus baccata*), which are colossal only in girth. At Emu Bay on the coast of Van Diemens Land, Mr. Backhouse found eucalyptus trunks with a circumference of 66 feet at the base and still 55 feet when measured five feet above the ground (Gould, *Birds of Australia*, vol. I, intro., p. XV).

Not to Malpighi, as is usually believed, but to the ingenious Michel Montaigne goes the credit for first having mentioned, in his *Voyage en Italie*, 1581, the relationship between tree rings and lifespan (Adrien de Jussieu, *Cours élémentaire de Botanique*, 1840, p. 61). An adept artist engaged in the construction of astronomical instruments, Montaigne brought attention to the importance of tree rings; he also maintained that the side of the trunk directed northward had thinner rings. Jean-Jacques Rousseau shared this belief, and his Emile, when he is lost in the woods, supposedly orients himself by the configuration of tree rings. New observations of plant anatomy teach us, however, that like the acceleration of vegetative growth, so too do the still periods (remissions) in the growth process, the greatly variegated generations of arborescent growth circles (annual rings) from the cambium cells, depend upon influences very different from position relative to geographical direction (Kunth, *Lehrbuch der Botanik*, vol. I, 1847, pp. 146 and 164; Lindley, *Introduction to Botany*, 2nd ed., p. 75).

Those trees among which single individuals attain a diameter of more than 20 feet and have a lifespan of several centuries belong to the broadest assortment of natural families. To wit: baobabs, dragon trees, varieties of eucalyptus, *Taxodium distichum* (Rich.), *Pinus lambertiana* (Douglas), *Hymenaea courbaril*, Caesalpinia, *Bombax, Swietenia mahogani*, the banyan tree (*Ficus religiosa*), *Liriodendron tulipifera* (?), *Platanus orientalis*, our lindens, oaks, and yews. That famous *Taxodium distichon* (*Cupressus disticha* [*Linn.*], *Schubertia disticha* [*Mirbel*]), the Ahuahuete of the Mexicans of Santa Maria del Tule in the state of Oaxaca, does not possess, as Decandolle states, a diameter of 57 Parisian feet, but rather one of 38 (Mühlenpfordt, *Versuch einer getreuen Schilderung der Republik Mexico*, vol. I, p. 153). The two beautiful Ahuahuetes near Chapoltepec that I have often seen (which are probably from an old landscaped garden of Montezuma) measure, according to accounts from Burkart's informative journey (vol. I, p. 268), only 34 and 36 feet in circumference, not diameter, as has often been erroneously asserted. The Buddhists of Ceylon venerate the gigantic trunk of the holy fig tree of Anuradhapura. The banyans, which sink roots from their branches, often achieve a thickness of 28 feet in diameter and, as Onesicritus once expressed in a true representation of Nature, build a roof of foliage like a many-pillared tent (Lassen, *Indische Alterthumskunde*, vol. I, p. 260). For the *Bombax ceiba*, see early notes from the voyages of Columbus in Bembo, *Historia Veneta*, 1551, fol. 83.

Among the oak trunks that have been precisely measured, the most immense one in Europe is probably the one in Saintes in the *Département de la Charente inférieure* on the way to Cozes. Standing 60 feet, the tree has a diameter of 27 feet, 8½ inches, near the ground, and even measures 21½ feet five feet higher up; it has a diameter of 6 feet at the point where the main limbs begin. In the dead portions of the trunk, a small chamber has been cut out, 10 to 12 feet wide and 9 feet high, with a semicircular bench carved into the green wood. A window provides light for the interior, allowing the walls of the chamber, which is closed in by a door, to be gracefully clothed by lichens and ferns. Considering the size of a small block of wood cut from over the door in which 200 rings were counted, the age of the Oak of Saintes was estimated to be 1,800 to 2,000 years (*Annales de la Société d'Agriculture de La Rochelle*, 1843, p. 380).

Regarding the so-called thousand-year-old rose tree (*Rosa canina*) by the crypt chapel of the cathedral of Hildesheim, carefully documented information (for which I am indebted to the kindness of the municipal court assessor Mr. Römer) indicates that only the rootstock is 800 years old. A legend connects the rose root with a vow made by the founder of the cathedral, Louis the Pious, and a document from the 11th century reports "that when Bishop Hezilo rebuilt the cathedral that had recently burned down, he surrounded the roots of the rose with an archway that had remained standing, ran the walls of the crypt chapel, newly consecrated in 1061, to this archway, and spread the branches of the rose along these walls." The current living stem, only two inches thick, is 25 feet tall; the shrub spreads out over about 30 feet on the outer wall of the eastern crypt and is certainly of a significantly old age and certainly worthy of the long-standing reputation that it has been afforded throughout Germany.

If enormity of size in organic development can be taken in general as evidence of a long lifespan, then among the thalassophytes of marine vegetation the kelp variety *Macrocystis pyrifera* Agardh (*Fucus giganteus*) is worthy of special attention. According to Captain Cook and George Forster, this sea plant reaches a length of 360 English or 338 Parisian feet and thus surpasses the height of the tallest conifers, even the *Sequoia gigantea* Endlicher (*Taxodium sempervirens*, Hooker and Arnott) of California (Darwin, *Journal of Researches into Natural History*, 1845, p. 239). Captain FitzRoy corroborated this figure (*Narrative of the Voyages of the "Adventure" and "Beagle,"* vol. II, p. 363). *Macrocystis pyrifera* grows from 64° south to 45° north latitude as far up as the San Francisco Bay on the northwestern coast of the New Continent. Joseph Hooker even believes that this variety of Fucus may be found as far north as the Kamchatka Peninsula. In the waters of the South Pole they may be seen swimming among the loose blocks of ice, or "pack ice" (Joseph Hooker, *Botany of the Antarctic Voyage under the Command of Sir James Ross*, 1844, pp. VIII, 1 and 178; Camille Montagne, *Botanique cryptogame du Voyage de la "Bonite,"* 1846, p. 36). The cellular, ribbon-and-string-shaped figures of the Macrocystis, which are attached to the seabed with a clawlike gripping organ, seem to be limited in their capacity for growth only by random or accidental destruction.

13. The phanerogamic plant varieties that have already been incorporated into the herbaria

One must carefully differentiate among three questions:

1. How many plant species are described in published works?
2. How many have already been discovered, that is, are present in the herbaria, but without having been described?
3. How many probably exist upon the surface of the Earth?

Murray's edition of the Linnaean system contains, including cryptogams, only 10,042 species. Willdenow had already described 17,457 species of phanerogams (*Monandria* to *Polygamia dioecia*) in his edition of the *Species Plantarum* of 1797 to 1807. If one adds to this 3,000 species of cryptogamic plants, then the contribution of Willdenow comprises 20,000 species. More recent investigations have shown how far this estimate of the number of described species and the others preserved in herbaria is below the actual figure. Robert Brown first counted (*General Remarks on the Botany of Terra Australis*, 1814, p. 4) over 37,000 phanerogams. I ventured at that time to present the geographical distribution of 44,000 phanerogams and cryptogams over the various parts of the Earth that had by that point been explored (Humboldt, *De distributione geographica Plantarum*, p. 23). Decandolle finds, in comparing Persoon's *Enchiridium* to his Universal System of 12 individual families, that between the publications of botanists and the herbaria of Europe one may safely presume there are more than 56,000 plant species (*Essai élémentaire de Géographie botanique*, 1820, p. 62). If one considers how many new species have been described by travelers since then (from my expedition alone there are 3,600 from the complete number of 5,800 species we collected in the equinoctial zone), and if one remembers that in all of the botanical gardens put together there are certainly more than 25,000 phanerogams being cultivated, then one will easily recognize how far below the true number Decandolle's assumption lies. In our complete ignorance of the interior of South America (Mato Grosso, Paraguay, the eastern slope of the Andes chain, Santa Cruz de la Sierra, all of the lands between the Orinoco, the Rio Negro, the Amazon, and the Purus), of Africa, Madagascar, Borneo, Inner and East Asia, the thought irresistibly arises that we are not familiar with a third, even a fifth, of the plants that exist on Earth! In South Africa alone, Drège collected 7,092 phanerogamic species (see *Meyers pflanzengeographische Documente*, pp. 5 and 12). He believes that the local flora there comprises more than 11,000 phanerogamic species, while on areas of the same size (12,000 square miles), Koch describes only 3,300 in Germany and Switzerland, and Decandolle only 3,645 in France. I must mention as well the new genera (in part tall forest trees) that are even now, and not far from large trading cities, being discovered in the Lesser Antilles, where Europeans have been visiting for the last 300 years. Such considerations, which I will develop more completely at the end of this annotation, to some extent substantiate the myth of the Zend-Avesta "as if the ancient Creative Force has called forth from the blood of the Sacred Bull 120,000 plant forms!"

If there is no direct scientific solution to the question of how many plant forms— including leafless cryptogams (water algae, mushrooms, and lichens), Characeae, liverworts and mosses, Marsileaceae, Lycopodiaceae, and ferns—exist upon dry land and upon the broad basin of the ocean in the current state of organic life on Earth, then all that is left to us is to attempt an approximating method showing certain figures to be the probable lower extremes (numerical values of minima). Since 1815, in making arithmetical observations of the geography of plants, I have first established a notion of the ratio of the sum of species of individual natural families to the entire mass of phanerogams in such countries where this second group has been sufficiently determined. Robert Brown, the greatest botanist among our contemporaries, had already determined before me the numerical relationship of the primary divisions: the acotyledons (agamae, cryptogams, or cellular plants) to the cotyledons (phanerogams or vascular plants), and the monocotyledons (endogenous plants) to the dicotyledons (exogenous plants). He finds the ratio of monocotyledons to dicotyledons in the tropical zone to be about 1:5, and in the frigid zone at the latitudes of 60 north and 55 south to be about 1:2.5 (Robert Brown, *General Remarks on the Botany of Terra Australis*, in Flinders's *Voyage*, vol. II, p. 338). By the method developed in that work, the absolute numbers of the species in three great divisions of the plant kingdom are compared to one another. I first moved beyond these primary divisions into the individual families and considered the number of species contained in each in relationship to the entirety of phanerogams within a given zone (cf. my work *De distribution geographica Plantarum secundum coeli temperiem et altitudinem montium*, 1817, pp. 24–44, and the further elaboration of numerical relationships that I have provided in the *Dictionnaire des Sciences naturelles*, vol. XVIII, 1820, pp. 422–436, and in the *Annales de Chimie et de Physique*, vol. XVI, 1821, pp. 267–292).

The numerical relationships among plant forms and the laws that may be observed in their distribution may be viewed in two very different ways. When one studies plants in their classification into natural families without considering their geographical distribution, one will ask: What are the basic forms, the structural types, that are exhibited by the greatest number of species? Are there more forms with glumes (Glumaceae) than there are Compositae on the Earth? Can it be that these two orders of plants make up one quarter of the phanerogams? What is the ratio of monocotyledons to dicotyledons?

These are questions of general phytology, the science that investigates the classification of plants and their mutual interconnection, indeed, the current state of all vegetation.

On the other hand, if one considers the plant species that have been grouped according to their analogous structures, not in an abstract manner but according to their climatic environment, that is, to their distribution over the planet, then these questions address a particularly different interest. One will then investigate the question, which plant families dominate the other phanerogams more in the torrid zone than they do nearer the arctic circle? One will ask, Are the Compositae, being at the same latitudes or between the same isothermal lines, more numerous in the New World than in the Old? Do the forms that cease to predominate moving from the equator toward the poles follow a similar law of decrease in climbing the equatorial mountains? Do the ratios of plant families to the entire mass of phanerogams at equal isothermal lines deviate between the temperate zone on this side of the equator and the one on the other side? These are questions of the actual geography of plants and are tied in to the most important challenges that meteorology and terrestrial physics can present. The character of a landscape, the very appearance of a desolate, gorgeous, laughing, or even majestic Nature, depends upon the predominance of certain plant families. The abundance of grasses that form upon the great savannas, the multitude of nourishing palms or socially growing conifers, have exerted a mighty effect upon the material situation of the nations of people, their customs and disposition, and the faster or slower rate of the advancement of their prosperity.

In the study of the geographical distribution of the plant forms, one may observe the species, genera, and natural families separately. Often a single plant species, especially among the social plants, will cover a broad expanse of land. Such is the behavior of fir or pine forests as well as heaths (*Ericeta*) in the North, of citrus groves in Spain, and of the populations of a single species of cactus, Croton, Brathys, or *Bambusa guadua* in tropical America. It is interesting to examine more closely these conditions of individual propagation and organic development. One might ask which species brings forth the most individuals within a given zone, or might simply name the families to which the dominant species in various climates belong. In an especially northerly region, where the Compositae and ferns stand in respective ratios of 1:13 and 1:25 to the sum of all phanerogams (that is, where one arrives at these ratios when the total number of all phanerogams is divided by the number of species belonging to the families of the Compositae or the ferns), a single species of fern can nevertheless cover ten times more land surface area than all of the species of Compositae combined. In this case, the fern plants surpass the Compositae in terms of their mass by virtue of their number of individuals belonging to the same species of *Pteris* or *Polypodium*; they are not predominant, however, if only the numbers of the different specific forms of *Filices* and of the Compositae are compared to the sum of all the phanerogams. Since reproduction then does not follow the same laws for all species, since not all species produce the same number of individuals, the quotients that represent the ratio of the species of a certain family to the sum of all phanerogams do not alone decide the defining characteristic of a given landscape's appearance or of the physiognomy of the various regions of the Earth. Should the traveling botanist be occupied with the profuse repetition of a single species, with its overall mass and the uniformity of vegetation that it produces, then his attention is even more captivated by the infrequency of many others, namely, those that are beneficial to mankind. In the tropical regions, where the forests are formed by Rubiaceae, myrtle shrubs, Leguminosae, or terebinths, it is astonishing how seldom one comes upon trunks of *Cinchona*, certain varieties of mahogany (*Swietenia*), *Haematoxylon*, *Styrax*, and the balsam-scented *Myroxylum*. I am reminded here of the lonely occurrences of the exquisite fever-bark trees (*Cinchona* species), which we had the opportunity to observe while descending the declivity of the high plateaus of Bogota and Popayan, as in the area around Loja, near the Amazon River and the unhealthy valley of the Catamayo. The quina hunters, *cazadores de cascarilla* (as the people of Loja call the Indians and mestizos who collect each year the most effective of all quina bark, that of the *Cinchona condaminea*, in the lonely mountains of Caxanuma, Uritusinga, and Rumisitana), climb at great risk to the tops of the tallest trees in the forest in order to obtain a distant view so that they can locate the slender and widely scattered cinchona trees by the reddish gleam of their large leaves. The average temperature of this important forest region (at 4° to 4.5° south lat.) at 6,000 to 7,500 feet absolute elevation is 12.5° to 16° Réaumur (Humboldt and Bonpland, *Plantes équinoxiales*, vol. I, p. 33, tbl. 10).

In observing the distribution of species, one can also, without regard to their individual propagation and mass, compare to one another the absolute number of species that belong to each family. A comparison of this sort is applied by Decandolle in the work *Regni vegetabilis Systema naturale* (vol. I, pp. 128, 396, 439, 464, 510). Kunth has carried it out on over 3,300 currently known Compositae. It does not indicate which families dominate in terms of mass of individuals or number of species relative to other remaining phanerogams, but rather, how many of the species of one single family are native to this or to that country or continent. The results of this method are on the whole more accurate, for they are reached through careful study of the individual families, without the necessity of being familiar with the entirety of phanerogams of every country. The greatest variety of fern forms, for example, is to be found in the tropics; in the moderate, humid, and shady mountain regions of the islands, every genus presents the most species. While there are fewer of these in the temperate zone than in the tropics, their absolute quantity diminishes even more when approaching the poles. While the frigid zone, Lapland for example, nurtures those species of a family that withstand the cold better than most phanerogams, ferns nevertheless predominate there over other plants more than they do in France or Germany, despite the low absolute number of northern species of fern in terms of the ratio of these species to all local phanerogams. In these other two lands, the quotients are 1:73 and 1:71 respectively; in Lapland the quotient is 1:25. I published these ratios (the entire mass of phanerogams divided by the species of each family) in 1817 in my *Prolegomenis de distributione geographica Plantarum* and in the later writings in French on the distribution of plants on land, amended according to the great works of Robert Brown. Progressing from the equator to the poles, they deviate, each according to its nature, from the ratios that arise from the comparison of the absolute number of species present in each family. One often sees the value of the fractions increase with the decrease of the denominator, even while the absolute number of the species has diminished. In this fractional method to which I adhere (it being more fruitful to the study of plant geography) there are two variables, for should one cross from one isothermal line into another, one sees the total sum of the phanerogams change, but not in the same proportion as the number of species of a particular family.

If we progress from the consideration of these species to the consideration of those divisions that the natural method delineates according to an ideal succession of abstractions, one can direct one's attention to categories or genera, to families, or to still higher classes. There are some genera, indeed some entire families, that belong exclusively to a single zone: not merely because they thrive only under a particular combination of climatic conditions, but also because they came into being only in very limited localities and have been greatly inhibited in their migrations. But there are a greater number of genera and families that have representatives in all countries and at all elevations. The earliest investigations into the distribution of plant forms were directed solely to genera. They may be found in the admirable work of Treviranus in his *Biology* (vol. II, pp. 47, 63, 83, and 129). This method is not so well suited to deliver general results, however, as is that which compares the number of species of each family or the great primary divisions (acotyledons, monocotyledons, and dicotyledons) with the number of all phanerogams. In the frigid zone, the multiplicity of forms does not affect the genus-value (i.e., the number of genera) to the same degree of diminution as it does that of the species; one finds proportionally more genera there with a smaller number of species (Decandolle, *Théorie élémentaire de la Botanique*, p. 190; Humboldt, *Nova genera et species Plantarum*, vol. I, pp. XVII and L). This holds true in virtually the same way on the peaks of tall mountains, which are the home of lone members of a great multitude of genera, of which one might be inclined to presume that they belong exclusively to the vegetation of the plains.

I felt myself obliged to indicate the various points of view from which one can consider the laws of the geographical distribution of plants. Only when one confuses these viewpoints with one another does one find contradictions, which are unjustly attributed to uncertainty of observation (*Jahrbücher der Gewächskunde*, vol. I, Berlin, 1818, pp. 18, 21, 30). When one employs the expressions "this form or this species dwindles upon approaching the frigid zone; its true home lies between this and that latitude; it is a southern form; it predominates in the temperate zone," then it must also be stated whether one is speaking of the absolute number of species and their absolute frequency of occurrence as it increases or decreases with latitude, or whether one intends to say that a family, compared to the entire number of phanerogams of a region of flora, predominates over other plant families. The sensory impression of predominance rests precisely upon the concept of relative quantity.

Like the solar system, geophysics has its numerical elements, and only with the concerted efforts of traveling botanists will we come gradually to a familiarity with its actual laws, the laws that determine the geographical and climatic distribution of plant forms. I have already mentioned that in the temperate zone of the Northern Hemisphere the Compositae (Synantherea) and the Glumaceae (with this term I include the three families of grasses, the Cyperoides, and the Juncaceae) constitute one-fourth of all phanerogamic vegetation. The following ratios are the result of my investigations of 7 large families of the plant kingdom within the same temperate zone:

Glumaceae	1/8	(grasses alone 1/12)
Compositae	1/8	
Leguminosae	1/18	
Labiatae	1/24	
Umbelliferae	1/40	

Amentaceae (Cupuliferae, Betulineae, and Salicineae)   1/45
Cruciferae   1/19

The forms of organic life are mutually dependent upon one another. The unity of Nature lies in the fact that these forms limit one another according to laws that are most likely bound to long periods of time. If one has an accurate knowledge of the number of species from one of the large families of Glumaceae, Leguminosae, or Compositae grow-ing upon a particular place on Earth, then one can with a certain probability approach or even determine the number of all phanerogams as well as the number of all other spe-cies from other plant families that are growing there. The number of Cyperoides deter-mines the number of Compositae, the number of Compositae that of the Leguminosae; indeed, these estimates place us in a position to recognize in which classes and orders the flora of a country are still incomplete; they teach us, as long as one is careful not to confuse greatly differing systems of vegetation with one another, what yields may yet be expected from particular families.

The comparison of the numerical relationships of the families in different well-explored zones led me to recognize the law by which the plant forms that make up a natural family numerically increase or decrease over the distance from the equator to the poles, when they are compared to the entire quantity of phanerogams belonging to each zone. Along with the direction of increase, its rapidity, i.e., the measure of increase, must also be observed. One will see the denominator of the fraction, which expresses the ratio, grow or diminish. Thus, for example, does the lovely family of the Legumi-nosae decrease in number moving from the equinoctial zone toward the North Pole. While one finds a proportion of 1/10 in the torrid zone (0° to 10° lat.), that section of the temperate zone that lies between 45° and 52° yields 1/18, and the frigid zone (67° to 70°) only 1/35. This same direction that the large family of Leguminosae follows (increas-ing toward the equator) is shared by the Rubiaceae, the Euphorbiaceae, and above all, the Malvaceae. Conversely, the grasses and Juncaceae (especially the latter) decrease to-ward the torrid zone, as do the Ericaceae and Amentaceae. The Compositae, Labiatae, Umbelliferae (umbel flora), and Cruciferae, moving from the temperate zone, decrease toward both the equator and the pole, the fastest rate of decrease occurring with the Umbelliferae and Cruciferae moving in the latter direction. Meanwhile, the Cruciferae appear in the temperate zone three times more frequently in Europe than in the United States of North America. In Greenland, the Labiatae decrease in number down to one species and the Umbelliferae down to two, though the entire number of phanerogams still reaches 315 species, according to Hornemann.

One cannot but note that the development of plants of different families and the distribution of forms is entirely dependent upon neither geographical nor isothermal latitude, but rather that the quotients upon one and the same isothermal line in the tem-perate zone are indeed not always the same, as in the case, for example, of the plains of America and those of the Old Continent. Within the circle of latitude there exists a very conspicuous difference between America, East India, and the western coasts of Africa. The distribution of organic life forms on the Earth depends not only upon extremely complex thermal and climatic conditions but also upon geological forces that remain almost wholly unknown to us, as they were the result of prehistoric conditions of the Earth and catastrophes that did not befall every part of our planet simultaneously. The great pachyderms are currently absent in the New World, while we can still find them in Asia and Africa in places of analogous climate. These differences must not distract us from seeking out the laws of Nature; indeed, they should entice us to study them in all their complexity.

The numerical laws of families, the often so conspicuous consistency of ratio values in those very places where the species that make up these families to a large extent differ from one another, lead into the mysterious darkness that covers all things connected to the classification of organic forms into animal and plant species, and all that proceeds from being to becoming. Let us take the examples of two long-investigated neighboring countries, France and Germany. Many varieties of grasses, of Umbelliferae and Cruciferae, of Compositae, Leguminosae, and Labiatae, which are among the most common species in Germany, are not to be found in France, yet the ratios of these six great families are nearly identical. Here are their comparative values:

Family	Germany	France
Gramineae	1/13	1/13
Umbelliferae	1/22	1/21
Cruciferae	1/18	1/19
Compositae	1/8	1/7
Leguminosae	1/18	1/16
Labiatae	1/26	1/24

This consistency in the ratio of the number of species of a particular family to the entire quantity of phanerogams in Germany and in France would by no means occur if the species absent from Germany were not compensated for by other species of the same family. Those who like to dream of the gradual alteration of species and who consider parrots respectively native to neighboring islands to be of one differently altered species will ascribe the remarkable equality of the ratios above to a migration of like species that, through climatic influences over the course of long millennia, have changed themselves and give this appearance of being different species that replace the original ones. Why then has our common heather, why have our oak trees, not pressed eastward from Europe, across the Urals and into Northern Asia? Why is there no species of the genus *Rosa* in the Southern Hemisphere, and virtually no *Calceolaria* in the Northern? Temperature requirements cannot explain this. Thermal conditions alone, like the hypothesis of plant migrations radiating from certain central points, are insufficient to make the current distribution of plant forms (or of defined forms of the organism) comprehensible. Thermal conditions hardly illuminate the individual cases of certain species that, on the plains approaching the poles or on mountain slopes at a certain vertical height, arrive at specific limits beyond which they do not proceed. The vegetative cycle of every species, however different their durations may be, requires a certain minimum temperature in order to thrive. (Playfair in the *Transactions of the Royal Society of Edinburgh*, vol. V, 1805, p. 202; Humboldt, *Über die Summe der Thermometergrade, welche ein Vegetations-Cyclus der Cerealien bedarf*, in *Mémoire sur les Lignes isothermes*, p. 96; Boussingault, *Économie rurale*, vol. II, pp. 659, 663, and 667; Alphonse Decandolle, *Sur les causes qui limitent les espèces végétales*, 1847, p. 8.) But all of the conditions of the existence of a plant in its natural distribution or culture (conditions of the geographical distance from the pole and the elevation of the location) are complicated by the difficulty of determining the beginning of the thermal vegetative cycle; they are also complicated by the influence that the inconsistent distribution of equal amounts of warmth over successive groups of days and nights exerts upon the plant's susceptibility to stimulation, upon its developmental progress, indeed, upon its entire life cycle; and finally, they are also complicated by the side effects of hygrometric and electrical atmospheric conditions.

My investigations of the numerical laws of the distribution of forms will eventually be able to be applied with some success to the various classes of vertebrates. The abundant collections of the Muséum d'Histoire naturelle in the Jardin des Plantes in Paris were estimated to have comprised, as of 1820, over 56,000 species of phanerogamic and cryptogamic plants in the herbaria, 44,000 species of insect (probably a low figure, but the one given to me by Latreille), 2,500 fish species, 700 reptiles, 4,000 birds, and 500 mammals. Europe is home to approximately 80 mammal species, 400 birds, and 30 reptiles; there are thus five times as many birds as mammals in the northern temperate zone (similar to how Europe has five times as many Compositae as Amentaceae and Coniferae, and five times as many Leguminosae as Orchidaceae and Euphorbiaceae). In the southern temperate zone there is an apparently corresponding ratio of mammals to birds at 1/4.3. The birds, and even more so the reptiles, increase at a greater rate in approaching the tropics than do the mammals. From Cuvier's research, one could well believe that the ratio was different at an earlier time and that more mammals than birds were wiped out by subsequent cataclysms. Latreille has shown which insect groups increase as they approach the poles and which increase when they approach the equator. Illiger has listed the native homes of 3,800 birds by continent, which is far more instructive than classifying them by climatic zone. It is possible to explain how it is that upon a given region of the planet, the individuals of a particular class of plant or animal limit one another in terms of number. and how, through struggles and long wandering through privations of nourishment and comfort, a state of equilibrium was achieved. But the causes that have spatially limited not the number of individuals of a form but rather the very forms themselves, and that have been the foundation of the forms' differences in type, lie beneath the impenetrable veil that continues to hide from our eyes everything that touches upon the beginning of all things and the first appearance of organic life.

As I remarked in the introduction to this annotation, should one wish to attempt to approximate the minimum number (French mathematicians say *le nombre limite*) below which the sum total of all phanerogams on the Earth may not be estimated, then the comparison of all currently recognized ratios of plant families to the number of species presently contained in our herbaria and being cultivated in great botanical gardens should be the surest indicator in this pursuit. As we have just noted above, as of 1820 the herbarium of the Jardin des Plantes in Paris was estimated to contain 56,000 species. I shall not allow myself any assumption as to what is contained in the herbaria of England; the great Parisian herbarium, however (which Benjamin Delessert, in an act of the noblest selflessness, established for free and general use), was said to contain at the time of his death some 86,000 species—nearly the same number that Lindley, even so late as 1835 (*Introduction to Botany*, 2nd ed., p. 504), conjectured to be the number of all species "upon the entire Earth." Few herbaria are numbered with particular care and categorized according to an established, strict, and consistently applied classification of the varieties. Additionally, the number of plants that are absent from the large, so-called general herbaria and yet may be found in single, less extensive ones is not small. Dr. Klotzsch, custodian of the large Royal Herbarium at Schöneberg near Berlin, estimates its total number of phanerogams at some 74,000 species.

Loudon's useful work (*Hortus britannicus*) presents an approximate overview of the species that are, or in not-too-distant times have been, cultivated in all of the collected gardens of England. Including native plants, the 1832 edition enumerates exactly 26,660 phanerogamic plants. This great number of formally and currently cultivated plants in all parts of Great Britain cannot compare to "what a single botanical garden at any given time" has to offer in terms of living plants. In this respect, the botanical garden near Berlin has long been held to be one of the richest in Europe. The reputation of this extraordinary treasure previously rested upon a merely approximated appraisal, and as my friend and colleague of many years Professor Kunth very correctly states (in a handwritten note presented to the *Gartenbau Verein* in December, 1846), "Only upon completion of a systematic catalog, based on a rigorous examination of the species, could an accurate count be conducted. This count yielded a figure of over 14,060 species; if one removes from these 375 cultivated ferns, there remain 13,685 phanerogams, among which 1,600 Compositae, 1,150 Leguminosae, 428 Labiatae, 370 Umbelliferae, 460 Orchidaceae, 60 palms, and 600 grasses and Cyperaceae may be found. If one should compare the figures above with the number of those described in recent works—approximately 10,000 Compositae (Decandolle and Walpers), 8,070 Leguminosae, 2,190 Labiatae (Bentham), 1,620 Umbelliferae, 3,544 grasses, and 2,000 Cyperaceae (Kunth, *Enumeratio Plantarum*)—then one will recognize that the Berlin botanical gardens, in terms of the very large families (Compositae, Leguminosae, and grasses), is currently cultivating only 1/7, 1/8, and 1/9 respectively, and in terms of the small families (Labiatae and Umbelliferae) perhaps 1/5 to 1/4 of the currently described species. Estimating therefore the number of all the various phanerogams cultivated at a given time in all of the botanical gardens in Europe to be around 20,000, one finds that these cultivated phanerogams appear to represent about one-eighth of those that have been described and can be found in the herbaria, a number that must be near to 160,000. This estimate may be fairly viewed as no exaggeration, for from many of the larger families (e.g., the Guttiferae, Malpighiaceae, Melastomeae, Myrtaceae, and Rubiaceae) barely one one-hundredth part belongs to our gardens." If we start with the number from Loudon's *Hortus britannicus* (26,660 species) as a basis, then the estimate of 160,000, in light of the well-founded conclusions borrowed from Professor Kunth's handwritten note, climbs to 213,000 species. And even this remains a very conservative estimate, considering that Heynold's *Nomenclator botanicus hortensis* (1846) puts the number of cultivated phanerogams at 35,600. Taken all in all (and at first sight, this conclusion alone is remarkable enough), more phanerogamic plant species have been identified than insects, thanks to the gardens, descriptions, and herbaria. By the median number of the figures given by several excellent entomologists with whom I was able to consult, the number of insects currently described, or contained in collections but not yet described, may be placed at 150,000 to 170,000 species. The splendid Berlin collection contains around 90,000, of which some 32,000 are beetles. There are some who have collected in distant regions a multitude of plants, without bringing with them the insects that live on or around them. If one restricts these numerical estimates to a specific part of the Earth where insects and plants have been studied most extensively, to Europe for example, then the ratio of these life forms, of phanerogamic plants to insects (since all of Europe hosts but seven to eight thousand phanerograms), changes in such a manner that the number of currently recognized insects of Europe enjoys a more than threefold predominance. According to the very interesting information provided to me by my friend Dohrn in Stettin, over 8,700 insects have already been collected from the abundant fauna of the region, even though many microlepidoptera remain uncollected. The number of phanerograms there barely exceeds 1,000. The insect fauna of Great Britain is estimated to be around 11,600. Such a predominance of animal forms must become less surprising if one considers that there are large categories of insects that feed upon only animal materials, and others that eat only agamic plants (fungi, above and beneath the ground). According to Ratzeburg, the pine-tree lappet alone (*Bombyx pini*), the most destructive of all forest insects, is host to 35 parasitic Ichneumonidae.

If these considerations have led us to the ratio in which the content of the gardens stands to the total of species that have already been described and are preserved in herbaria, then it remains to us to consider the ratio of these latter species to the presumably extant forms currently found upon the Earth, that is, to calculate the minimum estimate of these as based upon the ratios of the families—in other words, based upon hazardous multiples. Such a calculation gives such minuscule results for the lower limit that one may recognize just from this how even in the large families that appear in recent times to be the ones most greatly increased by the botanists' describing of plants, we have gained knowledge of but a small part of the trove that lies before us. Walpers's repertory serves to supplement Decandolle's *Prodromus* of 1825 up to the year 1846. This work lists the family of Leguminosae at 8,068 species. One may accept the ratio to be 1/21, since it is 1/10 in the tropics, 1/18 in the central temperate zone, and 1/33 in the northern frigid zone. The Leguminosae that have been described would thus lead us only to accept a figure of 169,400 existing phanerograms on the entire surface of the Earth, while the Compositae, as shown above, already indicate more than 160,000 phanerograms that are recognized (i.e., described and contained in herbaria). This contradiction is instructive and can be explained by the following analogous considerations.

The majority of Compositae, only 785 species of which were known to Linné and which have now grown to 12,000, seems to belong to the Old Continent; Decandolle, at least, described only 3,590 American species, compared to 5,093 European, Asiatic, and African. This abundance of Compositae in our plant systems is deceptive, however; it is only seemingly large, for the quotient of the family (1/15 in the tropics, 1/7 in the temperate zone, 1/13 in the frigid zone) allows us to recognize that a somewhat larger number of species of Compositae than of Leguminosae have as yet escaped the efforts of explorers, for multiplying by 12 will only yield the improbably low result of 144,000 phanerograms! The families of the grasses and the Cyperaceae produce even lower results, for, relatively speaking, still fewer species of these families have been described and collected. One need simply cast a glance at the map of South America and imagine the tremendous spaces that in botanical terms have been either incompletely explored or not explored at all: the grasslands of Venezuela, of Apure and Meta! And to the south of the forest region of the Amazon: in Chaco, in Eastern Tucuman, in the Pampas of Buenos Aires and Patagonia! Northern and Central Asia feature an almost equal area of steppes, where dicotyledonous plants (herbs), however, mingle in higher numbers with the Gramineae.

Assuming there is sufficient evidence to believe that half of the phanerogamic plants of our planet have been recognized, and if one stands by the notion that the number of these recognized species is either 160,000 or 213,000, then it must be that for grasses, the general ratio value of which seems to be 1/12, the first figure implies a minimum of 26,000, and the second a minimum of 35,000 different species, of which only 1/8 to 1/10 are recognized.

The hypothesis that we are already familiar with half of the phanerogams on Earth is contradicted by the following considerations. Several thousand mono- and dicotyledonous species, including among them some tall tree forms, are being discovered (I am thinking of my own expedition) in regions wherein a very considerable expanse of land has already been investigated by excellent botanists. The portions of the continents not yet visited by observers far, far exceed the size of the regions that have been (even superficially) explored. The greatest diversity of phanerogamic vegetation, that is, the greatest number of species within areas of a given size, are to be found between the tropical lines of latitude or in the subtropic zones. It is thus all the more important to remember how almost thoroughly unfamiliar we are with the New Continent north of the equator (with the flora of Oaxaca, Yucatan, Guatemala, Nicaragua, the Isthmus of Panama, El Choco, Antioquia, and the *Provincia de los Pastos*) and south of the equator (with the flora of the immeasurable forest country between the Ucayali, the Rio de la Madera, and the Tocantins, three mighty tributaries of the Amazon, and with the flora of Paraguay, and the *Provincia de las Misiones*). In Africa, except for the coasts, we are unfamiliar with the vegetation of the entire interior between 15° north and 20° south latitude. In Asia, we are unfamiliar with the flora of the regions south and southeast of Arabia, where highlands ascend to six thousand feet, with the flora of the regions between the Tien Shan, the Kunlun, and the Himalayas, with the flora of Western China and the greater portion of the Trans-Gangetic lands. Even less familiar to the botanist are the interiors of Borneo, New Guinea, and a portion of Australia. Farther to the south, the number of species diminishes dramatically, as Joseph Hooker so astutely shows in his own observations of Antarctic flora. The three islands that form New Zealand extend from 34.5° to 47.25° latitude and feature, along with snow peaks of over 8,300 feet, a considerable variety of climates. Only the northernmost island, since the voyage of Banks and Solander, as well as those of Lesson, the Cunningham brothers, and Colenso, has been more or less completely explored, and for more than 70 years now we know of fewer than 700 varieties of phanerogam among its native flora (Ernest Dieffenbach, *Travels in New Zealand*, 1843, vol. I, p. 419). This paucity of plant species correlates with a paucity of animal species. Joseph Hooker remarks that Iceland supports five times more phanerogamic species than Lord Auckland's and Campbell's Islands combined, though they are 8° to 10° closer to the equator in the Southern Hemisphere. This Antarctic flora is characterized both by uniformity and great luxuriance, due to the influence of an uninterruptedly cool and moist climate. In Southern Chile, in Patagonia, and even down to Tierra del Fuego, from 45° to 56° south latitude, this uniformity is conspicuous, not only on the plains but in the mountains as well, on whose slopes climb the same species. But if one were to compare the flora of Southern France (whose northern latitude matches the southern latitude of the Chonos archipelago off the coast of Chile) with the Scottish flora of Argyleshire (which is as far north of the equator as Cape Horn is south), how great is the variety of species! In the Southern Hemisphere, the same types of vegetation can spread over the course of many degrees of latitude. Approaching the North Pole, ten flowering phanerogams may be collected on Walden Island (80.5° north lat.), while if one moves toward the South Pole, even just to the South Shetland Islands at 63° south latitude, one may find but a single species of grass (Joseph Hooker, *Flora antarctica*, 1847, pp. 73–75). The ratios of plant distribution developed here testify to the great mass of the as yet unobserved, uncollected, and undescribed phanerogams belonging to the tropics and to the 12° to 15° of latitude that border them.

It did not strike me as unimportant, by means of this seldom-applied field of arithmetic botany, to expose the incomplete state of our knowledge and to more distinctly formulate numerical questions than would have previously been possible. With all conjecture regarding numerical relationships one must first concentrate on the possibility of establishing the lower limit: this is my approach in another work, where I address the question of the ratio of pressed gold and silver to the quantity of processed precious metals available; likewise in the question of how many stars of tenth to twelfth magnitude are scattered about the heavens—that is, how many of the smallest stars visible to a telescope might the Milky Way hold? (John Herschel, *Results of Astronomical Observations at the Cape of Good Hope*, 1847, p. 381.) If it were possible to research completely all of the species of one of the large phanerogamic families through observation, it is certain that one would in so doing come close to knowing the entire sum of all phanerogams on the planet, to being acquainted, as it were, with the embodiment of all families. The more the number of species in a large family is gradually exhausted through the progressive exploration of unknown regions, the higher the number of the lower limit rises, and (as the forms limit one another according to as yet uninterpreted laws of the planetary organism) the more closely we approach the solution of a great numerical life-problem. But is the number of organisms itself constant? Do not, after long periods of time, new forms of vegetation spring up from the ground, while others grow rarer and rarer and finally disappear? Geognosy, with its historical monuments to ancient life on Earth, affirms the second part of this question. "The primeval world," to employ the words of the ingenious Link (*Abhandlung der Akademie der Wissenschaft zu Berlin aus dem Jahre 1846*, p. 322), "compresses that which has gone away into wondrous forms, indicating a greater development and complexity in the world to follow."

14. Even the height of the atmosphere and its pressure

The pressure of the atmosphere has a decided influence upon the form and life of plants. Because of the abundance and importance of the leaf organs, equipped as they are with stomata, this life is for the most part facing outward. Plants live primarily on and through their outer surface, thus their dependence upon the surrounding medium. Animals are more subject to interior stimuli; they produce and maintain their own body temperature, through muscle motion their electrical currents and their chemical organic processes, which are dependent upon these currents and work upon them in return. A sort of skin-respiration is an active vital function of plants, and this respiration, inasmuch as it is the vaporization, inhalation, and exhalation of fluids, depends upon the pressure of the atmosphere. For this reason alpine plants are more aromatic, and also more hirsute, covered with numerous respiratory vessels. (See my work *Über die gereizte Muskel- und Nervenfaser*, vol. II, pp. 142–145.) For as I have demonstrated elsewhere, zoonomic experience indicates that the more easily the conditions for their functioning are fulfilled, the greater the number of organs that develop, and the more elegantly they construct themselves. Alpine plants therefore do not thrive easily on the plains, for the respiration of their outer coverings is disrupted by the increased barometric pressure.

Whether the ocean of air that surrounds our world has always exerted the same average pressure is thoroughly undecided. We do not even know for certain whether the average barometric level in any particular place has remained the same for a hundred years. According to Poleni's and Toaldo's observations, this pressure seemed to be variable. There has long been doubt surrounding the veracity of these observations. But the more recent research of the astronomer Carlini makes it seem probable that the average barometric level in Milan is decreasing. Perhaps the phenomenon is very localized and dependent upon the periodic changes of descending air currents.

15. Palms

It is remarkable that of this majestic plant form, the palm, some of which tower to twice the height of the Royal Palace in Berlin and which the Indian Amara Sinha quite characteristically dubbed the Queen of the Grasses, only 15 species had been described by the death of Linné. The Peruvian explorers Ruiz and Pavon added only eight to the list; Bonpland and I, having covered a larger piece of land from 12° south to 21° north latitude, described 20 new palm species and differentiated as many more, which we introduced by name without being able to procure complete blooms from them (Humboldt, *De distrib. geogr. Plantarum*, pp. 225-233). Today, 44 years after my return from Mexico, there are, including the East Indian varieties introduced by Griffith, over 440 methodically described species of palm from both continents. My friend Kunth's *Enumeratio Plantarum* (1841) alone contains 356 species.

Only a few palms belong, like our conifers, Quercineae, and Betulineae, to the group of socially growing plants; such are the Moriche palm (*Mauritia flexuosa*) and the two Chamaerops species, of which one (*Ch. humilis*) fills great tracts of land at the mouth of the Ebro and in Valencia, while the other, which we discovered on the Mexican shore of the Pacific (*Ch. mocini*), is completely without spines. Just as these shore palms are littoral plants, to which belong the coconut and the Chaemerops, so too are there a group of mountain palms in the tropics, which, if I am not mistaken, were quite unknown before my South American travels. Nearly all species of the palm family vegetate on the plains at an average temperature between 22° and 24° Réaum. These are seldom found above 1800 feet on the Andes chain, but the beautiful wax palm (*Ceroxylon andicola*), the palmetto of Azufral at the Quindio Pass (*Oreodoxa frigida*), and the reedlike *Kunthia montana* (*Caña de la Vibora*) of Pasto all grow at an elevation of between 6,000 and 9,000 feet above sea level, where the Réaumur thermometer often sinks to between 4.8° to 6° at night and where the average temperature barely reaches 11°. These alpine palms are clustered among nut trees, varieties of Podocarpus with needles like the yews, and oaks (*Quercus granatensis*). By taking exact barometric readings, I was able to ascertain the upper and lower limits of the wax palm. We first began to find this palm on the eastern slope of the Andes chain at Quindio at an elevation of 7,440 feet, but it continued to climb upward as far as Garita del Paramo and Los Volcanitos, to an altitude of 9,100 feet. Several years after my departure, the excellent botanist don José Caldas, who for a long time was our companion in the mountains of New Granada, and who later fell as a bloodied victim of Spanish partisan hatred, discovered in Paramo de Guanacos three more palm species very near to the line of perpetual snow, and thus probably at a height of more than 13,000 feet (*Semanario de Santa Fé de Bogotá*, 1809, no. 21, p. 163). Even outside of the tropical regions, at 28° latitude, *Chamaerops martiana* rises up in the foothills of the Himalayas (Wallich, *Plantae asiaticae*, vol. III, tbl. 211) at elevations reaching 5,000 English feet (4,690 Parisian feet).

If we observe the extreme geographical and thus also climatic limits of the palms in places that are not very high above sea level, we will see some forms (the date palm, the *Chamaerops humilis*, the *Ch. palmetto*, and the *Areca sapida* of New Zealand) far into the temperate zone of both hemispheres, penetrating into the regions where the average annual temperature just barely reaches 11.2° and 12.5° Réaumur. Placing the cultivated plants in the order of which ones demand the most warmth, they follow this succession (beginning with the maximum): cacao, indigo, pisang, coffee, cotton, date palm, citrus, olive, true chestnut, and wine grape. In Europe, the date palm and *Chamaerops humilis* both thrive up to the parallels of 43.5° to 44°, in the Genoese Riviera de Ponente for example, near Bordighera between Monaco and San Stefano (where there stands a grove of palms with more than 4,000 trunks), and in Dalmatia near Spalatro. Curiously, the *Chamaerops humilis* is present in large numbers in Nice and in Sardinia but is absent from the island of Corsica, which lies between them. On the New Continent, the *Chamaerops palmetto*, which can grow to 40 feet, extends to the north only as far as the 34° latitude, which can be explained by the curving of the isothermal lines. In the Southern Hemisphere, the palms in Australia, of which there are but very few (6 or 7 species), range only to 34°, according to Robert Brown (*General Remarks on the Botany of Terra Australis*, p. 45), and in New Zealand, where Sir Joseph Banks first saw the Areca, only as far as 38°. Africa, which quite contrary to old and still widespread belief is actually poor in palm species, features one species of palm, *Hyphaene coriacea*, which extends only to Port Natal at 30° south latitude. The mainland of South America presents almost the same limits. East of the Andes chain, in the Pampas of Buenos Aires and in the Cisplatina Province, the palms come as far as 34° and 35°, according to Auguste de St. Hilaire (*Voyage au Brésil*, p. 60). And according to Claude Gay, to the west of the Andes and almost exactly as far down may be found the *Coco de Chile* (our *Jubaea spectabilis?*), the only palm species in the entire country of Chile. (Cf. Darwin, *Journal*, ed. 1845, pp. 244 and 256.)

I would like to insert here some aphoristic remarks that I wrote down while still aboard the ship in March of 1801, at the moment when we left the palm-rich mouth of the Rio Sinu west of Darien to sail to Cartagena de Indias.

"In the last two years in South America we have seen 27 different species of palm. How many must Commerson, Thunberg, Banks, Solander, both Forsters, Adanson, and Sonnerat have observed on their long journeys! And yet our plant systems, in which I am recording this, still know of but 14 to 18 systematically described palm species. The difficulty of procuring palm blossoms, just to reach them, is in fact greater than one could ever imagine. We felt this all the more when we directed our attention primarily to palms, grasses, Cyperaceae, Juncaceae, Cryptogams, and all other objects that had until then been so neglected. Most palms bloom only once a year and indeed, near the equator, in January and February. Which traveler is likely to find himself in palm-rich regions in these specific months? And further, the blossoming time for many palms is of so few days' duration that one almost always arrives too late and sees the palm with swelling ovaries and no male blossoms. In land areas of 2,000 square miles one will often find only 3 or 4 species of palm. Who can simultaneously attend the blossoming months in the palm-rich missions of Rio Caroni, in the *Morichales* at the mouth of the Orinoco, in the valley of the Caura and the Erevato, on the banks of the Atabapo and the Rio Negro, or on the slopes of Duida? Added to this is the difficulty of acquiring the blooms when they hang atop 60 foot trunks armored with spines and standing amidst dense forest or on swampy shores, like those of the Temi and the Tuamini. When a traveler in Europe prepares for a natural history voyage, he is formulating dreams: of shears and curved knives that, when attached firmly to a shaft, could catch anything; of boys, their feet bound by a cord, who can climb the highest trees. These dreams, alas, almost all remain unfulfilled; the acquisition of the palm spathe, due to the great height, is not feasible. In the mission settlements along the river network of Guyana, one finds oneself amidst Indians who find their poverty, their stoicism, and their uncouth ways rich and free from want, such that neither money nor the offer of presents can persuade them to walk three steps off the footpath, assuming there is one. Such intractable coolness on the part of the Indians infuriates the European all the more upon seeing how members of this race of men climb everything with such incomprehensible ease, wherever their own inclinations take them, for example, to catch a parrot or an iguana, or perhaps a monkey that, having been wounded by an arrow, uses his curled tail to keep from falling. In the month of January in Havana, near the city on the public walk and in the nearby fields, all of the trunks of the *Palma Real* (our *Oreodoxa regia*) were crowned with snow-white blossoms. For days we offered the Negro boys that we encountered in the streets of Regla or Guanavacoa two piasters for a single spadix of the hermaphroditic blossoms, but to no avail! People in the tropics do not take on any strenuous work unless forced to it by the most extreme need. The botanists and painters of the Spanish Royal Commission on Natural History under the leadership of the Count of Jaruco y Mopox (Estevez, Boldo, Guio, Echeveria) have themselves admitted to us that for several years they had not been able to examine these palm blossoms, which were unattainable to them.

"After enumerating these difficulties, I am able to understand that which in Europe would have remained incomprehensible to me, namely that though we have in two years discovered over 20 different palm species, we have up to now been able to systematically describe no more than 12. What an interesting work an explorer could produce if in South America he were to concentrate exclusively on palms and depict the spathe, spadix, blossom parts, and fruits in actual size! [I wrote this many years before the Brazilian journey of Martius and Spix, thus before the publication of the first man's outstanding work on palms.]

"In the leaves there is much uniformity of shape: they are either feathered (*pinnata*) or fan-shaped (*palmo-digitata*); the leafstalk (*petiolus*) is sometimes free of spines, sometimes sharp-toothed (*serrato-spinosus*). The form of the leaves of *Caryota urens* and *Martinezia caryotifolia*, which we saw on the banks of the Orinoco and Atabapo, and later on the Andes pass of Quindiu at an elevation of over 3,000 feet, is nearly unique among the palms, much as the leaf form of the gingko is among the trees. In the habitus and physiognomy of palms in general there lies a great character that is difficult to put into words. The bole (*caudex*) is simple, very rarely parting, dracaenalike, into branches, as do *Cucifera thebaica* (the dum palm) and *Hyphaene coriacea*. It is sometimes bulky and thick (*corozo del Sinu*, our *Alfonsia oleifera*), sometimes reedlike and weak (*Piritu*,

Kunthia montana, and the Mexican *Corypha nana*), sometimes overly swollen at the bottom (coco), sometimes smooth and sometimes rough (*palma de covija ó de sombrero* in the Llanos), sometimes spiny (*corozo de Cumana* and *macanilla de Caripe*), the long spines regularly distributed in concentric rings.

"Characteristic differences can also be seen in the roots, which push forth at a height of about 1–1½ feet, thus forming either a scaffold that holds up the trunk or a sort of bulging base that grows around it. I have watched viverrines and even very small monkeys slipping through these root scaffolds of the Caryota. Often the bole is swollen only in the middle, and weaker above and below, as is the case of the *palma real* of the island of Cuba. The green of the leaves is sometimes dark and glossy (Mauritia, coco), sometimes a silvery white on the underside (as with the slender fan palm, *Corypha miraguama*, which we found near Trinidad Harbor in Cuba). Occasionally, the middle of the fan-shaped leaf is adorned with concentric yellow and bluish striping like a peacock's tail, as we found with the thorny Mauritia that Bonpland discovered on the bank of the Rio Atabapo.

"An even more important characteristic than the shape and color of the leaves lies in their orientation. The *foliola* are sometimes comblike, arranged close together across a surface, with stiff *parenchyma* (coco, *Phoenix*, whence derives the splendid reflection of the sun upon the upper surface of the leaves, which are of a fresher green in coco and a duller, ashen color in the date palm); sometimes the foliage appears to be woven like reeds of thinner, more flexible vessels, curling at the pointed end (jagua, *palma real del Sinu*, *palma real de Cuba*, *piritu del Orinoco*). Besides their axis itself (the trunk), it is primarily the orientation of its leaves that endows the palm with its impression of towering majesty. Part of the physiognomic beauty of a palm species is that not only in its earliest days (as with the date palm, the single variety that has been introduced in Europe) but throughout its entire life it has leaves that strive ever upward. The more acute the angle that the fronds create with the (upward) growth of its trunk, the more grand and dignified is its form. What a very different sight is created by the drooping leaves of the *palma de covija del Orinoco y de los Llanos de Calabozo* (*Corypha tectorum*), which has leaves that are closer to the horizontal line, or at least are less upright than those of the date and coconut palms, or the skyward-reaching fronds of the jagua, the cucurito, and the pirijao!

"Nature has gathered together all of the beautiful elements of form in the jagua palm, which together with the 80- to 100-foot-tall *cucurito* or *vadgihai* adorns the granite cliffs of the cataracts of Atures and Maypures, and was seen by us occasionally on the lonely banks of the Casiquiare. Their smooth and slender trunks lift themselves to a height of 60 to 70 feet, so that they rise above the deciduous forest like a colonnade of pillars. These lofty tops contrast wondrously with the thick-foliaged ceiba varieties, with the forests of Laurineae, Calophyllum, and Amyris species that surround them. Their fronds, which are few in number (perhaps 7 or 8), reach almost vertically upward for some 14 to 16 feet. The tips of the foliage are curled and feathery tufts. The individual leaves have a thin, grasslike *parenchyma* and flutter, light and airy, around the gently bobbing leafstalk. On all palms, the blossom structures burst forth below the collars where the fronds grow from the trunk. The nature of this bursting forth likewise modifies the physiognomic character. With a few (*corozo del Sinu*) the spathe stands vertically, and the fruits stand up in a configuration like a thyrse, similar to the fruits of the

Bromelia. With most, however, the spathes hang (some of them smooth, some prickly and rough) downward, and with a few the male blossom is brilliant white. The blooming spadix then can be seen gleaming at a great distance. On most palms, however, the male blossoms are yellowish, grouped densely together, and almost withered even as they emerge from the spathe.

"In palms with feathery leaves, either the leafstalks emerge (as with coco, *Phoenix, palma de real del Sinu*) from the dry, rough, and woody parts of the bole, or there is attached to the rough part of the bole, as with the *palma real de la Havana* admired by Columbus, a grass-green, smooth, thinner shaft from which the leafstalks emerge. On the fan palms (*foliis palmatis*), the leafy crown (*moriche, palma de sombrero de Havana*) often rests upon a bed of dry leaves, a trait that gives the plant an earnest, melancholy character. In some umbrella palms the crown consists of only a very few leaves growing out of slender stems (*miraguama*).

"In the form and color of the fruit, too, there is a much greater diversity than is believed in Europe. *Mauritia flexuosa* is ornamented with egg-shaped fruit, the flat, brown scaled surface of which gives them the appearance of young pine cones. How different from the immense three-sided coconut are the berry of the date palm and the stone-fruit of the corozo! But no palm fruit matches the fruit of the pirijao (pihiguao) of San Fernando de Atabapo and San Balthasar. Egg-shaped, golden, one side blushed with reddish purple, mealy and seedless apples, two to three inches thick, hang in bunches like grapes from the tops of the majestic palm trunks." (We have already mentioned this beautiful fruit, which grows in bunches of 70 to 80 and which, like bananas or potatoes, may be prepared in many ways, in the first volume of this work.)

The spathe of the palms, which encases the spadix, gives off in some species an audible sound when it suddenly bursts open. Richard Schomburgk (*Reisen in Britisch Guiana*, vol. I, p. 55) observed this phenomenon, as did I, with the *Oreodoxa oleracea*. The first blossoming of the palms being accompanied by sound reminds one of Pindar's Dithyrambus in praise of Spring, of that moment in Argive Nemea that "the first developing sprouts of the date palm announce the now bursting, aromatic spring" (*Kosmos*, vol. II, p. 10).

Three forms of exquisite beauty that are common to all tropical regions of the world are palms, pisang plants, and arborescent ferns. Where heat and moisture work in unison, the vegetation is the most luxuriant and the diversity of forms the greatest. For this reason, South America is the most beautiful part of the world of palms. The palm form is less common in Asia because, during the early natural revolutions of our planet, the greater portion of the Indian continent that lies below the equator was devastated and covered by the sea. Of the African palms between the Bight of Benin and the Ajan Coast we know almost nothing, and indeed we know very little, as I have already mentioned, of the African palm forms in general.

The palms, along with the conifers and some eucalyptus species of the family Myrtaceae, provide some examples of the tallest plant forms. Trunks of the Acai palm (*Areca oleracea*) measuring 150 to 160 feet have been observed (Augustin Saint Hilaire, *Morphologie végétale*, 1840, p. 176). The wax palm that we discovered on the ridge of the Andes between Ibague and Carthago in the *Montaña de Quindiu*, our *Ceroxylon andicola*, attains the tremendous height of 160 to 180 feet. I was able to measure accurately the felled trunks in the forest. After the wax palm, it seems to me that the next-highest of the American palms is *Oreodoxa sancona*, which we found in bloom near Roldanillo in the Cauca Valley, and which produces an excellent, very hard lumber. The fact that despite the tremendous amount of fruit that a single palm produces, the number of individuals of each species in the wild is not very considerable, may well be explained by the frequent abortive development of the fruit and the gluttonous feeding of enemies from every class of animals in the tropical world. And yet in the Orinoco River basin, entire tribes of natives live for many months of the year on palm fruits. "In the palmgroves planted with the pirijao palm, every single trunk annually produces nearly 400 pomiform fruit, and among the Brothers of St. Francis who live near the banks of the Orinoco and Guainia Rivers there is a worn-out saying: the bodies of the Indians grow wonderfully fat as often as the palms bring forth abundant fruit" (Humboldt, *De distrib. geogr. Plant.*, p. 240).

16. Since the earliest childhood of our culture

On all continents one will find in the tropical latitudes, for as far back as tradition and history can reach, the cultivation of pisangs. While it is certain that African slaves, over the course of centuries, brought varieties of banana fruit over to America, it is equally certain that even before Columbus's discovery, the natives there were already cultivating pisangs. The Guaiqueri Indians in Cumana assured us that on the Paria Coast, near the *Golfo Triste*, if one allows the pisang to ripen on the stem, it will produce germinating seeds. For just this reason, wild pisang plants can be found in the thicket of the forest, for the birds scatter the mature seeds. In Bordones near Cumana as well, fully formed seeds have been found now and then in pisang fruits. (Cf. my *Essai sur la Géographie des Plantes*, p. 29, and my *Relat. Hist.*, vol. I, pp. 104 and 587, vol. II, pp. 355 and 367.)

In another work (*Kosmos*, vol. II, p. 191) I have already cited that Onesicritus and other companions of the great Macedonian remember not the tall arborescent ferns but the umbrella palms with their fan-shaped leaves and the cultivated copses of pisang in their eternally fresh green. Among the Sanskrit names mentioned by Amara Sinha for the pisang (the *Musa* of the botanists) are *bhanu-phala* ("sun fruit"), *varana-buscha*, and *moko*. *Phala* means "fruit" in general. Lassen explains the words of Pliny (XII, 6) *arbori nomen palae, pomo arienae* by saying that "the Romans took the word *pala*, fruit, to be the name of the plant, and the word *varana* (*ourana* in the mouth of a Greek) was transformed to *ariena*. Out of *moko*, the Arabic word *mauza* may well have led to our *Musa*. The *bhanu* fruit is similar to the banana fruit." (Cf. Lassen, *Indische Alterthumskunde*, vol. I, p. 262, to my *Essai politique sur la Nouvelle-Espagne*, vol. II, p. 382, and *Rel. hist.*, vol. I, p. 491.)

17. Forms of the Malvaceae

Larger forms of mallow appear as soon as one crosses the Alps: at Nice and Dalmatia, *Lavatera arborea*, in Liguria, *L. olbia*. The dimensions of the baobab (the monkeybread tree) have been described above. Associated with the form of the Malvaceae are the botanically related families of the Buettneriaceae (*Sterculia, Hermannia*, and the large-leafed *Theobroma cacao*, whose blossoms break forth from the bark of the trunk as well as the root); the Bombaceae (*Adansonia, Helicteres*, and *Cheirostemon*); and finally, the Tiliaceae (*Sparmannia africana*). Among the glorious representatives of the mallow forms are our *Cavanillesia platanifolia* from Turbaco near Cartagena in South America, and the famed ochromalike hand-tree, the *Macpalxochiquahuitl* of the Mexicans (from *macpalli*, the flat hand), *arbol de las manitas* for the Spaniards, our *Cheirostemon platanoides*, with conjoined filaments that project like a hand (or claw) from the lovely purple-red blossoms. In all of the Mexican Free States there is only one single individual, one single ancient stem of this extraordinary species. It is believed to have been transplanted from a foreign origin by the kings of Toluca some 500 years ago. I found the place where the *arbol de las manitas* stands to be 8,280 feet above sea level. Why is there only one individual? From where did the kings of Toluca obtain the young tree or seed? Equally puzzling is that Montezuma did not possess it in his botanical gardens at Huaxtepec, Chapoltepec, and Iztapalapan, of which Hernandez, personal physician of Philip II, was still able to avail himself, and traces of which may still be found today; it is puzzling, too, that the hand-tree had not found a place among the natural historical illustrations that Nezahualcoyotl, king of Tezcuco, commissioned a half-century before the arrival of the Spanish. Some assert that the hand-tree grows wild in Guatemala (Humboldt and Bonpland, *Plantes équinoxiales*, vol. I, p. 82, pl. 24; *Essai polit. sur. la Nouv. Esp.* vol. I, p. 98). At the equator we saw two Malvaceae, *Sida philanthos* (Cavan.) and *Sida pichinchensis*, climb to the great heights of 12,600 and 14,136 feet on the sides of Antifana and the volcano Ruca-pichincha, respectively (see our *Plantes équin.*, vol. II, p. 113, pl. 116). The singular *Saxifraga Boussingaultii* (Brongn.) grows on the slopes of Chimborazo at six to seven hundred feet higher still.

18. The mimosa forms

The finely feathered leaves of the mimosas, acacias, schrankias, and desmanthus forms are truly typical of tropical vegetation. Yet we also can find representatives of this form outside of the tropical latitudes. In the Northern Hemisphere, and indeed only in Asia, can I point out but one low shrub: the *Acacia stephaniana*, first described by Marschall von Bieberstein. According to Kunth's recent investigations, it is a species of the genus *Proposis*. This socially growing plant covers the arid plains of the Shirvan Province on the Kura (or Cyrus) River near New Shamakhi to where the Kura meets the Aras. Olivier also found it near Baghdad. It is the *Acacia foliis bipinnatus*, as mentioned earlier by Buxbaum, and which extends northward to 42° latitude (*Tableau des Provinces situées sur la côte occidentale de la Mer Caspienne entre les fleuves Terek et Kour*, 1798, pp. 58 and 120). In Africa, *Acacia gummifera* (Willd.) presses northward as far as Mogodor, that is, up to 32° north latitude.

On the New Continent, the banks of the Mississippi and Tennessee, like the savannas of the Illinois, are decorated with *Acacia glandulosa* (*Michaux*) and *A. brachyloba* (Willd.). Michaux found that the *Schrankia uncinata* extended northward from Florida to Virginia, up to 37° north latitude. According to Barton, *Gleditsia triacanthos* grows as far up as 38° north latitude to the east of the Allegheny Mountains and up to 41° to the west. *Gleditsia monosperma* stops 2° to the south. These are the limits of the *Mimosae* in the Northern Hemisphere. In the Southern Hemisphere outside of the Tropic of Capricorn we find simple-leaved acacias as far as Van Diemen's Island; the *Acacia cavenia* described by Claude Gay grows in Chile between the 30th and 37th degrees of southern latitude (Molina, *Storia naturale de Chili*, 1782, p. 174). Chile has no true mimosas, but three species of the *Acacia* genus. The *Acacia cavenia* only attains a height of 12 feet, even in the northern part of Chile; in the south approaching the littoral, it barely rises a foot above the ground. The most excitable of the mimosas that we saw in the Northern Hemisphere of South America are (after the *Mimosa pudica*) *M. dormiens*, *M. somnians*, and *M. somniculosa*. The excitability of the African sensitive plant is mentioned as far back as Theophrastus (IV, 3) and Pliny (XIII, 10), but the first description of the South American sensitive (*dormideras*) I find in Herrera, *Décad.* II, lib. III, cap. 4. The plant first drew the attention of the Spaniards in 1518 in the savannas of the isthmus around Nombre de Dios: "parece como cosa sensible" (it seems a sensitive thing); and it was reported that the leaves ("de echura de una pluma de pajaros," the shape of a bird's feather) drew themselves together when someone touched them with a finger, but not when touched with a piece of wood. In the small marshes that surround the city of Mompox on the Magdalena River we discovered a beautiful swimming Mimosacea (*Desmanthus lacustris*). It is pictured in our *Plantes équinoctiales*, vol. I, p. 55, pl. 16. In the Andes chain of Caxamarca at 8,500 and 9,000 feet above the Pacific, we found two species of alpine mimosas, *Mimosa montana* and *Acacia revoluta*.

Up to now no true mimosa (as it is defined by Willdenow), nor even an inga, has been seen in the temperate zone. Among all the acacias, the *Acacia julibrissin* (which Forskål confused with the *Mimosa arborea*) can endure the greatest cold. In the botanic garden in Padua there stands a tall stem of considerable thickness in the open air, and the median temperature in Padua is under 10.5° Réaumur.

19. Heath plants

By no means do we include in these physiognomic observations under the name of heath plants the entire natural family of the Ericeae, which due to the similarity and analogousness of their floral parts includes *Rhododendrum*, *Befaria*, *Gautheria*, and *Escallonia*. We will restrict ourselves here to the so very consistent and characteristic form of the Erica species, including Calluna (*Erica vulgaris* [L.]).

"While *Erica carnea*, *E. tetralix*, *E. cinerea*, and *Calluna vulgaris* cover large expanses of land in Europe, from the German flatlands, from France and England, all the way to farthest Norway, it is Africa that offers the most variegated mixture of species. A single species, *Erica umbellata*, which in the Southern Hemisphere is native to the Cape of Good Hope, reappears in Northern Africa, Spain, and Portugal. *E. vagans* and *E. arborea* can also be found at the same time on the opposite coasts of the Mediterranean. The first-named is found in North Africa and around Marseille, in Sicily and Dalmatia, and even in England; the second in Spain, Istria, Italy, and on the Canary Islands." (Klotzsch, *Über die geographische Verbreitung der Erica-Arten mit bleibender Blumenkrone* [manuscr.]) The commom heath, *Calluna vulgaris* (Salisbury), a socially growing plant, forms great ranges from the mouth of the Scheldt to the western slope of the Urals. Beyond the Urals, oaks and heath simultaneously cease to be found. Both are absent from all of Northern Asia, from the entirety of Siberia, and all the way to the Pacific. Gmelin (*Flora Sibirica*, vol. IV, p. 129) and Pallas (*Flora Rossica*, vol. I, Pars 2, p. 53) both expressed their surprise at this disappearance of the *Calluna vulgaris*. On the eastern slope of the Urals it is even more decided, more sudden than one might wish to gather from the words of the latter of these great naturalists. Pallas merely says, "Beyond the Ural mountain range it gradually ceases. It very seldom appears on the plains of the Iset [i.e., the Siberian steppe], and it is entirely absent in remote Siberia." Chamisso, Adolph Erman, and Heinrich Kittlitz collected Andromeda heath in Kamchatka and on the northwest coast of America, but no Calluna. The accurate knowledge that we now possess regarding the mean temperature of the individual regions of Northern Asia and of the distribution of annual warmth in the different seasons in no way renders the halted progress of the heath explainable. Joseph Hooker, in a note to his *Flora antarctica*, presents the two contrasting phenomena of plant distribution ("uniformity of surface, accompanied by a similarity of vegetation," and "instances of a sudden change in the vegetation, unaccompanied with any diversity of geological and other feature") with a high degree of acuity (Joseph Hooker, *Botany of the Antarctic Voyage of the "Erebus" and "Terror,"* 1844, p. 210). Is there an *Erica* in Inner Asia? That which was described as *Erica vulgaris* by Saunders on Turner's journey to Tibet (*Philos. Transact.*, vol. LXXIX), alongside other European plants (*Vaccinium myrtillus* and *V. oxycoccus*), according to information from Robert Brown, is an Andromeda, probably the *Andromeda fastigiata* of Wallich. Just as conspicuous is the absence of *Calluna vulgaris* and all species of *Erica* in all of continental America, *Calluna* having been found in the Azores and Iceland. It has as yet not been found in Greenland but was discovered a few years ago in Newfoundland. The natural family of the Ericeae is also almost completely absent from Australia, where its place is taken by the Epacrideae. Linné described only 102 species of the genus Erica; after the editing done by Klotzsch, this genus comprises, if all variations are carefully excluded, 440 true species.

20. Cactus form

If the natural family of the Opuntiaceae are separated from the Grossulareae (*Ribes* species) and are compiled after the manner in which they were narrowed down by Kunth (*Handbuch der Botanik*, p. 609), then the entire family can be described as an exclusively American one. I am not unaware of the fact that Roxburgh, in the *Flora indica* (*inedita*), is introducing two cactus species that are said to be native to Southeast Asia, *Cactus indicus* and *C. chinensis*. Both are widespread, wild or have become wild, and different from *Cactus opuntia* and *C. coccinellifer*; it is conspicuous, however, that the Indian plant has no name in old Sanskrit. The so-called Chinese cactus has been introduced through cultivation to the island of St. Helena. Newer research, taking place at a time when, finally, interest in the original distribution of plants has been awakened, will give rise to the doubts that have many times been expressed regarding the existence of Asiatic Opuntiaceae. Isolated appearances of certain members of the animal kingdom have also been observed. How long the tapirs were held to be a creature characteristic of the New Continent! And yet the American tapir is identically reproduced in that of Malacca (*Tapirus indicus* [Cuv.]).

If the cactus species actually do belong to the tropics, some still have made their home in the New Continent in the temperate zone, on the Missouri and in Louisiana: the *Cactus missuriensis* and *C. vivipara*. It was with astonishment that Back, during his northern expedition, saw the shores of Rainy Lake under the latitude of 48°40′ (95.25° long.) covered with *C. opuntia*. South of the equator, cactus species do not extend any farther south than the Rio Itata (36° lat.) and Rio Biobio (37.25° lat.). In the portion of the Andes chain that lies between the tropic lines, I have seen cactus species (*C. sepium, C. chlorocarpus, C. bonplandii*) on high plateaus at elevations of nine to ten thousand feet, but in the temperate zone of Chile, a far more alpine character is displayed by *Opuntia ovallei*, the upper and lower limits of which were accurately determined by the learned botanist Claude Gay by means of barometric measurements. The yellow-blossomed *Opuntia ovallei* has a creeping stem, does not appear below 6,330 feet, extends upward to the line of eternal snow, and indeed even surpasses it in those places where individual jutting cliffs remain uncovered. The last little plants were collected at points that lie 12,820 feet above sea level (Claudio Gay, *Flora Chilensis*, 1848, p. 30). There are also some Echinocactus species that in Chile are true alpine plants. A counterpart to the sought-after fine-haired *Cactus senilis* is the wooly *C. (Cereus) lanatus*, known to the natives as *piscol* with its beautiful red fruit. We found this plant in Peru during our trip to the Amazon River near Huancabamba. The dimensions of the Cacti (a group over which much light was first spread by Prince von Salm-Dyck) present the most remarkable contrasts. *Echinocactus wislizeni*, at 4 feet in height and 7 feet in circumference, is still but third in size after *E. ingens* (Zucc.) and *E. platyceras* (Lem.) (Wislizenus, *Tour to Northern Mexico*, 1848, p. 97). The *Echinocactus stainesii* achieves a diameter of 2 to 2½ feet, *E. visnago* from Mexico, at 4 feet in height, reaches a diameter of 3 feet and can weigh between 700 and 2,000 lbs., while the *Cactus nanus*, which we collected near Sondorillo in the Jaén Province, is so small that, being shallowly rooted in the sand, it would stick between the toes of the dogs. Full of juice even in the driest seasons, the Melocacti, like the *ravenala* of Madagascar ("forest leaf" in the indigenous language: *rave* or *raven*, "leaf," and *ala*, whence comes the Javanese *halas*, "forest") are a spring in vegetable form. The wild horses and mules open them by stamping with their feet and in so doing are often injured (see vol. I). Over the last quarter-century, *Cactus opuntia* has spread in a most astonishing way throughout North Africa, Syria, Greece, and the whole of Southern Europe; indeed, proceeding from the coasts, the plant has penetrated deep into Africa, adding itself to the local plant life.

When one is accustomed to seeing cactus species only in our greenhouses, it is astonishing to see the density to which the wood fibers in old cactus stems will harden. The Indians know that cactus wood, being impervious to rot, is superb for the fashioning of oars and thresholds. To the newly arrived visitor, no plant physiognomy makes so strange and yet indelible an impression as a dry plain such as those at Cumana, New Barcelona, and Coro, or in the Province of Jaén de Bracamoros, which is thickly covered in the pillars or branching candelabra shapes of cactus stems.

21. Orchids

The occasionally almost animal-like form of orchid blooms is especially apparent in the *torito* (our *Anguloa grandiflora*), so celebrated in South America, in the *mosquito* (our *Restrepia antennifera*), in the *flor del Espiritu Santo* (also an *Anguola*, according to *Florae Peruvianae Prodrom.*, p. 118, tbl. 26), in the antlike flower of the *Chiloglottis cornuta* (Hooker, *Flora antarctica*, p. 69), in the Mexican *Bletia speciosa*, and in the entire wonderful tribe of our European Ophrys species: *O. muscifera, O. apifera, O. arinifera, O. arachnites*, etc. The popularity of this gorgeously flowering plant group has grown to such a degree that the number now being cultivated in Europe was estimated by the Loddiges brothers in 1848 to be 2,360 species, while in 1813 it was only 115, and in 1843 over 1,650. What a trove of gloriously blooming, as yet unknown orchids might the water-rich interior of Africa hold! In his fine work *The Genera and Species of Orchideous Plants* (1840), Lindley describes exactly 1,980 species; by the end of the year 1848, Klotzsch had counted 3,545 species.

While in the temperate and frigid zones grow only terrestrial orchids, confined to the ground, the beautiful lands of the tropics, on the other hand, feature both the terrestrial form and the parasitic, which grows on the trunks of trees. To the first group belong the tropical genera: *Neottia*, *Cranichis*, and most of the Habenaria. But we also found alpine examples of both forms on the declivity of the Andes chain near New Granada and Quito: the parasitic (*Epidendreae*) *Masdevallia uniflora* (9,600 ft.), *Cyrtochilum flexuosum* (9,480 ft.), and *Dendrobium aggregatum* (8,900 ft.), and the terrestrial *Altensteinia paleacea*, near Lloa Chiquito at the foot of the Pichincha volcano. Claude Gay believes that the orchids that people claim to have seen growing on trees on Juan Fernandez and even on Chiloe were probably just parasitic Pourretias, which extend to at least 40° south. In New Zealand, the tropical orchid form that hangs from trees may still be found as far down as 45° south latitude. The orchids of Auckland's and Campbell's Islands (*Chiloglottis*, *Thelymitra*, and *Acianthus*) grow in moss on level ground. In the animal world at least one tropical form goes even farther south. The Island of Macquarie (54°39' south lat.) has a native parrot that lives closer to the South Pole than Danzig is to the North Pole. (Cf. the section *Orchideae* in my book *De Distrib. geogr. Plant.*, pp. 241–247.)

22. Casuarineae

Acacias, upon which leaves are replaced by phyllodia, myrtles (*Eucalyptus*, *Metrosideros*, *Melaleuca*, *Leptospermum*), and Casuarineae uniformly characterize the flora of Australia (New Holland) and Tasmania (Van Diemen's Land). Due to the variations in the species, the Casuarineae, with their leafless, thin, threadlike, articulate branches, the rami covered in membranous serrated sheaths, are sometimes compared by travelers to the treelike Equisetaceae (horsetails) and sometimes to our Scotch firs (see Darwin, *Journal of Researches*, p. 449). In South America, I too received a peculiar impression of leaflessness from the small *Colletia* and *Ephedra* shrubs near the Peruvian coast. According to Labillardière, *Casuarina quadrivalvis* extends to 43° south in Tasmania. The melancholy form of the Casuarineae is also not unknown to East India and even the eastern coast of Africa.

23. Conifers

The family of conifers—including the genera *Dammara*, *Ephedra*, and the *Gnetum* of Java and New Guinea that essentially belong to this category, although they deviate in terms of the form and composition of their leaves—plays such a great role, due to the number of individuals of each species and to their geographical distribution, covering as socially growing plants in the northern temperate zone such vast expanses of land, that one cannot but be astounded by the relatively small number of its species. We are familiar with not even ¾ as many conifers as the number of palms that have been described; we know fewer conifers than Aroideae. In his articles on the morphology of conifers (*Abhandl. der mathem. Physikalisch. Classe der Akademie der Wiss. zu München*, vol. III, 1837–1843, p. 752), Zuccarini counts 216 species, of which 165 are in the Northern Hemisphere and 51 in the Southern. These ratios must now be adjusted in light of my research, for the *Pinus*, *Cupressus*, *Ephedra*, and *Podocarpus* species that we, Bonpland and I, discovered in the tropical regions of Peru, Quito, New Granada, and Mexico, raise the number of conifers vegetating between the tropical lines by 42. The outstanding most recent work by Endlicher, *Synopsis Coniferarum*, 1847, contains 312 species of currently living conifers as well as 178 species of prehistoric conifers, buried within anthracite formations, and colorful sandstone, and in the Keuper and Jura strata. The vegetation of the primeval world especially presents us with such objects that, in their simultaneous kinship with several families of the current world, remind us that along with themselves, many intermediate species have also been lost. These conifers, so common in the primeval world, are accompanied especially by palms and cycads; in the most recent layers of lignite or brown coal, however, we find conifers, our spruces and firs, accompanied once more by Cupuliferae, maples, and poplars. (*Kosmos*, vol. I, pp. 295–298 and 468–470.)

Were it not for the fact that portions of the Earth's surface in the tropics are lifted to great heights, the inhabitants of that region would have remained almost completely unaware of the distinctly characteristic form of the needle-bearing trees. Together with Bonpland, I took great pains in the Mexican high country to determine exactly the lower and upper limits of conifers and oaks. The heights at which both begin to grow (*los Pinales y Encinales*, *Pineta et Querceta*) are a welcome sight to those traveling from the seacoast, for they indicate a climate to which experience up to now indicates the deadly sickness of the black vomit (*vomito prieto*, a form of yellow fever) has not penetrated. For the oaks, especially for *Quercus xalapensis* (one of 22 Mexican oak species that we were the first to describe), the lower vegetation limit on the way from Veracruz to the Mexican capital was somewhat lower than the *Venta del Encero*, 2,860 feet above sea level. On the western slope of the high plateau between Mexico City and the Pacific, the bottom limit for oaks was somewhat lower; it begins at a cottage known as the *Venta de la Moxonera*, between Acapulco and Chilpanzingo, at an absolute elevation of 2,328 feet. I found a similar difference in the lower limit of the spruce forest. Near the Pacific at *Alto de los Caxones* north of Quaxiniquilapa, the limit for the *Pinus montezumae* (Lamb.), which we had at first mistaken for the *Pinus occidentalis* (Schwarz), lies at only 3,480 feet of elevation, while close to Veracruz on the *Cuesta del Soldado*, it is not reached until 5,610 feet. Both tree types, the mentioned oaks and spruces, grew at lower elevations on the Pacific side than on the side facing the Antillean Gulf. During my ascent of the *Cofre de Perote*, I found the upper limit of the oaks to be at 9,715 feet, and that of the *Pinus montezumae* to be at 12,138 feet (almost 2,000 feet higher than the summit of Aetna), where in February considerable masses of snow had already fallen.

The more significant the altitudes at which the Mexican conifers begin to show up, the more conspicuous it becomes that on the island of Cuba (where, admittedly, the air at the edge of the tropical zone is cooled 6.5° by the north winds), another *Pinus* species (*P. occidentalis* [Schwarz]) can be seen on the plain itself or on the low hills of the *Isla de Pinos* intermingled with palms and mahogany trees (*Swietenia*). In the journal of his first voyage (*Diario del 25 de Nov., 1492*), Columbus mentions a small fir forest (*pinal*) near Cayo de Moya in the Northeast of the Isle of Cuba. On Haiti (Santo Domingo) as well, *Pinus occidentalis* descends at Cape Samana from the mountains down to the littoral itself. The trunks of these spruce trees, carried by the Gulf Stream to Graciosa and Fayal in the Azores, were among the primary indicators of the existence of unknown lands in the West to the great explorer (see my *Examen crit.*, vol. II, pp. 246–259). Has it been established that on Jamaica, in spite of its high mountains, *Pinus occidentalis* is completely absent? One may also ask: what species of *Pinus* is found on the eastern littoral of Guatemala, since *P. tenuifolia* (Benth.) is probably limited to the mountains near Chinanta?

Taking a broad view of the plant species that define the upper limit of trees in the Northern Hemisphere from the frigid zone to the equator, I find for Lapland, in the Sulitelma Mountains (68° lat.), according to Wahlenberg, not needle-bearing trees but birches (*Betula alba*) far beyond the upper limit of the *Pinus sylvestris*; for the temperate zone in the Alps (45.75° lat.) *Pinus picea* (Du Roi), below which the birches end; in the Pyrenees (42.5° lat.) *Pinus uncinata* (Ram.) and *P. sylvestris var. rubra*; in the tropics of Mexico (19°–20° lat.) *Pinus montezumae* far above *Alnus toluccensis, Quercus spicata*, and *Q. crassipes*; in the snow-capped mountains of Quito at the equator *Escallonia myrtilloides, Arailia avicennifolia*, and *Drymis winteri*. This last species of tree, identical to *Drymis granatensis* (Mut.) and *Wintera aromatica* (Murray), provides, as has been shown by the younger Hooker (*Flora antarctica*, p. 229), the most extraordinary example of the uninterrupted dissemination of a single tree species, ranging from the southernmost parts of Tierra del Fuego and Hermit Island (where it was discovered by Drake's expedition in 1577) all the way to the northern highlands of Mexico, covering a meridional distance of 86 degrees of latitude, or 1,290 geographical miles. Where the tree boundaries of the highest mountains are formed not by the birches of the extreme North but rather by the conifers, as in the Swiss Alps and the Pyrenees, what follow just below in Europe and Western Asia, on the way to these same snow peaks that the trees so picturesquely enwreathe, are the alpine roses and *Rhododendra*; but on the Silla de Caracas and on the Peruvian Paramo de Saraguru, these blooming plants are replaced by the purple-red blossoms of another of the Ericeae, the graceful genus *Befaria*. In Lapland, *Rhododendron laponicum* follows after the conifers; in the Alps, *Rhododendron ferrugineum* and *R. hirsutum*; in the Pyrenees, *R. ferrugineum* alone, but which De Candolle also found in the Jura range (in the Creux du Vent) in an isolated space 5,600 feet lower, that is to say, at the low elevation of 3,100 to 3,600 feet; in the Caucasus, *R. caucasicum*. Should we wish to pursue this last zone of vegetation before the line of perpetual snow all the way to the tropics, by our own observations we would name in the Mexican tropics, *Cnicus nivalis* and *Chelone gentianoides*; in the cold mountainous region of New Granada, the wooly *Espeletia grandiflora, E. corymbosa*, and *E. argentea*; in the Andes of Quito, *Culcitium rufescens, C. ledifolium*, and *C. nivale*—yellow-blossomed Compositae, which here take the place of the somewhat more northerly hairy-leaved shrubs of New Granada that are so physiognomically similar to them, the Espeletia. Such replacement and repetition of similar, almost identical forms in regions separated by oceans or broad expanses of land is a wondrous law of Nature. It holds sway even over the most peculiar types of flora. In Robert Brown's family of Rafflesia, separate from the Cytineae, both of the Hydnorae described by Thunberg and Drege (*H. africana* and *H. triceps*) have their South American counterpart in *H. americana* (Hooker).

Far above the regions of alpine herbs, beyond the grasses and lichens, even beyond the boundary of the perpetual snow, seemingly alone and to the greatest astonishment of botanists, be they in the tropics or the temperate zone, a phanerogamic plant will now and then come wandering sporadically upward to blocks of stone that, warmed perhaps by open clefts, have remained free of snow. I have recalled above the *Saxifraga boussingaulti* that may be found at 14,800 feet on Chimborazo; in the Swiss Alps *Silene acaulis*, a Caryophyllea, has been seen at 10,680 feet. The first of these was vegetating at 600 and the second at 2,460 above the local snowline, as it was measured at the time the plants were found.

Among our European conifers the red and the white firs show great and unusual irregularities in their geographical distribution upon the mountain declivities. In the Swiss Alps, while the red fir (*Pinus picea* [Du Roi], *foliis compressotetragonis*, unfortunately called *Pinus abies* by Linné and most of the botanists of our time!) marks the last timberline at the moderate elevation of 5,520 feet, and only here and there the low mountain alder (*Alnus viridis* [Dec.], *Betula viridis* [Vill.]) presses on to the snowline, the white pine (*Pinus abies* [Du Roi], *Pinus picea* [Linn.], *foliis planis, pectinatodistichis, emarginatis*), according to Wahlenberg, remains about a thousand feet lower. The red fir does not appear at all in Southern Europe, in Spain, the Appennines, and Greece; it is seen on the slope of the northern Pyrenees, as Ramond notes, only at great altitudes, and is not found at all in the Caucasus. The red fir penetrates northward into Scandinavia farther than the white fir, which occurs in Greece (on the Parnassus, the Taygetus, and Oeta) as a long-needled variety, *foliis apice integris, breviter mucronatis*, the sharp-eyed Link's *Abies apollinis*. (Linnaeus, vol. XV, 1841, p. 529, and Endlicher, *Synopsis Coniferarum*, p. 96.)

In the Himalayas, the conifer form is distinguished by the powerful girth and height of the trunk as well as the length of the needles. Twelve to 13 feet thick and the jewel of the mountains is the cedar "Deodwara," Roxburgh's *Pinus deodara* (in Sanskrit actually *dèwa-dâru*, a lumber of the gods). In Nepal it climbs up to 11,000 feet above sea level. More than 2,000 years ago, the Deodwara cedar on the Behut (Hydaspes) River provided the material for Nearchus's fleet. In the Valley of Dudegaon north of the Dhunpur copper mine in Nepal, Dr. Hoffmeister, who was taken from science far too soon, found in a forest *Pinus longifolia* (Royle) (the Tschelu fir) growing alongside the tall trunks of a palm, the *Chamaerops martiana* (Wallach) (*Hoffmeisters Briefe aus Indien während der Expedition des Prinzen Waldemar von Preußen*, 1847, p. 351). Such an intermingling of *pineta* and *palmeta* had already astounded the companions of Columbus, as is reported by a contemporary and friend of the admiral's, Pietro Martire d'Anghiera (Dec. III, lib. 10, p. 68). I first saw this mix of firs and palms myself on the journey from Acapulco to Chilpanzingo. The Himalayas, like the Mexican highlands, have alongside the *Pinus* and cedar species forms of cypress (*Cupressus torulosa* [Don.]), of Taxus (*Taxus wallachiana* [Zucc.]), of Podocarpus (*P. nereifolia* [Rob. Br.]), and of junipers (*Juniperus squamata* [Don.] and *J. excelsa* [Bieberst.]; the latter species also found near Shipki in Tibet, in Asia Minor, Syria, and on the Greek islands); on the other hand, *Thuja*, *Taxodium*, *Larix*, and *Araucaria* are New Continent forms that are not to be found in the Himalayas.

Besides the 20 *Pinus* species in Mexico with which we are familiar, the United States of North America offer, within their current territory which extends to the Pacific Ocean, 45 described *Pinus* species, while Europe features but 15. This very difference between a richness and paucity of forms shows that the advantage also goes to the New Continent (a more consistent, meridionally oriented section of the planet's surface) in terms of the variety of oak species. The belief that many European *Pinus* species extend, through their broad distribution in Northern Asia, across to the Japanese islands, and there even intermingle with a true Mexican species, the Weymouth pine (*Pinus strobus* [L.]), as has been maintained by Thunberg, has quite recently been fully refuted by the extremely accurate investigations of Siebold and Zuccarini. What Thunberg considers to be European *Pinus* species are distinct species different from these. Thunberg's red fir (*Pinus abies* [Linn.]) is *P. polita* (Sieb.), often planted near Buddhist temples; his common northern (Scotch) pine (*Pinus sylvestris*) is *P. massoniana* (Lamb.); his *P. cembra*, the German and Siberian *Zwirbelnusskiefer* (Swiss pine), is *P. parviflora* (Sieb.); his common larch (*P. larix*) is *P. leptolepis* (Sieb.); his *Taxus baccata*, the fruits of which the people of the Japanese court enjoy as a precautionary agent for long ceremonies (Thunberg, *Flora Japonica*, p. 275) forms its own genus, *Cephalotaxus drupacea* (Sieb.). The Japanese islands, despite the proximity of the Asian mainland, have a vegetation of a very different character. Thunberg's Japanese Weymouth pine, which would constitute an important phenomenon, is a cultivated tree wholly different from the *Pinus* species of the New World. It is *P. koraiensis* (Sieb.), carried over to Nippon from the Korean Peninsula and Kamchatka.

Of the 114 currently known species of the genus *Pinus*, not a single one is to be found in the Southern Hemisphere, for the *Pinus merkusii* described by Junghuhn and De Vriese belongs to the portion of the Isle of Sumatra that lies north of the equator (the Battas district) and the *P. insularis* (Endl.) to the Philippines, though it was originally presented in the arboretum of Loudon as *P. timoriensis*. Excluded from the Southern Hemisphere along with the genus *Pinus* are also, according to our current understanding of the happily advancing field of plant geography, all species of *Cupressus, Salisburia* (ginkgo), *Cunninghamia* (*Pinus lanceolata* [Lamb.]), *Thuja*, from which one species (*Th. gigantea* [Nutt.]), growing along the Columbia River, measures 170 feet, *Juniperus*, and *Taxodium* (Mirbel's *Schubertia*). I am here able to introduce this last species all the more safely since a cape plant, Sprengel's *Schubertia capensis*, is no *Taxodium* but rather forms, in an entirely different division of conifers, a genus of its own, *Widringtonia* (Endl.).

This absence of the true Abietineae, Juniperineae, Cupressineae, and all Taxodineae, as well as the *Torreya*, the *Salisburia adiantifolia*, the *Cephalotaxus* of the Taxineae in the southern half of the globe, is vividly reminiscent of the puzzling and as yet unrevealed circumstances that determined the original distribution of plant forms and that, through similarities or differences in the ground, thermal conditions, or meteorological processes, can in no way be satisfactorily explained. I pointed out long ago that the Southern Hemisphere, for example, has many native plants of the Rosaceae family but not one single species of the genus *Rosa*. Claude Gay teaches us that the *Rosa chilensis* described by Meyer is a feral variety of *Rosa centifolia* (Linn.), which became a European plant several thousand years ago. Such varieties, turned feral in Chile, spread out across large area of land near Valdivia and Osorno (Gay, *Flora Chilensis*, p. 340). And in the entire tropical region of the Northern Hemisphere we have only a single native rose, our *Rosa montezumae*, which was found in the Mexican highlands near Moran at 8,760 feet of elevation. Among the curious phenomena in connection to the distribution of plants is the fact that along with palms, pourretias, and many species of cactus, there are no agaves in Chile, even though *A. americana* vegetates luxuriantly in Roussillon, near Nice and Botzen, and in Istria, where it has probably been imported from the New Continent since the end of the 16th century, and despite the fact that starting in Northern Mexico, it forms a contiguous plant migration across the Isthmus of Panama and down to Southern Peru. I have long believed that the Calceolaria, like the roses, were to be found exclusively to the north of the equator. As it turned out, none of the 22 species that we brought back with us had been collected north of Quito and the Pinchincha volcano; my friend Professor Kunth, however, remarks that *Calceolaria perfoliata*, which Boussingault and Captain Hall found near Quito, also presses on to New Granada and that this species, like *C. integrifolia*, had been sent by Mutis from Santa Fé de Bogota to the great Linné.

The *Pinus* species that are so numerous in the purely tropical Antilles as well as in the mountainous parts of Mexico do not extend across the Isthmus of Panama, and remain foreign to the equally mountainous portions of the tropics of South America that lie to the north of the equator, foreign to the high plains of New Granada, Pasto, and Quito. I have been on the plains and in the mountains near the Rio Sinu close to the Isthmus of Panama down to 12° south latitude, and in this stretch of nearly 400 geographical miles, the only coniferous forms I saw were a Taxus-like, 60-foot Podocarpus in the Andean pass of Quindiu at 4°26′ north latitude and in the Paramo de Saraguru at 3°40′ south, and an Ephedra (*E. americana*) near Guallabamba, north of Quito.

Those members of the group of conifers that are equally common to the Northern and Southern Hemispheres are *Taxus, Gnetum, Ephedra*, and *Podocarpus*. Long before l'Héritier, on the 25th of November, 1492, Columbus already knew to distinguish this last genus from *Pinus*; he writes of *pinales en la Serrania de Haiti que no llevan piñas, pero frutos que parecen azeytunos del Axarafe de Sevilla* (pine groves in the mountains of Haiti that bear no cones, but rather fruits like those of the olive trees of the plain of Seville; see my *Examen crit.*, vol. III, p. 24). Species of *Taxus* are spread from the Cape of Good Hope to 61° north in Scandinavia, i.e., over 95 degrees of latitude; nearly as widespread are *Podocarpus* and *Ephedra*. Even from the Cupuliferae, those oak species that we generally call a northern form (which indeed in South America do not extend below the equator) appear once again on the Indian Archipelago in the Southern Hemisphere, on the Isle of Java. Exclusively indigenous to this latter hemisphere are ten genera of conifer, of which we shall here mention only the most prevalent: *Auracaria, Dammara (Agathis* [Sal.]), *Frenela* (of which there are around 18 Australian species), *Dacrydium*, and *Lybocedrus*, being in both Australia and the Strait of Magellan. New Zealand has a species of the genus *Dammara* (*D. australis*) and no *Auracaria*. In Australia, by a strange contrast, the opposite is true.

In the form of the needle-bearing trees, Nature provides the most greatly extended longitudinal axis of all arborescent plants. I say "arborescent plants" for, as mentioned above, in the genus of the Laminaria (oceanic algae), the *Macrocystis pyfera* between the California coast and 68° south latitude often reaches 370 to 400 feet in length. Excluding the 6 Araucaria of Brazil, Chile, Australia, the Norfolk Islands, and New Caledonia, the tallest of the conifers are those indigenous to the northern temperate zone. Similar to how we found the most gigantic member of the palm family (*Ceroxylon andicola*, which stands over 180 feet tall) growing in the temperate alpine climate of the Andes, so too do the tallest conifers belong to the temperate northwestern coast of America and in the Rocky Mountains (40°–52° north lat.) of the Northern Hemisphere, and in the southern half of the globe to New Zealand, Tasmania (or Van Diemen's Land), to Southern Chile or Patagonia (again 43°–50° latitude, but south). The most gigantic forms are of the genera *Pinus, Sequoia* (Endl.), *Araucaria*, and *Dacrydium*. I am naming here only those species that not only reach but often surpass 200 feet. In order also to present comparative measurements, one must be reminded that in Europe, the tallest red and white firs (especially the latter) attain approximately 150–160 feet, and that, for example, the famed spruce of the Lampersdorfer Forest near Frankenstein in Silesia already enjoys a great reputation, despite measuring, at a circumference of 16 feet, only 153 Prussian feet (148 Parisian ft) in height (cf. Ratzeburg, *Forstreisen*, 1844, p. 287). Here are some reliable figures, the English units reduced to Old French feet:

Pinus grandis (Dougl.), in New California, reaches 190–210 feet;

Pinus fremontiana (Endl.), same place, probably reaches the same height (Torrey and Frémont, *Report of the Exploring Expedition to the Rocky Mountains in 1844*, p. 319);

Dacrydium cupressinum (Solander), in New Zealand, over 200 feet;

Pinus lambertiana (Dougl.), Northwest America, 210–220 feet;

Araucaria excelsa (R. Brown), the *Cupressus columnaris* (Forster), on Norfolk Island and the surrounding emerging rock cliffs, 170–210 feet. According to Endlicher, the 6 Araucaria known up to now fall into two groups:

A. The American (Brazil and Chile, *A. brasiliensis* [Rich.], between 15° and 25° south lat., and *A. imbricata* [Pavon], between 35° and 50° south lat., the latter species reaching 220–244 ft.);

B. The Australian (*A. bidwilli* [Hook.] and *A. cunninghami* [Ait.], on the eastern side of Australia, *A. excelsa* of Norfolk Island, and *A. cookii* [R. Brown]of New Caledonia). Corda, Presl, Göppert, and Endlicher have up to now discovered 5 prehistoric Araucaria in the Lias, in the chalk and brown coal (Endlicher, *Coniferae fossiles*, p. 301).

Pinus douglasii (Sab.), in the valleys of the Rocky Mountains and along the Columbia River—43°-52° north lat. (The very worthy Scottish botanist for whom the tree is named, having traveled from New California to the Sandwich Islands, suffered in 1833 a martyr's ghastly death while collecting samples there. He carelessly plunged into a pitfall into which one of the country's feral and always ferocious bulls had fallen before him.) With careful measurements, this traveler described a tree trunk that stood 230 Parisian feet (245 English ft.) tall and had, three feet above the ground, a circumference of 54 Parisian feet. (Cf. *Journal of the Royal Institution*, 1826, p. 325.)

Pinus trigona (Rafinesque), from the western slope of the Rocky Mountains, described in Lewis and Clark's *Travels to the Source of the Missouri River, and across the American Continent to the Pacific Ocean (1804–06)*, 1814, p. 456. These "gigantic firs" were painstakingly measured; the circumference of the trunk six feet above the ground was often 36 to 42 feet. One tree had a height of 282 feet (300 Engl. ft), and the first 180 feet was completely without branches.

Pinus strobus (in the eastern part of the United States of North America, especially east of the Mississippi, but again in the Rocky Mountains from the source of the Columbia to Mt. Hood, from 43° to 54° north latitude), called Lord Weymouth's pine in Europe, white pine in North America, usually only 150 to 180 feet, but in New Hampshire, several of 235 to 250 feet have been observed (Dwight, *Travels*, vol. I, p. 36, and Emerson, *Report on the Trees and Shrubs Growing Naturally in the Forests of Massachusetts*, 1846, pp. 60–66).

Sequoia gigantea (Endl.) (*Condylocarpus* [Sal.]), of New California, like *Pinus trigona*, over 280 feet tall.

The composition of the soil and the conditions of temperature and humidity, upon all of which the nourishment of plants depends, certainly promote the thriving and increase in number of the individuals that make up a species; the tremendous height, however, to which the trunks of only a few ascend while in the midst of many closely related species of the same genus is not determined by soil and climate but rather, in the plant and animal kingdoms, by a specific organization, by innate natural qualities. As the greatest contrast to *Araucaria imbricata* of Chile, to *Pinus douglasii* on the Columbia River, and to the *Sequoia gigantea*, I do not name the two-inch-tall willow (*Salix arctica*), stunted by cold or the mountain altitude, but rather, a small phanerogam from the lovely climate of the southern tropics, from the Brazilian Province of Goias. The moss-like *Tristicha hypnoides* of the monocotyledonous family of the Podostemeae reaches a height of barely 3 lines. "While crossing the Rio Claro in the Province of Goyaz," says the outstanding observer Auguste de St. Hilaire, "I saw, on a rock, a plant whose stem was no taller than three lines high and which I at first took for moss. It was, however, a phanerogamic plant, the *Tristicha hypnoïdes*, equipped with sexual organs like our oaks and the gigantic trees that stood round raising their majestic tops" (Auguste de Saint-Hilaire, *Morphologie végétale*, 1840, p. 98).

Along with the height of the trunk, the length, width, and position of the leaves and the fruit, the almost umbrellalike spread of the horizontal or uplifted branches, and the graduated differences in color from a vivid green or a silvery grayish-green to a blackish-brown give the conifers a singular physiognomic character. The needles of *Pinus lambertiana* (Dougl.) of the American Northwest are five inches long, those of the *P. excelsa* (Wallich) on the southern slope of the Himalayas near Katmandu seven, those of the *P. longifolia* (Roxb.) in the mountains of Kashmir over twelve. In the very same manner, the needles too vary in the most conspicuous ways due to the influences of soil and airborne nourishment and elevation above sea level. Over a distance of 80 degrees of longitude (more than 760 geographical miles), from the mouth of the Scheldt across Europe and Northern Asia to Bogoslowsk in the Northern Urals and Barnaul on the far side of the Ob River, I found these changes in the length of the needles of our common pine (*Pinus sylvestris*) to be so great that there are those who occasionally believe, misled by the shortness and stiffness of the needles, that they have suddenly found a new *Pinus* species related to the mountain spruce, *P. rotundata* (Link) (*Pinus uncinata* [Ram.]). Those are, as Link has already correctly observed (*Linnaea*, vol. XV, 1841, p. 489), transitions to Ledebour's *P. siberica* of the Altai.

On the high plains of Mexico, I found the tender, pleasantly green, but deciduous foliage of the Ahuahuete (*Taxodium distichum* [Rich.], *Cupressus disticha* [Linn.]) especially uplifting. This tree, which swells to great thickness and whose Aztec name means "water drum" (from the Aztec *atl*, water, and *huehuetl*, drum), thrives in this tropical region, between 5,400 and 7,200 feet above sea level, while in the United States of North America it descends into the lowlands to the marshy region (the cypress swamps) of Louisiana at 43° latitude. In the southern states of North America, as on the high plains of Mexico, *Taxodium distichum* (*Cyprès chauve*) attains at 120 feet in height the monstrous girth of 30 to 37 feet in diameter, measured near the ground (Emerson, *Report on the Forests*, pp. 49 and 101). The roots in these cases feature the remarkable phenomenon of woody outgrowths that, being sometimes conical and rounded, sometimes tablelike, stand 3 to 4½ feet above the ground. Travelers have compared these root outgrowths, in the places where they are especially numerous, to the headstones in Jewish cemeteries. Auguste de Saint-Hilaire astutely remarks, "These outgrowths of the bald cypress, resembling boundary stones, can be regarded as exostoses, and since they live high up in the air, adventitious buds would no doubt burst if the nature of the coniferous plants' tissue did not prevent the development of the hidden sprouts that give rise to these sorts of buds" (*Morphologie végétale*, p. 91). In the roots of the conifers, by the way, a remarkably persistent life-force manifests itself in the phenomenon known as *Umwallen* or *Überwallung* (surging; boiling over), which has in many cases attracted the attention of plant physiologists and which is seldom seen with other dicotyledons. The portion of the trunks of felled white firs that have been left behind (stumps or stubs) will put forth for many years, without the development of shoots, twigs, and leaves, new layers of wood, continuing to grow in girth. The commendable Göppert believes that this occurs only by means of a nourishing of the roots that the end of the trunk (the stump) receives from a living tree of the same species standing nearby. The roots of the foliated individual, he maintains, have organically fused with those of the felled one. (Göppert, *Beobachtungen über das sogenannte Umwallen der Tannenstöcke*, 1842, p. 12.) In his excellent new textbook on botany, Kunth speaks against this explanation of the phenomenon, which was known, albeit imperfectly, even to Theophrastus (*Hist. Plant.*, lib. III, cap. 7, pp. 59 and 60, Schneider). According to him, this *Überwallung*, this "boiling over," is quite analogous to the processes by which sheets of metal, nails, letters carved into the wood, and even antlers can be found enveloped in the wood. "The *cambium*, that is, the thin-walled web of cells that move the granular, mucilaginous sap from which all new growth issues, irrespective of the buds (without regard to them at all), continues to lay down new layers of wood upon the outermost layer of the trunk." (Part I, pp. 143 and 166.)

The relationship mentioned above, between the absolute elevation of the ground and position in relation to latitudinal and isothermal lines, often reveals itself when one compares the arborescent vegetation of the tropical portions of the Andes chain with the vegetation of the northwestern coast of America or that by the shores of Canadian lakes. Darwin and Claude Gay made the same observation in the Southern Hemisphere, as they pressed onward from the high plain of Chile to Eastern Patagonia and to the archipelago of Tierra del Fuego, where *Drymis winteri*, along with forests of *Fagus antarctica* and *Fagus forsteri* growing in long stretches oriented north-south, cover all uniformly all the way to the lowlands. Minor exceptions to the law of constant positional relationships between mountain elevation and geographical latitude, dependent upon local causes as yet not satisfactorily explained, may be found even in Europe itself. I call to mind the altitudinal boundary of the birch trees and the common pine in a portion of the Swiss Alps at the Grimsel Pass. The pines (*Pinus sylvestris*) grow there as high as 5,940 feet, the birches (*Betula alba*) up to 6,480; above the birches there lies yet another level of Swiss pines (*Pinus cembra*), the upper boundary of which is 6,890 feet. Thus the birch stands flanked by two zones of conifers. According to the excellent observation of Leopold von Buch and the latest from Martins, who also visited Spitzbergen, the boundaries of geographical distribution in high Northern Scandinavia (in Lapland) are as follows: the pine reaches 70°, *Betula alba* 70°40′, *B. nana* all the way to 71°, and *Pinus cembra* is completely absent from Lapland. (Cf. Unger, *Über den Einfluß des Bodens auf die Vertheilung der Gewächse*, p. 200; Lindblom, *Adnot. In geographicam plantarum intra Sueciam distributionem,* p. 89; Martins in *Annales des Sciences naturelles,* vol. XVIII, 1842, p. 195.)

As the length and configuration of the needles determine the physiognomic character of the conifers, this determination is even more a function of the specific differences in the widths of the needles and in the parenchymatous development of the appendicular organs. Several species of Ephedra may almost be considered leafless, but in *Taxus, Araucaria, Dammara (Agathis),* and the *Salisburia adiantifolia* (Smith) (*Gingko biloba* [Linn.]), the width of the needle increases incrementally. I have arranged the genera here morphologically. The names first chosen by the botanists reflect such an order themselves. *Dammara orientalis* of Borneo and Java, often 10 feet in diameter, was originally called *loranthifolia; Dammara australis* (Lamb.) from New Zealand, growing up to 140 feet, was first *zamaefolia.* Both plants do not have needles, but *folia alterna oblongolanceolata, opposita, in arbore aldultiore saepe alterna, enervia, striata.* The underside of the leaf is packed with rows of slitlike openings. These transitions of the appendicular system from the greatest degree of contraction to a broad leaf surface are, as is ever the case in a progression from the simple to the complex, of both a morphological and a physiognomical interest (Link, *Urwelt,* vol. I, 1834, pp. 201–211). The short-stemmed, broad, split leaf of the *Salisburia* (Kämpfer's gingko) also has the breathing slits only on the underside. The original native land of this tree remains unknown. In connection with Buddhist congregations, it migrated in ancient times from the temple gardens of China to those of Japan.

During a trip from a harbor on the Pacific, across Mexico, and on to Europe, I was an eyewitness to the strange and frightful impression that the first sight of a pine forest near Chilpanzingo made upon one of our companions who, having been born in Quito at the equator, had never before seen conifers and *folia acerosa.* To him the trees seemed leafless, and since we were traveling toward the frigid North, he believed that he was already perceiving, in this ultimate contraction of the organs, the impoverishing influence of the pole. The traveler, whose impressions I describe here and whose name Bonpland and I cannot utter without a pang of sorrow, was a fine young man, the son of the Marques de Selvalegre, don Carlos Montufar, who, not many years later, courageously met a violent but honorable death in the war of independence that rose from the Spanish colonies' noble and impassioned love of freedom.

24. Varieties of Pothos, Aroideae

Caladium and *Pothos* are exclusively tropical forms, while *Arum* species belong more to the temperate zone. *Arum italicum, A. dracunculus,* and *A. tenuifolium* advance as far as Istria and Friuli. No Pothos has yet been discovered in Africa. East India has some species of this genus (*P. scandens* and *P. pinnata*), which in their physiognomy are less attractive and grow less luxuriantly than the American Pothos plants. We discovered a beautiful, truly treelike aroid (*Caladium arboreum*) with a 15- to 20-foot stem not far from the Caripe Cloister east of Cumana. Beauvois found an unusual Caladium (*Culcasia scandens*) in the kingdom of Benin (Palisot de Beauvois, *Flore d'Oware et de Benin,* vol. I, 1804, p. 4, pl. III). In the Pothos form, the parenchyma extends itself so much that the leaf surface becomes perforated, as in *Calla pertusa* (Kunth) (*Dracontium pertusum* [Jacquin]), which we collected in the forests around Cumana. The Aroideae first led us to the remarkable phenomenon of the fever-heat that certain plants register on the thermometer during the development of their blossoms and that is connected to a large, temporary increase in oxygen absorbtion from the atmosphere. Lamarck noted the temperature increase in *Arum italicum* in 1789. According respectively to Bory de St. Vincent and Hubert, the vital warmth of *Arum cordifolium* in Ile de France climbed to 39° or 35°, while the surrounding air temperature was only 15.2°. Even in Europe, Becquerel and Breschet found a temperature difference of up to 17.5°. Dutrochet noted a paroxysm, a rhythmic increase and decrease of internal temperature, which seemed by day to reach a double maximum. Théodore de Saussure observed analogous increases in heat in other plant families such as *Bignonia radicans* and *Cucurbita pepo*, but only of 0.5° to 0.8° Réaumur. In the latter, the male plant exhibits a greater temperature increase than the female, as measured with an extremely sensitive thermoscopic instrument. Dutrochet, taken away too soon and so worthy in the realms of medical physics and plant physiology, also found (*Comptes rendus de l'Institut*, vol. VIII, 1839, p. 454, vol. IX, pp. 614 and 781), through the use of thermomagnetic multiplicators, a vital warmth of 0.1° to 0.3° Réaumur among many young plants (*Euphorbia lathyris*, *Lilium candidum*, *Papaver somniferum*), even among mushrooms of the *Agaricus* and *Lycoperdon* species. This vital warmth disappeared by night but not by day, even if the plants were set in a dark place.

The physiognomic contrast the Casuarineae, the conifers, and the nearly leafless Peruvian Colletia present to the Pothos plants (Aroideae) becomes even more marked when those forms with the greatest contraction in the foliate form are compared to the Nymphaeaceae and Nelumboneae. Here again we find, as with the Aroideae, the stretched cellular web of the leaf surface upon long, fleshy succulent leafstalks; this is true of *Nymphaea alba*, *N. lutea*, *N. thermalis* (once called *N. lotus*, from the Pecze hot spring in Hungary near Großwardein), the Nelumbo species, *Euryale amazonica* (Pöppig), and also of the *Victoria Regina*, discovered in 1837 by Sir Robert Schomburgk in the Berbice River of English Guyana and related to the prickly Euryale, but according to Lindley very different in genus. The round leaves of this magnificent water plant have a diameter of 5–6 Parisian feet and are surrounded by an upright border of 3–5 inches that is light green on the inside but a bright carmoisine red on the outside. The delicately scented blossoms, of which 20–30 can appear within a small space, have a diameter of 14 inches, are white and pink, and have several hundred petals (Robt. Schomburgk, *Reisen in Guiana und am Orinoko*, 1841, p. 233). Pöppig too gives a diameter of 5 ft 8 in for the leaves of his *Euryale amazonica*, which he found near Tefé (Pöppig, *Reisen in Chile, Peru, und auf dem Amazonenstrome*, vol. II, 1836, p. 432). If Euryale and Victoria are the genera that present the greatest parenchymatous extension of the leaf form in all dimensions, then a parasitic Cytinea discovered by Dr. Arnold in 1818 in Sumatra presents the most gigantic development of the flower. *Rafflesia arnoldi* (R. Brown) has a stemless flower nearly 3 feet in diameter, surrounded by large, leaflike scales. It has a fungoid odor of bad beef.

25. Lianas, climbing plants (Spanish: vejucos)

By Kunth's classification of the Bauhiniae, the actual genus *Bauhinia* belongs to the New Continent. The African Bauhinia (*B. rufescens* [Lam.]) is a *Pauletia* (Cav.), a genus of which we also found some species in South America. So too are the Banisteria, of the Malpighiaceae, actually an American form; two species are native to East India, and one, the *B. Leona* described by Cavanilles, is native to West Africa. Various members of the tendrillar and sinuous climbing plants belong to the tropics and the Southern Hemisphere, making the forests there so impenetrable for human beings and so accessible and homey for the races of monkeys and apes (of all four-handers), to the Cercolepts (Procyonidae), and the small tiger-cats. The rapid ascent of tall trees, the crossing from one tree to the next, even over streams, is made easier for hosts of coexisting animals by the lianas.

As in the South of Europe and in North America the hops of the Urticeae and the *Vitis* species of the Ampelideae belong to the group of Lianas, so too are there in the tropics twining and climbing grasses. We have seen a Bambusacea, our *Chusquea scandens* (related to *Nastus*) on the high plateau of Bogota, in the Andean pass of Quindiu, and in the cinchona forests of Loja, wrap itself around powerful tree trunks resplendent with blooming orchids. Also, the *Bambusa scandens* (Tjankorreh), which Blume found in Java, probably belongs to Nastus, or to the grass genus Chusquea, the *carrizo* of the Spanish settlers. In the fir forests of Mexico, it seemed to me that the climbing plants were completely absent; in New Zealand, however, next to the Smilacea (*Ripogonum parviflorum* [Robt. Brown]) that renders the forest nearly impassable, an aromatic Pandanea, *Freycinetia banksii*, twines around an enormous 200-foot conifer, *Podocarpus dacryoides* (Rich.), or *kakikatea* in the native tongue (Ernest Dieffenbach, *Travels in New Zealand*, 1843, vol. I, p. 426).

Contrasting by means of their splendid, colorful flowers with the tendrillar grasses and tendrillar Pandaneae are the Passiflora (one of which, a treelike, upright species, *Passiflora glauca*, we ourselves found in the Andes of Popayan at an elevation of 9,840 feet), the Bignoniaceae, Mutisteae, Alstroemeria, Urvillea, and Aristolochiaea. Of the last-mentioned, our *Aristolochia cordata* has a colorful (purple-red) calyx that measures a full 16 inches in diameter! "*Flores gigantei, pueris mitrae instar inservientes.*" Many of these twining plants, due to the four-sided form of their stalks, to the flat spots that exert no outer pressure, and to a linear, wavelike bending back and forth, have a physiognomic appearance that is theirs alone. The slanting courses of the Bignonias and Banisterias create criss-cross or mosaiclike patterns with the furrows they press into the wood and the splitting that results from the deep penetration of the bark. (See very accurate pictures of this phenomenon in Adrien de Jussieu, *Cours de Botanique*, pp. 77–79, figs. 105–108.)

26. Aloe plants

Belonging to this group of plants which are all so similarly characterized are *Yucca aloifolia*, ranging northward to Florida and South Carolina; *Y. augustifolia* (Nutt.), extending up to the banks of the Missouri; *Aletris arborea*; the dragon tree of the Canary Islands and two other dragons of New Zealand; arborescent Euphorbiae; and *Aloe dichotoma* (Linn.) (once of the genus *Rhipidodendrum* of Willdenow)—the celebrated kokerboom, with a trunk 20 feet tall and four feet thick, and a crown with a circumference sometimes reaching 400 feet (Patterson, *Reisen in das Land der Hottentotten und der Kaffern*, 1790, p. 55). The forms combined here come from greatly differing families: the Liliaceae, Asphodeleae, Pandaneae, Amaryllideae, and Euphorbiaceae—all but the last, however, belonging to the great division of monocotyledons. One of the Pandaneae, *Phytelephas macrocarpa* (Ruiz), which we found in New Granada on the bank of the

Magdalena River, looks with its feathered leaves very much like a small palm tree. The *tagua* (its Indian name) is also, as Kunth remarks, the New Continent's only Pandanea. The peculiar agavelike yet long-stemmed *Doryanthus excelsa* of New South Wales, which was first described by the astute Correa de Serra, is an Amaryllidea, like our low-growing narcissus and polyanthus.

In the candelabra form of the aloe plants, one must not confuse the arms of the trunk with flower stalks. It is these latter that, in the American aloe (*Agave americana, maguey de Cocuiza*, which is completely absent from Chile) as in the *Yucca acaulis* (*maguey de Cocui*), present a candelabralike arrangement of the flowers with the rapid and enormous development of the inflorescence—a well-known, if far too transitory phenomenon. In some arborescent Euphorbeae the characterizing physiognomy lies in the branches and their distribution. In his *Reisen im südlichen Afrika* (vol. I, p. 370), Lichtenstein gives a vivid description of the impression made upon him by a *Euphorbia officinarum* that he found in Chamtoos Revier on the Cape. The form of the tree was so symmetrical that a candelabra shape was repeated in miniature on each branch, even up to a height of 30 feet. Every branch was covered with sharp spines.

Palms, yuccas and aloe plants, tall-stemmed ferns, some Araliae, and the *Theophrasta*, in places where I have seen them growing abundantly, present to the eye through the nakedness (absence of branches) of their stems and the ornamentation of their crowns a certain physiognomic similarity of natural character, however different the composition of their floral structures. The *Melanoselinum decipiens* (Hofm.), which was introduced to our gardens from Madera and sometimes reaches a height of 10 to 12 feet, belongs to a distinct group of arborescent Umbelliferae to which the Araliaceae are at any rate related, and which, in due time, other as yet undiscovered species will join. Certainly *Ferula, Heracleum*, and *Thapsia* likewise attain a considerable height, but there are also herbaceous shrubs. *Melanoselinum* is almost completely alone as an umbelliferous tree; *Bupleurum (Tenoria) fruticosum* (Linn.) from the shores of the Mediterranean, *Bubon galbanum* from the Cape of Good Hope, *Crithmum maritimum* from our own shores are but shrublike. The tropical regions, on the plains of which, according to the old and quite correct assertion of Adanson's, the Umbelliferae and Cruciferae are almost totally absent, showed us on the other hand, on the high mountain ridges of the South American and Mexican Andes, the most dwarflike of all the Umbelliferae. Among the 38 species that we collected at high elevations with a mean temperature below 10° Réaumur, there can be found, vegetating like moss and seemingly conjoined to the rocks and the frozen earth at 12,600 feet above the sea, *Myrrhis andicola, Fragosa arctioides*, and *Pectophytum pedunculare*, mingled with an equally dwarflike alpine Draba. The only Umbelliferae of the tropics that we observed on the plains of the New Continent were two species of *Hydrocotyle* (*H. umbellata* and *H. leptostachya*) between Havana and Batabano, that is, at the outermost boundary of the torrid zone.

27. The grass form

The group of arborescent grasses, which Kunth classifies under the name *Bambusaceae* in his excellent work on the plants that Bonpland and I collected, belongs among the most splendid jewels of the tropical plant world. (*Bambu*, also *mambu*, is a word in the Malay language, but according to Buschmann appears only in isolated cases, the much more common expression being *buluh*; on Java and Madagascar *wuluh* or *voulou* is the only name for this reed species.) Thanks to the efforts of explorers, the number of genera and species that make up the group has increased extraordinarily. It is recognized as fact that the genus *Bambusa* is completely absent from the New Continent, while the gigantic 50- to 60-foot Guadua that we discovered is exclusively indigenous to it, as is the Chusquea; also, that *Arundinaria* (Rich.) appears on both continents, while the specifically differing *Bambusa* and *Beesha* appear in India and on the Indian Archipelago, and *Nastus* is found on Madagascar and Ile Bourbon. With the exception of the high-climbing Chusquea, these are forms that morphologically replace one another in the different parts of the world. In the Northern Hemisphere, it is the joy of the traveler to encounter, in the Mississippi River Valley far outside of the torrid zone, the *Arundinaria macrosperma*, formerly called *Miegia* and *Ludolfia*. In the Southern Hemisphere, Gay discovered a 20-foot Bambusacea (a nonclimbing, undescribed Chusquea that stood upright like a tree) in Southern Chile between the latitudes of 37° and 42°, in a place dominated by a uniform forest of *Fagus obliqua* mixed with *Drymis chilensis*.

While the Bambusa bloom in East India to such a degree that the people of Mysore and Orissa enjoy eating the seed kernels like rice, mixing them with honey (Buchanan, *Journey through Mysore*, vol. II, p.341, and Stirling, in the *Asiat. Res.* vol. XV, p. 205), the Guadua in South America blooms so unusually seldom that, in four years, we were able to collect blossoms for ourselves only twice: once on the lonely banks of the Cassiquiare, the arm by which the Orinoco connects with the Rio Negro and the Amazon, and later in Popayan Province between Buga and Quilichao. It is truly remarkable how certain plants, while growing mightily, do not bloom in certain locations: such is the case in the tropics with the European olive trees cultivated for centuries around Quito, 9,000 feet above sea level; so too with the walnuts, hazel shrubs, and, again, with the lovely olive trees (*Olea europaea*) in Ile de France (see Bojer, *Hortus Mauritianus*, 1837, p. 201).

Just as some of the Bambusaceae (arborescent grasses) extend into the temperate zone, they also do not suffer in the more temperate climate of the mountains within the torrid zone. Certainly they are more luxuriant than the socially growing plants between the shoreline and 2,400 feet of elevation, for example, in Esmeraldas Province west of the Pichincha volcano, where *Guadua augustifolia* (*Bambusa guadua* in our *Plantes équinoxiales*, vol. I, tbl. XX) produces in its interior a great deal of the siliceous tabashir (Sanskrit *tvakkschîra*, cow's milk). In the Andean pass of Quindiu we saw the Guadua growing, according to barometric measurements, at up to 5,400 feet above the level of the Pacific. *Nastus borbonicus* is regularly referred to by Bory de St. Vincent as an alpine plant. By his account, it does not descend on the Ile Bourbon any lower than 3,600 feet from the slope of the volcano to the plain. This occurrence, such a repetition at great heights of certain forms of the warm plains, is reminiscent of the group of mountain palms I discuss above (*Kunthia montana, Ceroxylon andicola, Oreodoxa frigida*), and of a clump of 15-foot Musaceae (*Heliconia*, possibly *Maranta*) that I found isolated on the Silla de Caracas at an elevation of 6,600 feet (*Relation hist.*, vol. I, pp. 605–606). If the grass form, excepting a few individual herbaceous dicotyledons, makes up the highest phanerogamic zone on the snowy peaks, so too does the vegetational region of the phanerogams, extending in the horizontal direction approaching the northern and southern polar regions, end with the grasses.

To my young friend Joseph Hooker, who, having only just returned with Sir James

Ross from the frigid austral countries, now forges onward into the Tibetan Himalayas, the geography of plants is grateful not only for a great mass of important material but also for outstanding general data. He points out that phanerogamic flowering plants (grasses) approach the North Pole 17.5° closer than they do the South Pole. In the Falkland Islands (Malvinas), beside the dense clumps of tussock-grass (*Dactylis caespitosa* [Forster], a *Festuca* according to Kunth), in Tierra del Fuego in the shadow of the birch-leaved *Fagus antarctica*, there vegetates the same *Trisetum subspicatum* that stretches out over the spines of the Peruvian cordilleras and over the Rocky Mountains, and on to Melville Island, Greenland, and Iceland, and is also found in the Swiss and Tyrolean Alps, in the Altai, in Kamchatka and on Campbell Island south of New Zealand: in short, from 54° south to 72°50′ north latitude, a latitudinal difference of 127°. ("Few grasses," says Joseph Hooker in the *Flora antarctica*, p. 97, "have so wide a range as *Trisetum subspicatum* [Beauv.], nor am I acquainted with any other Arctic species that is equally an inhabitant of the opposite polar regions.") The South Shetland Islands, which separate the Bransfield Strait from d'Urville's "Terre de Louis-Philippe" and the 6,612-Par.-ft-high volcanic Mount Haddington (64°12′ south lat.), have been most recently visited by a botanist from the United States of North America, Dr. Eights. There he found (probably at 62° or 62.25° south latitude) a small grass, *Aira antarctica* (Hooker, *Icon. plant.*, vol. II, tbl. 150), "the most Antarctic flowering plant hitherto discovered."

At Deception Island in the same group, 62°50′, one already finds only lichens, and no more grass species; similarly, on Cockburn Island farther to the southeast (64°12′) near Palmer Land, only Lecanorae, Lecideae, and five mosses have been collected, one of which is our German *Bryum argenteum*. "That seems to be the *ultima Thule* of Antarctic vegetation"; farther south even the land cryptogams are absent. In the large gulf formed by Victoria Land, on a small island opposite Mount Herschel (71°49′) and on Franklin Island, 23 geographical miles north of the 11,603-Parisian-foot-tall volcano, Mount Erebus (i.e., 76°7′ south lat.), Hooker found no more traces of plant life. Very different is the distribution of even the higher organic forms in the far North. Phanerogams there come 18.5° closer to the pole than they do in the Southern Hemisphere. Walden Island (80.5° north lat.) still has 10 species of phanerogam. The Antarctic phanerogamic vegetation is poorer in species at the same distance from the pole (Iceland has five times as many phanerogams as the southern group of Lord Auckland's and Campbell's Islands), but the more uniform Antarctic vegetation is more succulent and luxuriant, due to climatic causes. (Cf. Hooker, *Flora antarctica*, pp. VII, 74 and 215, with Sir James Ross, *Voyage in the Southern and Antarctic Regions, 1839–1843*, vol. II, pp. 335–342.)

28. Ferns

If one estimates (with the aid of Dr. Klotsch, a profoundly knowledgeable expert in agamic vegetation) the total number of cryptogams thus far described to be about 19,000 species, then mushrooms number about 8,000 (of which the *Agarici* make up 1/8); the lichens, according to J. von Flotow in Hirschberg and Hampe in Blankenberg, at least 1,400; the algae about 2,580; the mosses and liverworts, according to Carl Müller in Halle and Dr. Gottsche in Hamburg, about 3,800; the ferns about 3,250. For this last important figure, we are indebted to the groundbreaking research into this group of plants by Herr Professor Kunze in Leipzig. It is worth noting that of the entire number of described *Filices*, the family of Polypodiaceae alone comprises 2,165 species, while other forms, even the Lycopodiaceae and Hymenophyllaceae, comprise but 350 and 200 respectively. Still, there are nearly as many described ferns as described grasses.

It is remarkable that in the works of the classical authors of antiquity—Theophrastus, Dioscorides, Pliny—the beautiful treelike form of the fern is not mentioned, while in the information disseminated by the companions of Alexander—Aristobulus, Megasthenes, and Nearchus—there is mention of the Bambusae, "which, once their internodes had been split apart, sailing vessels in turn used to convey on a regular basis," of the trees of India, "whose leaves are not smaller than a shield," of the fig tree that sinks roots with its branches, and of the palms "of such great height that they are unable to shoot arrows over the top of them." (Humboldt, *de distrib. geogr. Plant.*, pp. 178 and 213.) I find the first description of arborescent ferns in Oviedo, *Historia de las Indias*, 1535, fol. XC: "Among the many ferns," says the widely traveled man, appointed by Ferdinand the Catholic to the post of director of gold placer mining in Haiti, "there are also such ones as I count among the trees, for they are as thick and tall as firs [*Helechos que yo cuento por arboles, tan gruesos como grandes pinos y muy altos*]. They grow mostly in the mountains and where there is much water." The value for the height is exaggerated. In the dense forests around Caripe, even our *Cyathea speciosa* reaches only 30 to 35 feet, and an excellent observer, Ernst Dieffenbach, saw no stems of *Cyathea dealbata* on the northernmost of the three islands of New Zealand in excess of 40 feet. In the *Cyathea speciosa* and the *Meniscium* of the missions of the Chaymas we observed, in the midst of the most shadowy ancient forest, the scaly trunks of very healthy, lushly growing individuals covered in a gleaming powder like coal dust. It appeared to be a peculiar decomposition of the fibrous parts of the old leafstalks (Humboldt, *Rel. hist.*, vol. I, p. 437).

Between the tropical circles, where the climates on the slopes of the cordilleras overlap one another in stages, the true zone of the tree ferns lies between three and five thousand feet above sea level. In South America and in the Mexican high country they seldom descend lower than 1,200 feet above the hot plains. The median temperature of this fortunate zone falls between 17° and 14.5° Réaumur. It reaches the cloud layer, which first looms over the ocean and then the plain, and thus enjoys, along with a great consistency of thermal conditions, an uninterruptedly high level of humidity (Robert Brown in *Exped. to Congo*, app., p. 423). The inhabitants of Spanish descent call this zone *tierra templada de los helechos* (the temperate land of the ferns). The Arabic term is *feledschun, felix*, fern, the *f* being changed to *h* in the Spanish custom; connected perhaps to the verb *faladscha*, "he cuts apart," referring to the so finely incised frond (Abu Zacaria Ebn el Awam, *Libro de Agricultura, traducido por J. A. Banqueri*, vol. II, Madr. 1802, p. 736).

The conditions of the mild temperature of an atmosphere impregnated with water vapor and a great consistency of humidity and warmth come together on the declivity of the mountains, in the valleys of the Andes chain, and above all in the mild and moist Southern Hemisphere, where arborescent ferns extend not only to New Zealand and Van Diemen's Land (Tasmania) but to the Strait of Magellan and Campbell Island, that is, to a southern latitude that is nearly the equal of the northern latitude of Berlin. Of the tree ferns, *Dicksonia squarrosa* grows abundantly at 46° south latitude in Dusty Bay (New Zealand), Labillardière's *D. antarctica* in Tasmania, a *Thyrsopteris* in Juan Fernandez, an undescribed *Dicksonia* with a 12- to 15-foot stem in Southern Chile not far from Valdivia, and a somewhat smaller *Lomaria* near the Magellanic Strait. Campbell Island lies still closer to the South Pole at 52.5° lat., and even there the leafless stem of *Aspidium venustum* rises to a height of four feet.

The climatic conditions under which the ferns (*Filices*) generally thrive become apparent in the numerical laws of their distribution quotients. In the level regions of of the large continents, this quotient in the tropics is, according to Robert Brown and more recent research, 1/20 of all phanerogams; in the mountainous portions of the large continents, 1/6 to 1/8. Quite different is the relationship on the small islands scattered about the wide ocean. The portion of ferns in relation to the totality of phanerogams increases there to the extent that this quotient in the island groups in the tropical zone of the Pacific rises to ¼; indeed, ferns account for almost half of the total phanerogamic vegetation of the sporadic islands of St. Helena and Ascension. (See an excellent discussion by d'Urville, *Distribution géographique des fougères sur la surface du Globe* in the *Annales des Sciences Nat.*, vol. VI, 1825, pp. 51, 66, and 73.) Moving away from the tropics (d'Urville accepts the numeric ratio of the large continents as a whole to be 1/20), one will observe a rapid decline in the relative frequency of ferns in the temperate zone. The quotient for North America and the British Isles is 1/35, for France 1/58, for Germany 1/52, for the arid parts of Southern Italy 1/74, for Greece 1/84. Moving toward the frigid North, the relative frequency grows again considerably. The family of the ferns itself decreases much more slowly in total number of species there than does the number for the phanerogamic plants. The abundantly growing quantity of individuals of every species increases the deceptive impression of absolute frequency. According to Wahlenberg's and Hornemann's catalogs, the ratios of *Filices* are 1/25 for Lapland, 1/18 for Iceland, and 1/12 for Greenland.

Those are, to our current knowledge, the natural laws that become apparent in the distribution of the graceful form of the fern. But it seems that we have recently gained ground in the pursuit of another law of Nature, the morphological law of the reproduction of the family of ferns, which for so long has been held to be cryptogamic. Count Leszczyc-Suminski, who brings together in a happy combination the capacity of microscopic research with a truly distinguished artistic talent, has discovered the system that facilitates fertilization in the prothallium of the fern. He differentiates two sexual apparatuses: a female, in hollow ovoid cells located in the middle of the prothallium, and a male, in the ciliary antheridia or the organs that produce spiral filaments, which have already been investigated by Nägeli. The fertilization is said to take place not through pollen tubes but through mobile spiral filaments with cilia. (Count Suminski, *Zur Entwicklungsgeschichte der Farrnkräuter*, 1848, pp. 10–14.) By this view, the ferns would be, as Ehrenberg expresses it (*Monatliche Berichte der Akad. zu Berlin*, January 1848, p. 20), the product of a microscopic fertilization that occurs on the prothallium, which serves as a starting pot, and throughout the whole course of their often treelike development, they remain bulbil-forming plants with neither blossom nor fruit. The spores, which lie in clusters (sori) on the underside of the fern frond, are not seeds but buds.

29. Lily plants

The primary home of this form is Africa: there may be found the greatest diversity of lily plants; there they form masses and determine the natural character of the region. The New Continent also has splendid species of *Alstroemeria, Pancratium, Haeman-*

thus, and *Crinum*, and the first-named of these genera we increased by 9 species, the second by 3. But these American lily plants are scattered, less social that the European Irideae.

30. Willow form

Of the primary representative of the form, that is, of the willow itself, over 150 different species are already known. They bedeck the northern half of the world from the equator to Lapland. Their number and diversity of forms increase between the 46th and 70th degrees of latitude, especially in that portion of Northern Europe that was so fantastically furrowed by early planetary cataclysms. Of willows as tropical plants I am familiar with 10 or 12 species which, like the willows of the Southern Hemisphere, are worthy of special attention. As it seems that Nature pleases to have in all zones a tremendous variety of certain animal forms, such as the Anatidae (Lamellirostrae) and the doves, so too are willows, varieties of *Pinus*, and oaks equally widespread: the latter always producing similar fruits, though widely varying in leaf forms. Among willows of the most contrasting climates, the similarity of the foliage, the branching, and the whole physiognomic formation is exceedingly great, perhaps greater even than that of the conifers. In the southern portion of the temperate zone north of the equator, the number of willow species decreases considerably, and yet (according to the *Flora atlantica* of Desfontaines), Tunis still has an indigenous species, similar to the *Salix caprea*, and Egypt, according to Forskål, has 5 species whose male catkins will produce through distillation the medicament often used in the Orient, the *Moie chalaf* (*aqua salicis*). The willow I saw on the Canary Islands, according to Leopold von Buch and Christian Smith, is likewise an indigenous species (*S. canariensis*), but common both to these islands and to Madera. Wallich's plant catalog of Nepal and the Himalayas currently lists 13 species from the subtropical zone of East India, some of which have been described by Don, Roxburgh, and Lindley. Japan has indigenous willows, one of which, *S. japonica* (Thunb.), also occurs in Nepal as a mountain plant.

Before my expedition there was, to my knowledge, still no other known species besides the Indian *S. tetrasperma* in the torrid zone between the tropical parallels. We collected seven new species, three of them coming from the Mexican high plains up to 8,000 feet. Still higher, on mountain plateaus that we often visited, for example, between twelve and fourteen thousand feet, nothing presented itself to us in the Andes of Mexico, Quito, and Peru that might remind us of the small, creeping alpine willows of the Pyrenees, the Alps, or Lapland (*S. herbacea*, *S. lanata*, and *S. reticulata*). In Spitzbergen, whose meteorological conditions are so analogous to the Swiss and Scandinavian snow peaks, Martins described two dwarf willows whose woody little stems and twigs, pressed to the earth, lie so well hidden in the turf moor that only with effort does one find its tiny leaves under the moss. The species that I discovered at 4°12′ south latitude in Peru near Loja at the beginnings of the cinchona forests, which was described by Willdenow as *Salix humboldtiana*, is the most widespread willow in the western part of South America. A beach willow, *S. falcata*, which we found on the sandy Pacific coast near Trujillo is probably just a variety of it, according to Kunth. Equally likely to be identical to it is the attractive, often pyramidal willow that accompanied us along the banks of Magdalena River from Mahates to Bojorque, and which according to the residents of the region has only in the last few years become so widespread. At the confluence of the Magdalena with the Rio Opon, we found all of the islands covered with willows, of which many, with stems of 60 feet, had diameters of a bare 8–10 inches (Humboldt and Kunth, *Nova Gen. Plant.*, vol. II, p. 22, tbl. 99). From Senegal, that is, from the African equinoctial zone, Lindley (*Introduction to the Natural System of Botany*, p. 99) has made known a species of *Salix*. On Java, Blume has also found two species of willow near the equator: the wild species indigenous to the island (*S. tetrasperma*) and a cultivated willow (*S. sieboldiana*). From the southern temperate zone, I am familiar with only two willows, already described by Thunberg (*S. hirsuta* and *S. mucronata*). They vegetate alongside the *Protea argentea*, which itself has the physiognomy of the willow; its leaves and young twigs are the food of the hippopotamuses of the Orange River. In Australia and on the nearby islands, the willow genus is completely absent.

31. Myrtle plants

A delicate form with stiff, gleaming, densely packed, mostly nonserrated, small, and foraminous leaves. Myrtle plants give a distinct character to three regions of the Earth's surface: Southern Europe, especially the islands (limestone cliffs and trachyte rocks) that rise up from the basin of the Mediterranean; the continent of Australia, adorned with *Eucalyptus*, *Metrosideros*, and *Leptospermum*; and a section of land that lies between the lines of the tropics, which is partly low and flat, and partly elevated nine to ten thousand feet above sea level—the high comb of the Andes in South America. This mountainous region, which in Quito is called the *Paramos*, is covered everywhere with trees of myrtlelike appearance, even if not all are actually members of the natural family of the Myrtaceae. Growing at this altitude are *Escallonia myrtilloides*, *E. tubar*, *Symplocos alstonia*, species of *Myrica*, and the beautiful *Myrtus microphylla*, of which we had a depiction included in *Plantes équinoxiales*, vol. I, p. 21, pl. IV, and which vegetates at up to 9,400 feet on the mica schist near Vinayacu and Alto de Pulla on the Paramo de Saraguru, which is arrayed in so many gracefully blooming alpine plants. *M. myrsinoides* can even be found on the Paramo de Guamani as high as 10,500 feet. Out of 40 species of the genus *Myrtus* that we collected in the equinoctial zone, and of which 37 were undescribed, the majority by far belong to the plains and foothills. From the mild tropical mountain climate of Mexico we brought back only one single species (*M. xalapensis*), but the *tierra templada* in the proximity of the Orizaba volcano certainly contains many more. We found *M. maritima* near Acapulco on the shore of the Pacific itself.

Together with the European and American alpine roses (*Rhododendrum* and *Befaria*), and with *Clethra*, *Andromeda*, and *Gaylussacia buxifolia*, the Escalloniae, among which *E. myrtilloides*, *E. tubar*, and *E. floribunda*, the jewels of the Paramos, are physiognomically so very reminiscent of the myrtle form, formerly made up the family of the Ericeae. Robert Brown, however (see the addenda to Franklin's *Narrative of a Journey to the shores of the Polar Sea*, 1823, p. 765), has elevated them to the status of being a family of their own, which Kunth places between the Philadelpheae and the Hamamelideae. The *Escallonia floribunda* presents, in its geographical distribution, one of the most conspicuous examples of the relationship between distance from the equator and vertical elevation of position above sea level. I can support myself here once again with the testimony of my perspicacious friend August de Saint-Hilaire (*Morphologie végétale*,

1840, p. 52): "Messrs. von Humbolt and Bonpland discovered on their expedition the *Escallonia floribunda* at 1,400 toises, 4° southern latitude. I found it at 21° in Brazil in the high country, yet infinitely lower than the Andes of Peru: it is common between 24°50′ and 25°55′ in the Campos Geraes. Finally, I see it at the Rio de la Plata around 35°, at sea level."

The group of the Myrtaceae, to which belong *Melaleuca*, *Metrosideros*, and *Eucalyptus* (which have been included under the general name of Leptospermeae), create to some degree, either where true leaves are replaced by phyllodes (petiole leaves) or by the positioning of the leaves, i.e., their direction turning back on the unexpanded petiole, a distribution of streaks of light and shadow which we do not see in the shade trees of our forests. Even the earliest explorers to visit Australia as botanists were moved to astonishment by the peculiarity of this impression. Robert Brown was the first to demonstrate how this phenomenon arises from the vertical direction of the outspread petioles (the phyllodes of *Acacia longifolia* and *A. suaveolens*) and from the fact that the light, instead of striking horizontally oriented surfaces, falls down between vertically oriented ones (Adrien de Jussieu, *Cours de Bot.*, pp. 106, 120, and 700; Darwin, *Journal of Researches*, 1845, p. 433). Morphological laws in the development of the leaf-organ determine the particular character of illumination and the limitations of light and shadow. "Phyllodes," says Kunth, "can, in my view, occur only in families that have bunched, feather-like leaves; in fact, up to now they have been found only in the Leguminosae (Acacias). In *Eucalyptus*, *Metrosideros*, and *Melaleuca* the leaves are simple (*simplicia*), and their postion, on average, is achieved by a half-turn of the leaf stalk (*petiolus*); it is notable here that both surfaces of the leaf are of identical construction." In the shade-poor forests of Australia, the optical effects touched on here are all the more frequent since two groups of Myrtaceae and Leguminosae, species of *Eucalyptus* and *Acacia*, constitute almost half of the entire, gray-green, arborescent vegetation. Additionally, *Melaleuca* forms easily removed membranes between the bast layers that come to the surface and, with their white color, are reminiscent of the bark of our birches.

The sphere of distribution of the Myrtaceae is quite dissimilar in the Old and New Continents. In the New Continent this family extends, especially in the western part, barely above the 26th parallel of northern latitude, according to Joseph Hooker (*Flora antarctica*, p. 12). In contrast, in the Southern Hemisphere there are, according to Claude Gay, 10 species of *Myrtus* and 22 of *Eugenia* in Chile. There they form forests, mingling with Proteaceae (*Embothrium*, *Lomatia*) and the *Fagus obliqua*. The Myrtaceae become more numerous from 38° south latitude on: on Chiloe Island (where a *Metrosideros*-like species, *Myrtus stipularis*, known there as *tepuales*, forms virtually impenetrable thickets), in Patagonia, and down to the remotest point of Tierra del Fuego at 56° south latitude. While the Myrtaceae in Europe spread to the north only as far as 46°, in Australia, Tasmania, New Zealand, and Lord Auckland's Islands they penetrate to 50.5° south latitude.

32. Melastomae

This group comprises the genera *Melastoma* (*Fothergilla* and *Tococa* [Aubl.]) and *Rhexia* (*Meriana*, *Osbeckia*), of which we collected on both sides of the equator, in tropical America alone, over 60 new species. Bonpland has published a magnificent work on the Melastomaceae with colored drawings in two volumes. There are species of *Rhexia* and *Melastoma* that, as alpine or *Paramo* shrubs of the Andes chain, ascend to 9,000 to 10,500 feet, such as *Rhexia cernua, R. stricta, Melastoma obscurum, M. aspergillare,* and *M. lutescens.*

33. Laurel forms
Belonging to this group are *Laurus* and *Persea,* as well as the Ocoteae, which are so numerous in South America, and from the Guttiferae, due to physiognomic similarity, the *Calophyllum* and the impressively burgeoning *Mammea.*

34. How interesting and instructive to the landscape painter would be a work that depicts all of the primary forms of vegetation!
To more distinctly delineate that which here has been alluded to only briefly, I hope I might be allowed to insert the following considerations from my sketch of a history of the art of landscape painting and of the graphic depiction of the physiognomy of plants (*Kosmos*, vol. II, pp. 88–90):

"Everything that is connected to the expression of the passions, and to the beauty of the human form, reached its highest level of perfection in the northern temperate zone under the Greek and Hesperian skies; from the depths of his mind as from the thoughtful examination of humankind, the artist, at once representational and creatively free, calls forth the types of historical depiction. Landscape painting, which is just as far from being merely imitative, has a more material substratum, a drive that is more earthly. It requires a great depth and diversity of immediate, thoughtful examination that the mind should absorb and, fructified by a force of its own, present to the senses as a free work of art. The great style of the heroic landscape is the result of a profound understanding of Nature and of this inner, spiritual process.

"Certainly, Nature in every corner of the Earth is but a reflection of the whole. The shapes of the organism repeat themselves in other and still other connections. The icy North too enjoys for months the green-covered earth, large-blossomed alpine plants, and the gentle blues of the heavens. Up to now, being familiar with only the simpler forms of local flora, but for all that not without a depth of feeling or a fullness of creative imagination, landscape painting has completed among us its graceful work. Lingering with those elements of the plant kingdom that are native to the homeland or have naturalized over time, landscape painting has traveled a narrower circle, but even here, highly talented artists—the Carraccis, Gaspard Poussin, Claude Lorrain, and Ruysdael—found room enough, through the variation of tree forms and of illumination, to bring forth magically the happiest and most manifold creations. What art has yet to expect from more lively interaction with the tropical world, from the atmosphere that a glorious Nature, rich in the variety of forms, breathes into the creative imagination—to which I am compelled to refer, as a reminder of the old bond that the lore of Nature shares with poetry and artistic sensibility—will in no way diminish the fame of those masterworks. For in landscape painting, and in every other branch of art, one must differentiate between that which, in a more restricted way, is created by sensory examination and immediate observation, and that which arises unrestricted from the depths of feeling and the strength of idealizing intellect. The greatness for which landscape art, as a more or less inspired form of Nature poetry, is indebted to this creative intellect (see for example the graduated progression of the tree forms from Ruysdael and Everdingen, through Claude Lorrain, and on to Poussin and Hannibal Carracci) is, like the individual endowed with fantasy, not something that is shackled to the ground. Among the first masters of the art, limitations of place are not perceptible, but broadening of the horizon of perception and familiarity with nobler and grander forms in Nature, and with the torrid zone's luxuriant fecundity of life, all provide the advantage that they not only effect an enrichment of the material substratum of landscape painting, but also that they have the effect of more vitally animating the sensitivity of less gifted artists and thus increasing their creative energy."

35. From the dense, rough barks of the Crescentia *and Gustavia*

In the *Crescentia cujete* (the tutuma or calabash tree), the rinds of which are so indispensable to the households of the natives, in the *Cynometra* or cacao tree (*Theobroma*), and in the Pirigara (*Gustavia* [Linn.]), the tender bloom organs break forth through the half-carbonized bark. When children enjoy the fruit of the *Pirigara speciosa* (the chupo), their entire bodies turn yellow: it is a jaundice that lasts from 24 to 36 hours and then disappears on its own, with no medical intervention.

Unforgettable to me has been the impression of the abundant vegetational energy of the tropical world that I received upon first seeing, after a wet night on a cacao plantation (*cacahual*) in the *Valles de Aragua*, large blossoms bursting forth from a *Theobroma* root, deeply submerged in black earth and far away from the trunk. It is in the tropics that the actions of the driving forces within the organism most obviously reveal themselves. The people of the North speak of the "awakening of Nature by the first airs of spring." Such an expression contrasts with the imaginative lament of the Stagirite, who in plants recognizes objects "that lie in a quiet slumber beyond waking, free of the desires that drive them to move." (Aristotle, *De generat. Animal. V, 1,* p. 778 and *De somno et vigil.*, cap 1, p. 455, Bekker.)

36. Pull over their heads

The blossoms of our *Aristolochia cordata*, which are mentioned in note 25, above. Besides the Compositae (the Mexican *Helianthus annuus*), the largest blossoms in the world are carried by *Rafflesia arnoldi, Aristolochia, Datura, Barringtonia, Gustavia, Carolinea, Lecythis, Nymphaea, Nelumbium, Victoria regina, Magnolia, Cactus,* the orchids, and the lily plants.

37. The dome of the heavens from pole to pole hides from him none of its luminous worlds

For the inhabitants of Europe, the most magnificent portion of the southern heavens, where the Centaur, the ship Argo, and the Southern Cross gleam, where the Magellanic Clouds orbit, will remain forever hidden. At the equator alone does a person enjoy the unique and beautiful vantage point from which to see of all of the stars of the Southern and Northern Hemispheres at once. When viewed from there, some of our northern constellations (Ursa Major and Minor, for example), due to their low position over the horizon, appear to be of astonishing, almost frightening size. Much as the inhabitant of the tropics can see all the stars, so too has Nature, in that place where plains, deep valleys, and mountains come in turns, surrounded him with representatives of all of the forms of plants.

In the preceding sketch of a physiognomy of plants I have stressed three primary objects of concentration that are closely related to one another: the absolute differences in the forms, their numerical ratios (i.e., their localized level of predominance amidst the collective total of phanerogamic flora), and their geographical and climatic distribution. If one wishes to rise to a general understanding of the views concerning the various forms of life, then my ideas regarding physiognomy, the study of numeric ratios (the arithmetic of botany), and the geography of plants (the study of the spatial zones of distributions) cannot be separated from one another. The physiognomy of plants should not remain exclusively concerned with the noticeable contrasts in form that the large organisms present when considered individually; it should dare to come to a recognition of the laws that determine the physiognomy of Nature in general, the scenic vegetational character of the entire surface of the Earth, and the vital impression evoked by the aggregation of contrasting forms in various zones of latitude and elevation. Only when examined from this point of view does it become clear wherein lies the close, interior interlinking of the material dealt with in the preceding pages. We have been led here into a field that has been little examined until now. I have ventured to pursue the method that first appears so brilliantly in the zoological works of Aristotle and that is particularly suited to establish scientific confidence: the method in which the most extraordinary of phenomena are penetrated by the introduction of specific examples, along with unceasing efforts toward the generalization of concepts.

The listing of forms by physiognomic dissimilarity is by its very nature not capable of strict classification. Here, as ever in the consideration of external shape, there are certain primary forms whose contrasting points are especially obvious: the groups of arborescent grasses, of aloe plants and cactus species, of palms, conifers, Mimosaceae, and bananas. Even sparsely dispersed individuals of these groups determine the character of a region, leaving with the unscientific but receptive observer a lasting impression. A perhaps larger, predominant number of other forms, however, do not characteristically stand out: not through the shape or position of the foliage, nor through the relationship of the trunk to the branches; not through powerful luxuriance or blithe gracefulness, nor through a melancholy stunting of the appendicular organs.

If a physiognomic classification, an assignment into groups according to external *facies*, may thus not be applied to the entire kingdom of plants, it is also true that the basis for division in plant physiognomy is a very different one from that by which our all-inclusive system of natural plant families are so happily arranged. Physiognomy bases its divisions, the choice of its types, upon everything that has mass: upon the stem, the branches and appendicular organs (leaf form, leaf position, leaf size, composition, and the luster of the parenchyma), that is, upon the now so suitably named "vegetational organs," the organs upon which the preservation (nourishment, development) of the individual depends; systematic botany, on the other hand, bases the arrangement into natural families upon the reproductive organs, those organs upon which the preservation of the species depends (Kunth, *Lehrbuch der Botanik*, 1847, vol. I, p. 511; Schleiden, *Die Pflanze und ihr Leben*, 1848, p. 100). In the school of Aristotle (*Probl.* 20, 7) it was already taught that procreation is the ultimate goal of the existence and life of plants. The developmental process in the fertilization organs has become since Caspar Friedrich Wolff (*Theoria Generationis*, §5–9) and since our own great poet, the morphological foundation of all systematic botany.

This approach and plant physiognomy proceed (I repeat it here) from two divergent views: the first from a consistency in inflorescence, in the reproduction of the delicate sex organs, the other from the formation of the axillary parts (the stem and the branches), and from the morphological group of the leaves, which depends primarily upon the distribution of the vascular bundle. Since axil and appendicular organs are also predominant in terms of volume and mass, they serve to dictate and strengthen the impression that we receive; they individualize the physiognomic character of the formation, as they do the character of the landscape and of a given zone in which distinctive forms appear. Consistency and relatedness in the characteristics that lie in the vegetative, i.e., nutritive organs dictate the law here. In all of the colonies of Europeans, similarities of physiognomy (*habitus, facies*) have led the immigrants to attach the names of trees native to their country of origin to certain tropical trees that bear blossoms and fruits completely different from those of the plant genera of the mother country to which these names were originally given. All over, in both hemispheres, northern settlers have believed that they were seeing alders and poplars, apple and olive trees. It was primarily the form of the leaves and the orientation of the branches that misled them. Sweet memories of the forms of their native lands reinforced the deception, and European plant names are passed from generation to generation, enriched in the slave colonies by appellations from the Negro languages.

The contrast so frequently presented by a conspicuous similarity of physiognomy together with the greatest difference in the blossom and fruit structures, the contrast between external formation determined by the appendicular or leaf system and the reproductive organs that are the foundation of the groupings of natural plant systems, is a wondrous phenomenon. One might be inclined to believe that the morphological group of the exclusively (so-called) vegetational organs (e.g., the leaves) must be less independent of the structure of the reproductive organs, but such a dependency is revealed only in a small number of families: amongst ferns, grasses, Umbelliferae, and Aroideae. In the case of the Leguminosae, consistency of the physiognomic character and of the inflorescence may for the most part be recognized only if they are divided into specific groups (Papilionaceae, Caesalpineae, and Mimosae). Some types that display an external physiognomic consistency but very divergent blossom and fruit structures when compared to one another are palms and cycads, the latter being most closely related to the conifers; *Cuscuta* (a Convolvulacea) and the leafless *Cassytha* (a parasitic Laurinea); *Equisetum* (from the division of the cryptogams) and *Ephedra* (a conifer). The gooseberries (*Ribes*) are in terms of inflorescence so closely related to the cactus (that is, the family Opuntiaceae) that they were only recently separated from them! United in one and the same family (Asphodeleae) are the gigantic tree *Dracaena draco*, the common asparagus, and the colorfully blooming *Aletris*. Not only do simple and compound leaves often belong to the same family, they can even be found in one and the same genus. In the high plains of Peru and New Granada, we found among 12 new species of *Weinmannia 5 foliis simplicibus*, the others having feathered leaves. The genus *Aralia* displays an even greater independence in leaf forms: *folia simplicia, integra, vel lobata, digitata*, and *pinnata*. (Cf. Kunth, *Synopsis Plantarum, quas in itinere collegerunt Al. de Humboldt et Am. Bonpland*, vol. III, pp. 87 and 360.)

It seems to me that feathered leaves belong primarily to the families that stand at the highest level of organic development, namely the Polypetalea—among the perigynous plants, to the Leguminosae, Rosaceae, terebinths, and Juglandeae; among the hypogynous plants, to the Auranteaceae, Cedrelaceae, and Sapindaceae. The beautiful, doubly feathered leaves, one of the main adornments of the torrid zone, are most frequently found among the Leguminosae; among the Mimosae they are also found in some Caesalpineae, Colteriae, and Gleditsiae; never, as Kunth observes, among the Papilionaceae. *Folia pinnata* and indeed any *folia composita* are foreign to the Gentianeae, the Rubiaceae, and the myrtles. In the morphological development presented by the plenitude and variety of forms of the appendicular organs of the dicotyledons, there are but a small number of general laws to be recognized.

6

Concerning the Structure and Action of Volcanoes in Various Regions of the Earth

If one considers the influence that expeditions to different regions and increased knowledge of earth sciences have exerted for centuries on the study of Nature, one soon sees how this influence varies depending on whether the investigations concentrate on the forms of the organic world or on the formations of inanimate earth—the varieties of stone, their relative ages, and their origins. Different forms of plants and animals inhabit the Earth in every zone, be it in the oceanlike plain where the heat of the atmosphere moves according to latitude and the manifold twists and turns of the isothermal lines, or places where the heat moves almost vertically along the steep slopes of mountain chains. Organic Nature gives to each stretch of land a physiognomic character of its own—not so with inorganic Nature, in places where the hard crust of the planet is exposed from under the cover of plants. The same sorts of stone, seeming to attract and repel one another in groups, occur in both hemispheres from the equator to the poles. On a distant isle, surrounded by strange vegetation and under a sky where the old stars no longer shine, the seafarer, joyfully astonished, will recognize the clay slate of home, the comfortably familiar type of rock he knows from his fatherland.

This independence of geognosy from the current constitution of the climate does not diminish the beneficial influence that numerous observations made in foreign parts of the world have exerted upon the progress of mineralogy and physical geology; rather, it gives to these sciences their own particular orientation. Each expedition enriches natural history with new plant and animal genera. Sometimes they are organic forms that may be added to long-familiar types, displaying to us the regularly woven, though often seemingly interrupted, network of animated natural creation in its original perfection; sometimes they are forms that appear in isolation, as escaped remnants of

(This treatise was read before the public assembly of the Academy at Berlin on the 24th of January, 1823.)

extinct genera or, arousing our expectations, as unknown members of groups yet to be discovered. Such diversity is clearly not offered by analysis of the hard crust of the Earth. This reveals rather a consistency in the constituent minerals, the stratification of various masses, and their periodic reappearance, which excites the wonderment of the geognost. In the Andes chain, as in the central range of Europe, one formation seems to some degree to call forth the other. Correspondent masses take the shape of similar forms: basalt and dolerite into twin peaks; dolomite, sandstone, and porphyry as looming cliff walls; trachyte rich in feldspar into bell shapes or high-vaulted domes. In the most distant zones, crystals of the same rock separate themselves as if by inner development from the dense weave of the groundmass, envelope one another, come together in secondary formations, and, as such, announce the proximity of a new and independent formation. Thus every mountainous region of considerable extent reflects, with greater or lesser clarity, the entire inorganic world; yet to recognize completely the important phenomena of the composition, the relative age, and the emergence of the various types of rock, observations from the most disparate regions of the planet must be compared to one another. Problems that have long seemed puzzling to the geognost in his northern homeland find their solution at the equator. If the distant zones, as has often been noted, present to us no new types of rock, i.e., no unknown combinations of basic materials, then they teach us rather how to unmask the great laws that are the same everywhere, the laws by which the layers of the Earth's crust alternately support one another, break apart into channels, or are lifted by elastic forces.

In light of the profit just described that investigation gives to geognostic knowledge, it should not surprise us that one class of phenomena, upon which I will concentrate here, has long been considered all the more one-sidedly as the points of comparison were more difficult, one might even say more painful, to discover. By the close of the previous century, everything that was thought to be known regarding the structure of volcanoes and the action of their subterranean forces was gathered from two mountains of Southern Italy: Vesuvius and Aetna. Since the first is more accessible and (like nearly all short volcanoes) erupts more frequently, it came about that a hill, as it were, came to serve as the generic model by which an entire distant world, the mighty strings of volcanoes of Mexico, South America, and the Asiatic islands, was pictured in the mind. Such a thought process understandably calls to mind Virgil's deluded shepherd who perceived in his narrow hut the model of the regal eternal city of Rome.

Certainly a more painstaking investigation of the entire Mediterranean, especially the eastern islands and coastal lands where humanity first awakened to intellectual civilization and nobler feelings, could have eradicated such a one-sided view of Nature. From the depths of the seafloor here, trachyte cliffs heaved themselves up amidst the Sporades to form islands—similar to what periodically occurred in the Azores, where it happened three times over the course of three centuries and at nearly equal intervals. Between Epidaurus and Troezen, near Methoni, the Peloponnese features a *Monte nuovo*, which was described by Strabo and seen again by Dodwell: taller than the *Monte nuovo* of the Phlegraean Fields of Baiae, perhaps even taller than the new volcano of Jorullo on the Mexican plains, which I found encircled by several thousand small basalt cones that had been pushed out through the surface of the ground and were, at the time, still smoking.

In the Mediterranean basin too, the volcanic fire does not break forth only from permanent craters, isolated mountains with an abiding connection to the interior of the Earth, such as Stromboli, Vesuvius, and Aetna. On Mount Epomeo on the island of Ischia and, according to the reports of the ancients, on the Lelantine Plain near Chalcis, lava has flowed from suddenly opening fissures in the ground. Along with these phenomena (which fall in the realm of historical time, in that narrow region of assured tradition, and which will be collected and elucidated by Carl Ritter in his masterful geographical work), the coasts of the Mediterranean contain manifold remnants of the effects of more ancient fire. Southern France shows us its own enclosed system of volcanoes ordered in rows in the Auvergne region: trachyte domes alternating with volcanic cones from which pour forth ribbonlike streams of lava. The plain of Lombardy, which lies even with the sea and forms the innermost bay of the Adriatic, encloses the trachyte of the Euganean Hills, where domes of granular trachyte, of obsidian, and of perlite rise up: three masses growing out of one another which break through the chalk and nummulite lime below but have never flowed in narrow streams. Similar evidence of ancient revolutions of the Earth can be found in many parts of continental Greece and Western Asia, countries that will offer to the geognost rich material for investigation, once the light returns to the place from which it first shone forth over the West, when tormented humanity is no longer persecuted by the wild barbarism of the Ottomans.

I mention the geographical proximity of these diverse phenomena in order to demonstrate that the basin of the Mediterranean, with its strings of islands, could have provided the attentive observer with everything that has recently been discovered in many various forms and structures in South America, on Tenerife, or in the Aleutians close to the polar region. These objects of observation were certainly all concentrated closely together. But travels to distant climes and the comparison of larger sections of the Earth's surface both within and outside of Europe were necessary in order to recognize the commonality of volcanic phenomena and their dependence upon one another.

Conventional language (which often lends longevity and credibility to the earliest errors, but also often instinctively indicates the truth) describes as "volcanic" all eruptions of underground fire and molten materials: columns of smoke and vapor that sporadically rise up out of the rocks, as in Colares after the great earthquake of Lisbon; salses, that is, clay cones that spew forth wet mud, bitumen, and hydrogen, like those near Girgenti in Sicily and Tur-baco in South America; hot geysers that rise up under the pressure of elas-tic vapors—in general, all effects of those wild forces of Nature that originate deep in the interior of our planet. In Central America (Guatemala) and in the Philippine Islands, the natives differentiate (and indeed formally) between water and fire volcanoes, *volcanes de agua y de fuego*. With the first of these names, they indicate mountains from which now and then, upon the occasion of sizable seismic shocks, a muffled crack is accompanied by the expulsion of subterranean water.

Without denying the connectedness of the aforementioned phenomena, it still seems advisable to give to the oryctognostic areas of geognosy a more specific language, and not to use the word *volcano* to indicate in one context a mountain that terminates with a permanent mouth of fire and in another any sort of subterranean cause of volcanic phenomena. In the current situa-tion of the Earth, in every region of the planet, the isolated conical mountain (of Vesuvius, Aetna, the peak of Tenerife, of Tungurahua and Cotopaxi) is clearly the usual form of volcanoes; I have seen them in sizes from the lowest hills up to 18,000 feet above sea level. But beside these cone-shaped peaks one also finds permanent chasms of fire, enduring communications with the interior of the Earth upon long-abiding jagged ridges, and not always in the middle but at the lower end of their wall-like tops, near the slope. This is the form of Pichincha, which rises between the Pacific and the city of Quito, and which was first made famous by Bouguer's barometric formulas; so too the volcanoes that rise from the ten-thousand-foot-high steppe of los Pastos. All of these peaks of widely diverse shapes consist of trachyte, once called trap-porphyry: a granular, fissured stone composed of varieties of feldspar (labradorite, oligoclase, albite), augite, hornblende, and occasionally admixed mica or even quartz. Where the evidence of the initial eruption, or I might say where the old structure has remained intact, a high wall of rock, a mantle of ac-cumulated layers surrounds the isolated cone in a circular fashion. Such walls or ring-shaped enclosures are called elevation craters, a large and important phenomenon about which the first geognost, Leopold von Buch, from whose writings I have borrowed many of the views in this essay, presented a truly memorable essay to our academy five years ago.

Volcanoes that communicate with the atmosphere by means of fiery mouths, conical basalt hills, and domed, craterless trachyte mountains (these last being sometimes low, like Sarcouy, sometimes tall, like Chimborazo) form many diverse groups. Sometimes comparative geography shows us small archipelagos, closed mountain systems as it were, with craters and lava flows in the Canary Islands and the Azores, and without craters or actual lava flows in the Euganean Hills and the Siebengebirge near Bonn; other times it describes for us volcanoes lined up in simple or double chains, ranges of several hundred miles, either parallel to the primary direction of the mountains, as in Guatemala, Peru, and Java, or else perpendicularly bisecting the axis of the mountains, as in tropical Mexico. In this land of the Aztecs, fire-spewing trachyte mountains alone attain the height of the snow line, and, probably breaking out along a fault, follow a line of latitude that extends for 105 geographical miles and cuts across the entire continent from the Pacific to the Atlantic.

This concentration of volcanoes, sometimes in separated, somewhat circular groups, and sometimes in double ranges, provides the most convincing evidence that volcanic effects are not dependent upon trivial causes close to the surface but are immense phenomena with deep-rooted causes. The entire metal-poor eastern region of the American continent, in its present condition, is without fiery chasms and trachyte masses, perhaps even without basalt or olivine. All American volcanoes are in the regions that lie across from Asia, in the longitudinally extending 1,800-geographical-mile-long chain of the Andes.

Indeed, the entire plateau of Quito, whose summit is formed by Pichincha, Cotopaxi, and Tungurahua, is a single volcanic hearth. The subterranean fire breaks forth now from one of these openings, now from another, the three of them having been viewed by habit as separate volcanoes. The progressive movement of the fire here has been directed from north to south over the course of the last three centuries. Even the earthquakes that have so devastated this part of the world give peculiar evidence of subterranean connections: not only between countries without volcanoes, which have long been known, but also between mouths of fire that lie at great distances from one another. Thus did the volcano of Pasto, east of the Guaitara River, uninterruptedly pour out a tall column of smoke for three months in 1797, the column then disappearing at the very moment when, 60 miles away, the great earthquake of Riobamba and the mud-eruption, or *Moya*, killed thirty to forty thousand Indians.

The sudden appearance of Sabrina Island in the Azores on 30 January 1811 was the harbinger of the terrible earthquakes far to the west that from May of 1811 to June of 1813 almost unceasingly shook first the Antilles, then the plain of the Ohio and Mississippi, and finally the coastline of Venezuela that lies across from this plain. Thirty days after the total destruction of Caracas, the beautiful capital city of this country, the eruption of the long-dormant volcano of St. Vincent occurred in the nearby Antilles. An odd natural phenomenon accompanied this eruption. In the same moment that this explosion occurred, on 30 April 1812, a terrible subterranean noise was heard in South America throughout an area of 2,200 square geographical miles. The people living on the Apure, near the influx of the Rio Nula, compared the sound to the roar of heavy artillery, as did the most distant inhabitants of the Venezuelan coast. From the confluence of the Rio Nula and the Apure, by way of which I came to the Orinoco, the distance to the volcano of St. Vincent following a straight line measures 157 geographical miles. This noise, which was certainly not propagated through the air, must have had a deep subterranean origin. Its intensity was hardly more extreme on the coasts of the Caribbean, closer to the erupting volcano, than in the country's interior, in the basin of the Apure and Orinoco.

It would serve no purpose to increase the number of such examples as I have collected; but to think of an event that was more historically significant for Europe I need only mention the famous earthquake of Lisbon. Simultaneous to it, on 1 November 1755, not only were the Swiss lakes and the sea on the coast of Sweden violently disturbed, but even in the eastern Antilles around Martinique, Antigua, and Barbados, where the tide never rises above 28 inches, it suddenly rose to twenty feet. All of these phenomena demonstrate that subterranean forces manifest themselves either dynamically, intensely, and convulsively, or else in the productive and chemically altering manner of volcanoes. They also demonstrate that these forces do not exert their influence superficially, coming only from the thin crust of the Earth, but instead deeply, coming from the interior of our planet through clefts and unfilled passages, rising simultaneously to the most distant points on the Earth's surface.

The more complex the structure of volcanoes, that is, the elevated points that contain the channel through which the molten masses of the Earth's interior rise to the surface, the more important it is to fathom this structure by means of exact measurements. The interest in these measurements, which were a particular object of my investigations on another continent, is heightened by the consideration that that which is to be measured is at many points a variable quantity. Natural philosophy is at pains, in this variability of phenomena, to connect the present with the past.

To explore a periodic recurrence or even the laws of progressive alterations in Nature, one needs certain firm points, carefully executed observations that, attached to specific epochs, can be the basis for numerical comparisons. If over the course of millennia the mean temperature of the atmosphere or the average level of the barometer at sea level could have been determined, then we would know to what extent the heat of the climate has risen or decreased, or whether the height of the atmosphere has undergone changes. Just such points of comparison as these are needed for the declination and inclination of the magnetic needle, and for the intensity of the electromagnetic forces, over which two outstanding physicists, Seebeck and Erman, have shed so much light within the halls of this academy. If it is a laudable occupation of learned societies to pursue tenaciously the cosmic changes in heat, air pressure, magnetic direction, and electrical charge, it is on the other hand the duty of the traveling geognost, while determining the irregularities of the Earth's surface, to consider primarily the changeable heights of volcanoes. That which I attempted in the past in the Mexican mountains, on the *volcan de Toluca*, on Popocatepetl, on the *Cofre de Perote* or Naucampatepetl, and on the Jorullo as well as in the Andes of Quito on Pichincha, I have had the opportunity since my return to Europe to repeat at different points in time on Vesuvius. Where complete trigonometric or barometric measurements are not available, carefully taken elevation angles from specifically determined points may be substituted for them. The comparison of such elevation angles measured at different points in time can often be preferable to the complication of complete operations.

Saussure had measured Vesuvius in 1773, at a time when both rims of the crater, the northwest and the southeast, seemed to him to be of equal height. He found their height to be 609 toises or 3,654 Parisian feet. The eruption of 1794 caused a cave-in on the south side, creating the unevenness of the crater rim which the untrained eye can discern even from a great distance. Leopold von Buch, Gay-Lussac, and I measured Vesuvius three times in 1805 and found the north rim, which stands opposite the Somma, *la Rocca del Palo*, to be the same as Saussure had found it, but we found the south rim to be 75 toises lower than it was in 1773. The entire height of the volcano at that time in the direction of Torre del Greco (toward a side on which the fire has been working more or less primarily for 30 years) had decreased by 1/8. The ratio of the cinder cone, relative to the entire height of the mountain, is 1 to 3 on Vesuvius, 1 to 10 on Pichincha, 1 to 22 on the peak of Tenerife. Vesuvius thus has the proportionately tallest cinder cone of these three volcanoes, possibly because it is a low volcano and thus most of its action was from its peak.

A few months ago I was fortunate enough not only to repeat my earlier barometric measurements on Vesuvius but also, in three ascents of the mountain, to undertake a more complete determination of all crater rims.[1] This work is perhaps worthy of attention because it covers the long period of large eruptions between 1805 and 1822, and because it is the only measurement on any volcano yet presented to the public that is comparable in all its parts. It proves that the rims of the craters, not only in cases (such as the peak of Tenerife and all of the volcanoes of the Andes chain) where they are visibly composed of trachyte, but rather in general, are a much more constant phenomenon than was believed after the fleeting observations made before now. According to my latest determinations, the northwestern rim of Vesuvius had since Saussure (i.e., over the last 49 years) seemingly not changed at all, and the southeastern rim, facing Boscotrecase, which became 400 feet lower in 1794, had barely changed by 10 toises (60 feet).

If in the descriptions of great eruptions in published works one finds the completely changed shape of Vesuvius mentioned so often; if one believes these assertions to be substantiated by the picturesque views of the mountain that are sketched in Naples; then the cause of the error lies in the fact that the outline of the crater rim is confused with the outline of the spatter cone, which takes shape in the middle of the crater, resting on the floor around the vent, which has been raised by vapors. Such a spatter cone, made of loosely piled *rapilli* (lapilli) and slag, gradually became visible over the southeastern crater rim in the years 1816 and 1818. The eruption of February 1822 increased its size to the extent that it even grew to be 100 to 110 feet higher than the northwestern crater rim (the *Rocca del Palo*). On the occasion of the last eruption on the night of 22 October, this peculiar cone, which the people of Naples had come to regard as the actual peak of Vesuvius, caved in with a fearful cracking, such that the floor of the crater, which had been uninterruptedly accessible since 1811, presently lies 750 feet deeper than the north rim of the volcano and 200 feet lower than the south rim. The changeable shape and relative position of the spatter cone, the openings of which must not, as so often occurs, be confused with the crater of the volcano, gives to Vesuvius at different times a particular physiognomy; the historian of the volcano, by means of the outline of the mountaintop, indeed by merely looking at the landscapes by Hackert in the Palace of Portici, would be able to guess the year in which the artist made the sketch study for his painting according to whether the north or south side of the mountain is represented as higher.

One day after the collapse of this 400-foot cone of slag, once the small but numerous lava streams had flowed off, the fiery expulsion of ash and *rapilli* began in the night of the 23rd to the 24th of October. This went on uninterrupted for 12 days, though it was heaviest for the first 4 days. During this time, the detonations inside the volcano were so strong that merely the shocks to the air (there was no trace whatever of earthquake) cracked the ceilings of the Palace of Portici. There appeared in the close-lying villages of Resina, Torre del Greco, Torre dell'Annunziata, and Boscotrecase a curious phenomenon. The atmosphere was so filled with ash that the entire region, in the middle of the day, was wrapped for several hours in the deepest darkness. People on the streets walked with lanterns, as so often happens in Quito at times when Pichincha erupts. Never was there a more universal flight of the inhabitants. They fear lava flows less than eruptions of ash: an event that, to such an extreme, is unknown to us here and that, through the dark legends of the manner in which Herculaneum, Pompeii, and Stabiae were destroyed, fills the human imagination with visions of terror.

The hot steam that rose out of the crater during the eruption and poured into the atmosphere formed in cooling a thick bank of cloud around the 9,000-foot column of ash and fire. Such a sudden condensation of vapors and, as Gay-Lussac demonstrated, the formation of the clouds themselves, increased the electrical tension. Lightning snaked in all directions from around the pillar of ash, and one could clearly differentiate the rolling of the thunder from the inner rumbling of the volcano. During no other eruption had the play of the electrical forces been so apparent.

On the morning of the 26th of October, the uncanny news spread: a river of boiling water had poured forth from the crater and come crashing down from the cinder cone. Monticelli, the assiduous and learned observer of the volcano, soon recognized that an optical illusion had been the cause of this erroneous rumor. The supposed river was a terrific amount of dry ash that had shot forth like quicksand from a cleft in the highest rim of the crater. After a drought that had rendered the fields desolate had preceded the eruption of Vesuvius, the volcanic storm just described brought on, near the end of the eruption, a pouring rain of long duration. In all zones, such an event is characteristic of the end of an eruption. Since the cinder cone is usually wrapped in clouds during the eruption, and since the rainfall is generally heaviest in its vicinity, one will see streams of mud pouring down it on all sides. The shocked landsman perceives these as water that climbs up from the interior of the volcano and pours out from the crater; the deceived geognost believes them to be seawater or mudlike products of the volcano, the so-called *éruptions boueuses*, or, in the language of old French systematists, products of a conflagrant-aqueous liquefaction.

When the summits of the volcanoes (and this is usually the case in the Andes) reach higher than the snow line, or even to twice the height of Aetna, the inundations described above, due to the melting and penetrating snow, become extremely frequent and devastating. These events have a meteorological connection to the eruption of volcanoes, and they are modified in many ways by the height of the mountains, the girth of their eternally snow-covered peaks, and the warming of the walls of the cinder cone, but they cannot be viewed as actual volcanic phenomena. In broad caves, sometimes on the slope of the volcano, sometimes at the foot, underground lakes form that communicate in many directions with the alpine streams. When earthquakes, which precede all fiery eruptions in the Andes chain, mightily shake the entire mass of the volcano, the subterranean vaults break open and water, fish, and tufflike mud burst forth from them all at once. This is the strange phenomenon that provides the catfish of the Cyclops (*Pimelodes cyclopum*) that the inhabitants of the highlands of Quito call *preñadilla* and that I described shortly after my return. When the peak of the 18,000-foot Carihuairazo north of Chimborazo caved in on the night of 19 to 20 June 1698, all the fields around for some two square miles were covered in mud and fish. Similarly, seven years previous to this, the putrid fever in the city of Ibarra was ascribed to a fish eruption of the volcano Imbaburu.

I call these facts to mind because they shed some light upon the difference between the eruption of dry ash and that of mudlike mixtures of wood, coals, and shells within alluvia of tuff and trass. The amount of ash that Vesuvius recently ejected has been, like all things associated with volcanoes and other terrifying natural phenomena, greatly exaggerated in the published papers; indeed, two Neapolitan chemists, Vicenzo Pepe and Giuseppe di Nobili, even ascribed to the ash, despite contradiction from Monticelli and Covelli, gold and silver content. According to my investigations, the ash layer that fell over 12 days on the slope of the *conus* in the direction of Boscotrecase, where there were also *rapilli* mixed in, reached a thickness of only three feet, and down on the plain, 15 to 18 inches at most. Measurements such as these must not be made in locations where ash, like snow or sand, is blown into heaps by the wind or flooded with water into a sort of slurry. The times are past when, in the way of the ancients, one sought in volcanic phenomena only the wondrous, when someone like Ktesias might describe the ash of Aetna as flying down as far as the Indian Peninsula. Admittedly, a portion of the Mexican lodes of gold and silver are to be found in trachitic porphyry, but in the ash of Vesuvius that I brought with me and that an excellent chemist, Heinrich Rose, examined at my request, there was no trace of gold or silver to be seen.

Though the results that I am developing here (and which Monticelli's more exact observations substantiate) diverge greatly from those that have been disseminated in the last few months, the ash eruption of Vesuvius from 24 to 28 October remains the most memorable event of its kind (of which there is a reliable report) since the death of Pliny the Elder. The amount of ash is perhaps three times more than has ever been seen to fall since volcanic phenomena have been attentively observed in Italy. A layer of 15 to 18 inches seems at first glance to be insignificant compared to the mass that we find covering Pompeii. But without thinking of the rainfall and alluvia that might certainly have increased this mass over the centuries, without reviving the lively debate which has been carried on with great skepticism on the other side of the Alps regarding the causes of the destruction of the Campanian cities, one may call to mind here that the eruptions of a volcano in epochs widely separated in time may in no way be compared to one another in terms of their intensity. All conclusions based on analogy are insufficient when they are related to quantitative circumstances, to the amounts of lava and ash, to the height of the column of smoke, or to the strength of the detonation.

From the description by Strabo and the judgment of Vitruvius regarding the volcanic origin of pumice stone, one may conclude that until the year of Vespasian's death, that is, until the eruption that covered Pompeii, Vesuvius resembled more a burnt-out volcano than a solfatara. When, after a long quiet period, the forces beneath the ground suddenly opened new passages for themselves, when they again broke through layers of primordial stone and trachyte, powerful events must have been unleashed to which subsequent ones could offer no comparison. From the well-known letters to Tacitus in which Pliny the Younger reports the death of his uncle one can clearly discern that the renewal of eruptions, one could say the reanimation of the slumbering volcano, began with the eruption of ash. This very thing was observed in September 1759 with Jorullo, when the new volcano, penetrating strata of syenite and trachyte, suddenly lifted itself up from the plain. The country people fled, because upon their huts they found ash that had been flung from the many cracks in the ground. In the usual periodic actions of volcanoes, on the other hand, each partial eruption usually ends with the rain of ash. Moreover, the letter of the younger Pliny contains a passage which clearly shows that right at the beginning and without the influence of alluvial deposits, the dry ash that fell from the air reached a height of 4 to 5 feet. "The court," the narrative continues, "through which one came to the room in which Pliny took his afternoon rest was so filled with ash and pumice that, had the sleeper hesitated any longer, he would have found the exit blocked." In the enclosed space of a court, the effect of winds causing the ash to drift can well have been negligible.

I have interrupted my comparative overview of volcanoes with specific observations made at Vesuvius partly because of the great interest evoked by the last eruption, but also partly because every strong ashfall reminds us almost involuntarily of the classical ground of Pompeii and Herculaneum. In a supplemental text, the reading of which is not really suited to this collection, I have condensed all elements of the barometric measurements that I had the opportunity to make at the end of the past year on Vesuvius and in the Phlegraean Fields.

Thus far we have considered the form and actions of those volcanoes that stand in a state of enduring connection with the interior of the Earth by means of a crater. The peaks of such volcanoes are elevated masses of trachyte and lavas cut through by manifold vents. The permanence of their effects indicates a very complex structure. They have, so to speak, an individual character that remains the same for long periods. Mountains of the same sort situated nearby mostly produce greatly differing materials: leucite and feldspar lavas, obsidian with pumice, basaltlike masses containing olivine. They belong among the newer phenomena of the Earth, they break through most all of the layers of the floetz stone, and their eruptions and lava flows are of later origin than our valleys. Their life, if one may employ this figurative expression, depends upon the type and durability of their connections with the planet's interior. They often rest for centuries, suddenly reignite, and then end up emitting steam and various gases and acids as solfataras; sometimes, however, as was seen in the case of the peak of Tenerife, their top has already become a factory for regenerated sulfur—and yet from the sides of the mountain, powerful streams of lava still pour out, of a type that is like basalt down low, of an obsidian sort with pumice stone above, where the pressure is less.[2]

Independent of these volcanoes with permanent craters, there is another type of volcanic phenomenon that is less frequently observed but that, in a manner especially instructive to the field of geognosy, is reminiscent of the primordial world, that is, of the earliest revolutions of our planet. Trachyte mountains suddenly open, eject lava and ash, and then close themselves up again, possibly forever. So it was with the mighty Antisana in the Andes chain; so too with Epomaeus on Ischia in the year 1302. Such an eruption occasionally occurs even on the plains, as on the plateau of Quito, in Iceland far from the volcano Hekla, and on Euboea in the Lelantine Fields. Many of the elevated islands are the results of these transitory phenomena. The connection with the planet's interior is in these cases not permanent; the effect ceases as soon as the cleft, the communicating canal, is closed again. Channels of basalt, dolerite, and porphyry, which in various parts of the world cut through almost all formations; syenite, augite-porphyry, and amygdaloid masses, which all characterize the newest layers of transition rock and the oldest layer of the floetz strata: all are probably formed in a similar way. In the youthful days of our planet, the materials of the interior that had remained fluid forced their way through the cracks that were everywhere in the Earth's crust; sometimes hardening as granular dike rock, at other times spreading in layers atop one another. What the primordial world has passed down to us as exclusively so-called volcanic rock types did not flow forth ribbonlike, as do the lavas of our isolated cone-shaped mountains. The masses of augite, ilmenite, feldspar, and hornblende may have been the same at different epochs, now more like basalt, now more like trachyte; the chemical materials in mixtures of certain ratios may have aligned themselves in a crystalline fashion (as can be learned from Mitscherlich's important work and the analogy of artificial products of ignition): time and again we recognize that similarly structured materials have come to the surface of the ground in greatly different ways, either lifted up or forced out of temporary fissures, and that, breaking through the older stone strata (that is to say the parts of the Earth's crust that oxidized earlier), they finally pour forth as lava flows from conical peaks that possess a permanent crater. Confusing these actually quite different phenomena leads the geognosy of volcanoes back into the darkness from which a great number of comparative experiences have gradually begun to wrest it away.

The question often arises: what is burning in the volcanoes, what creates the heat under which melting earth and metals mingle? Recent chemistry has attempted to answer: "That which burns there are the earth, the metals, and the alkalis themselves; it is the metalloids of these materials. The firm, already oxidized crust of the Earth separates the surrounding oxygen-rich atmosphere from the combustible, unoxidized materials of the interior of our planet. Upon the contact of these metalloids with the oxygen pressing in, the release of heat occurs." The celebrated and ingenious chemist who proposed this explanation of volcanic phenomena would himself abandon it shortly thereafter. The experiences gathered in mines and caves in every zone of the Earth, and which Arago and I collected and presented in a separate treatise, demonstrate that even at shallow depths, the heat of the terrestrial body is significantly higher than the mean temperature of the atmosphere at the same location. Such a remarkable and generally established fact goes along with that which volcanic phenomena teach us. The depth at which the planetary body may be considered a molten mass has been calculated. The primitive cause of this subterranean heat is, as with all planets, the process of formation itself, the coalescence of a sphere-forming mass from a vaporous cosmic fluid, the cooling of the layers of the Earth at different depths through radiation. All volcanic phenomena are probably the result of a stable or temporary connection between the interior and exterior of our planet. Elastic vapors push the molten, oxidizing materials upward through deep fissures. The volcanoes are thus intermediary earth-springs; the liquid mixtures of metals, alkalis, and earth that settle into lava streams flow softly and quietly when, lifted up, they find an egress. In a similar way, the ancients (according to Plato's *Phaeton*) imagined all volcanic streams of fire as runoffs of the Pyriphlegethon.

I might be permitted, I hope, to add to these considerations another, more daring, observation. Might not the cause of one of the most extraordinary phenomena in the study of petrifaction also lie in the interior heat of the planetary body, a heat that is indicated by the observation of volcanoes and by thermometric experiments on springs[3] that rise from various depths? Tropical animal forms, arborescent ferns, palms, and bamboo plants lie buried in the frigid North. Everywhere, the primeval world shows us a distribution of organic forms that the current conditions of climate contradict. As a solution to such an important problem, several hypotheses have been conceived: the near approach of a comet, a change in the inclination of the ecliptic, an increased intensity of sunlight. None of these has been able to satisfy the astronomers, physicists, and geognosts all at once. I am happy to leave unchanged the axis of the Earth or the light of the sun's disc, from the spots of which a famed stellar expert explained the fertility and failure of crops; I believe I can recognize, however, that in any sort of planet, regardless of its relationship to a central body or of its astronomical position, there are manifold causes for the release of heat: through oxidation processes, precipitation, and chemically changed capacity of the planetary body, through the increase in electromagnetic charge, and through opened communication between the interior and exterior parts.

In the places where the deeply fissured crust of the prehistoric world radiated heat from its clefts, there, for perhaps centuries and over vast stretches of land, palms, arborescent ferns, and all the animals of the torrid zone could thrive. By this view of things, which I have already suggested in the recently released work *Geognostischer Versuch über die Lagerung der Gebirgsarten in beiden Hemisphären*, the temperature of volcanoes would be that of the interior of the terrestrial body itself; and the same cause that now brings about such frightful devastation may once, upon the newly oxidized crust of the Earth and on the deeply riven layers of rock, have been able to bring forth in any zone the most luxuriant of vegetation.

Even if one is inclined, by way of explaining the astonishing distribution of tropical forms among their resting places, to accept that the longhaired elephantine creatures now encased in ice floes were originally native to the northern climates, and that similar forms belonging to the same primary type as lions and lynxes could live simultaneously in very different climates, such a means of explaining things would still probably not extend to plant products. For reasons involving plant physiology, palms, pisang trees, and arborescent monocotyledons cannot survive having their appendicular organs stripped away by the northern cold, and in the geognostic problem that we are touching upon here, it seems difficult to me to separate the plant and animal forms from one another. The same line of explanation must apply to both forms.

At the close of this treatise, I have followed the facts that have been collected in the most widely varying regions of the world with uncertain hypothetical assumptions. Natural philosophy transcends a mere description of Nature. It does not consist in a sterile accumulation of facts. It is the privilege of the curious and active mind of humanity to occasionally drift out of the present and into the darkness of prehistory, to gain a sense of what cannot yet be clearly discerned, and thus to take delight in the ancient myths of geognosy in their many recurring forms.

Annotations and Additions

1. A more complete determination of the crater rims of Vesuvius

Oltmanns, my astronomical collaborator who was unfortunately taken away from science so soon, once more calculated the barometric measurements I made on Vesuvius (from the 22nd and 25th of November and the 1st of December, 1822) that are mentioned here, and compared the results with the results garnered from the handwritten measurements passed on to me from Lord Minto, Visconti, Monticelli, Brioschi, and Poulett Scrope. (All measurements in toises.)

A) Rocca del Palo, highest northern crater rim of Vesuvius:

Saussure, calculated 1773, barometric, probably according to Deluc's formula	609
Poli, 1794, barometric	606
Breislak, 1794, barometric (but, as with Poli, uncertain by which barometric formula)	613
Gay-Lussac, Leopold von Buch, and Humboldt, 1805, barometric, calculated according to the Laplace formula (as in all following barometric results)	603
Brioschi, 1810, trigonometric	638
Visconti, 1816, trigonometric	622
Lord Minto, 1822 (often repeated), barometric	621
Poulett Scrope, 1822, (somewhat uncertain due to unknown ratio of diameters of tubes and cistern)	604
Monticelli and Covelli, 1822	624
Humboldt, 1822	629

Probable end result: 317 toises above the settlement or 625 above sea level.

B) The lowest, southeastern crater rim, across from Boscotrecase:

After the eruption of 1794, this rim became 400 feet lower than the Rocca del Palo; assuming the latter to be 625 toises	559
Gay-Lussac, Leopold von Buch, and Humboldt, 1805, barometric	534
Humboldt, 1822	546

C) Height of spatter cone within crater, and which collapsed on 22 October, 1822:

Lord Minto, barometric	650
Brioschi, trigonometric, by various combinations, either	636 or 641

Probable end result for the height of the spatter cone that collapsed in 1822: 646 toises.

D) Punta Nasone, highest peak of Somma:

Shuckburgh, 1794, barometric, probably according to his own formula	584
Humboldt, 1822, barometric, according to Laplace formula	586

E) Plain of Atrio del Cavallo:

Humboldt, 1822, barometric	403

F) Foot of cinder cone:

Gay-Lussac, Leopold von Buch, and Humboldt, 1805, barometric	370
Humboldt, 1822 barometric	388

G) Settlement of Salvatore:

Gay-Lussac, Leopold von Buch, and Humboldt, 1805, barometric	300
Lord Minto, 1822, barometric	307.9
Humboldt, 1822, again barometric	308.7

A portion of my measurements is printed in Monticelli's *Storia de' fenomeni de Vesuvio, avvenuti negli anni 1821–1823*, p. 115, but the neglected correction there of the mercury level in the cistern barometer has somewhat distorted the values of elevation. If one considers that the results in the table above were achieved using barometers of widely varying construction at different times of day, with winds coming from all parts of the globe, and on the unevenly warmed slope of a volcano in a locality in which the reduction in air temperature deviates greatly from that implied by our barometric formulas, they will find the agreement of the figures to be perfectly sufficient.

My measurements of 1822, at the time of the Congress of Verona, when I accompanied the late king to Naples, were accomplished with greater care and under more favorable circumstances than were those of 1805. Differences in elevation, furthermore, are preferable to absolute elevations. These differences demonstrate, however, that since 1794 the relationship in height of the rim on the Rocca del Palo and the one facing

Boscotrecase has remained almost the same. I found in 1805 exactly 69 toises, in 1822 almost 82 toises. An excellent geognost, Mr. Poulett Scrope, found 74 toises, although the absolute elevations that he ascribes to the two crater rims seem a bit too low. Such a small degree of variability in the crater over a period of 28 years, and accompanied by such violent agitation within, is certainly a remarkable phenomenon.

The height to which the spatter cones on Vesuvius climb from the floor of the crater is also worthy of attention. In 1776, Shuckburgh found one such cone to have reached 615 toises above the level of the Mediterranean; by the measurements of Lord Minto (a very exact observer), the spatter cone that collapsed on October 22, 1822, was all of 650 toises high. Both times, the spatter cones inside the crater surpassed the maximum of the crater rim. If one compares the measurements of the Rocca del Palo from 1773 to 1822, one arrives almost involuntarily at the bold supposition that the northern rim of the crater may have been pushed gradually upward by subterranean forces. The agreement of the three measurements between 1773 and 1805 is almost as remarkable as that of the ones between 1816 and 1822. In the last period, the heights of 621 to 629 toises are beyond doubt. Should the measurements that 30 to 40 years earlier yielded results of only 606 to 609 toises be less certain? Only after longer periods will we be able to decide what to attribute to errors in measurement or to the rising of the crater rim. Accumulation of loose materials on top is not taking place here. If the solid trachytelike lava layers of the Rocca del Palo are truly climbing, then one must assume that they are being lifted from below by volcanic forces.

My friend Oltmanns, who was learned, diligent, and indefatigable in performing calculations, presented to the public in detail the particulars of all of the measurements mentioned here, accompanied by a painstaking critique, in the *Abhandlungen der königl. Akademie der Wissenschaften zu Berlin* (for the years 1822 and 1823, pp. 3–20). May this work stimulate the geognosts over the course of centuries to conduct frequent hypsometric monitoring of the hill-like and, after Stromboli, most accessible of all European volcanoes, Vesuvius.

2 . *Where the pressure is less*
Cf. Leopold Von Buch regarding the peak of Tenerife in his *Physikalische Beschreibung der canarischen Inseln*, 1825, p. 213, and in *Abhandlungen der königl. Akademie zu Berlin* for the years 1820–1821, p. 99.

3. *Springs that rise from various depths*
Cf. Arago in the *Annuaire du Bureau des Longitude pour 1835*, p. 234. The heat increase in our latitude is 1° Réaumur for every 113 Parisian feet. In the artesian borehole at Neu-Salzwerk (Bad Oeynhausen's) not far from Minden, which has reached the greatest depth below sea level now known, the temperature of the water at a depth of 2,094.5 Parisian feet is a full 26.2° Réaumur, while the average temperature of the air at ground level measures 7.7° Réaum. It is truly remarkable that St. Patricius, who was bishop of Pertusa, was led in the third century by the hot springs emerging at Carthage to a very correct assessment of the cause of such increased heat. (*Acta S. Patricii*, p. 555, Ruinart ed.; *Kosmos*, vol. I, p. 231.)

7

The Life Force, or The Rhodian Genius: A Tale

The Syracusans, like the Athenians, had their Stoa Poikile. Depictions of gods and heroes, works of Greek and Italian art, graced the colorful walls of the portico. The people could be seen there, streaming in unceasingly: the young warriors to revel in the deeds of their forefathers, the artists to do the same in the brushstrokes of the great masters. Among the countless paintings that the Syracusans had collected with assiduous diligence from their mother country, there was but one that had for a full century attracted the attention of all passersby. Even when the Olympian Jupiter, or Cecrops, founder of Athens, or the valor of Harmodius and Aristogieton wanted for admirers, still throngs of people stood crowded around this picture. Whence came this affection? Was it a rescued work of Apelles, or had it come perhaps from the school of Callimachus? No; dignity and grace indeed radiated from the picture, but in the blending of colors, in character and style as a whole, it could not measure up to several of the other works in the Poikile.

The people gaze in admiration of what they do not understand, and this characteristic of the public encompasses many classes. A century ago the picture was installed, and though Syracuse contained within its narrow walls more artistic genius than was to be found in all the rest of sea-bound Sicily, the meaning of the picture remained a mystery. No one even knew in what temple it had originally stood, for it had been saved from a stranded ship, and only the ship's other cargo hinted that the picture had come from Rhodes.

In the foreground of the painting, youths and maidens were to be seen, crowded together in a group. They were unclad and well formed, but not with the slender figures admired in the statues of Praxiteles and Alcamenes. The strong build of their limbs, which bore the signs of hard labors, and their human expressions of longing and sorrow—all seemed to strip them of anything heavenly or godlike and bind them to their earthly homeland. Their hair was adorned simply with leaves and flowers of the field. Imploringly they stretched their arms out to one another, but their solemn, mirthless gaze was directed to a *Genius* that, surrounded by a bright shimmer, floated in their midst. A butterfly sat upon the spirit's shoulder, and in his right hand he held aloft a blazing torch. The shape of his body was rounded like that of a child, his countenance divinely animated. Commandingly, he gazed down upon the youths and maidens at his feet. Further characteristic elements of the painting were not distinguishable, though some thought they discerned the letters ζ and ς, which (for the antiquarians in those days were no less bold than today) they most unfortunately put together to indicate the name of an artist called "Zenodorus," that is, having the same name as the builder of the second Colossus.

The Rhodian *Genius*, as the puzzling picture was called, was not without its interpreters in Syracuse. Art experts, especially the youngest, having returned from a brief trip to Corinth or Athens, might have thought themselves compelled to abandon all claims of talent, were they not immediately to come forward with a new explanation. Some held the *Genius* to be the expression of spiritual love that forbids the enjoyment of sensuous pleasures; others believed that he indicated the mastery of reason over the desires. The wiser among them remained silent and had a notion of something more sublime as they stood in the Poikile and delighted in the simple composition of the group.

And thus the matter remained, ever undecided. The picture was copied with many additions and sent off to Greece, without anyone ever receiving even an explanation of its origins. When finally the early rising of the Pleiades announced that shipping in the Aegean Sea was again opened, ships from Rhodes entered the harbor of Syracuse. They carried a trove of statues, altars, candelabras, and paintings that the love of art had inspired the Dionysians in Greece to collect. Among the paintings was one that was immediately recognized as a companion piece to the Rhodian *Genius*. It was of the same size and similar coloring, though the hues were better preserved. Again the *Genius* stood in the middle, but without the butterfly and with his head bowed, the extinguished torch pointed earthward. The circle of youths and maidens were now almost falling over him in many an embrace; their glance was no longer solemn and obedient, but bespoke a condition of wild abandon, the fulfillment of a long-nurtured desire.

Immediately, as the antiquarians of Syracuse sought to revise their earlier interpretations of the Rhodian *Genius* such that their ideas might also apply to this work of art, the Tyrant commanded that it be brought to the house of Epicharmus. This philosopher, from the school of Pythagoras, lived in an outlying part of Syracuse that was called Tyche. He seldom visited the house of the Dionysians, not as if excellent men of all of the Greek colonial cities did not gather there, but because such close proximity to princes robs even the most intellectual of men of both their intellect and their freedom. He occupied himself constantly with the nature of things and the forces within them, with the development of plants and animals, with the harmonic laws that govern how celestial bodies in vast spaces and snowflakes and hailstones in small ones take the form of spheres. As he was exceedingly old, he would daily take a ride to the Poikile and from there to Nasos at the harbor, where before the broad ocean his eye, as he put it, was afforded a view of the unlimited, the infinite, after which the mind strives in vain. He was held in high esteem not only by the common people but by the Tyrant as well. This man he avoided, as surely as he encountered others joyously and often helpfully.

Epicharmus now lay fatigued upon his bed of ease when the command of Dionysus sent the work of art to him. Care had been taken to supply to him a faithful copy of the Rhodian *Genius* as well, and the philosopher had both of them set up side by side before him. His gaze dwelt long upon them; then he called together his pupils and lifted his voice with emotion, saying, "Push back the curtain from the window, that I may once again revel in the sight of the animate Earth, teeming with life! For sixty years I have pondered the hidden wheels that drive Nature onward, the differences in the forms of matter, and today at last, the Rhodian *Genius* lets me see more clearly that which, until now, I only suspected. As the difference between the sexes beneficially and fruitfully links living beings to one another, so too is the raw material of inorganic Nature moved by the same drives. Even in the darkness of chaos, materials accumulated or repelled one another as friendship or enmity attracted or repelled them. The heavenly fire follows the metals, the magnet follows the iron, a rubbed piece of amber will move light substances, earth mixes with earth, cooking salt condenses from seawater, and the sour moisture of *stypteria* (στυπτηρία ὑγρά; "alum"), like the woolly-haired salt halotrichite, loves the loam of Milos. Everything in inanimate nature rushes to join its kind. No earthly material (does anyone dare to count light as one?) is thus to be found anywhere in a simple and pure, virgin state. All things strive from their inception onward toward new compounds, and only the discriminating art of humanity can depict separately that which you seek in vain within the interior of the Earth and in the moving oceans of water and air. In dead, inorganic material there is a listless quiescence, so long as the bonds of interrelatedness are not loosed, so long as a third material does not force its way in to join the others. But then, this disturbance too is followed by fruitless quiescence.

"But different is the mixing of like materials in the bodies of animals and plants. Here the Life Force imperiously asserts its rights; it cares not for the friendship or enmity of atoms described by Democritus, it unifies materials that in inanimate Nature forever flee from one another, and separates those that in that same world seek one another without ceasing.

"Come closer around me, my pupils, and see in the Rhodian *Genius*, in the expression of his youthful strength, in the butterfly on his shoulder, in the regal glance of his eye, the symbol of the Life Force and how it animates every particle of organic creation. The earthly elements at his feet seem to strive to follow their own desires and to join together. Commandingly, the *Genius* threatens them with the upraised, blazing torch and forces them, with no thought for their ancient right, to follow his laws.

"Consider now the new work that the Tyrant has sent to me for interpretation; direct your eyes from the image of life to the image of death. Away has flown the butterfly, the inverted torch burnt out, and the head of the youthful spirit hangs down. The spirit has escaped to other spheres, the Life Force has died. Now the hands of the youths and maidens are happily outstretched to one another. The shackles now loosened, they wildly pursue, after long deprivation, their drives to unite; the day of death is for them a wedding day. —Thus does dead material, animated by the Life Force, go through an innumerable succession of generations; the same material in which once a puny worm momentarily enjoyed his existence may well have encased the godlike spirit of Pythagoras.

"Go, Polycles, and tell the Tyrant what you have heard! And you, my dear Euryphamos, Lysis, and Skopas, come closer, still closer to me! I feel that the weak Life Force within me, too, will not much longer reign over the earthly material. This material is demanding its freedom once more. Take me once again to the Poikile, and from there to the open shore. Soon you will gather my ashes!"[1]

Commentary and Addendum

1. In the forewords to the second and third editions of *Views of Nature*, I have already mentioned the reappearance of the preceding composition, which was first published in Schiller's *Die Horen* (1795, No. 5, pp. 90–96). It contains, in half-mythical garb, the development of a physiological idea. In 1793, in my Latin aphorisms from the chemical physiology of plants which I attached to my work on subterranean flora, I had defined the Life Force as the unknown cause that hinders the elements from following their natural attractive forces. The first of my aphorisms read thus:

"If you should examine the whole nature of matter, you will discern a division, great and lasting, which exists among the elements. Half of them seem obedient to the laws of affinity; the other half, once their bonds have been set free, seem compounded in sundry different ways. Certainly this division by no means has been situated in the elements themselves or in their innate nature, since it appears that it must be sought in solely the distribution of the individual elements. We call that matter sluggish, dull, and inanimate whose filaments have been united according to the laws of chemical affinity. We designate those particles in preference to all else as animate and organic, which although they constantly strive to be changed into new shapes are kept in check by some internal force so that they do not forsake their ancient and innate shape.

"We call vital the internal force that loosens the bonds of chemical affinity as well as prevents the elements of the particles from freely compounding. And thus no surer criterion of death is given than putrefaction, by which the first principles or the filaments of matter obey the laws of affinity after the ancient bonds have been revoked. For inanimate particles there can be no putrefaction." (See *Aphorismi ex doctrina Physiologiae chemicae Plantarum* in Humboldt, *Flora Fribergensis subterranea*, 1793, pp. 133–136.)

These tenets, about which the sharp-eyed Vicq-d'Azyr already gave warning in his *Traité d'Anatomie et de Physiologie*, vol. I, p. 5, and yet which even today are shared by many well-known men with whom I am friendly, I placed in the mouth of Epicharmus. Consideration and further studies in the area of physiology and chemistry have profoundly shaken my earlier belief in a so-called Life Force. By 1797, in the conclusion of my *Versuche über die gereizte Muskel- und Nervenfaser, nebst Vermuthungen über den chemischen Prozeβ des Lebens in der Thier- und Pflanzenwelt* (vol. II, pp. 430–436), I asserted that by no means do I consider the existence of such intrinsic life forces to be proven. Since then, I no longer use the term *intrinsic forces* for that which is perhaps only actuated by the interaction of long-familiar individual materials and material forces. One may, however, deduce from the chemical properties of elements a definition of animate and inanimate materials that is more certain than the criteria that are derived when one speaks of voluntary movement, of the circulation of fluid parts within solid ones, of internal acquisition and the fibrous organization of elements. I apply the term *animate* to any material "whose voluntarily separated parts will, after the separation and under the previous external conditions, change the condition of their composition." This definition is merely the statement of a fact. The equilibrium of elements maintains itself in living material by virtue of the elements being part of a whole. One organ dictates to another, one gives to the other, as it were, the temperature, the disposition in which these and no other affinities have effects. Thus is everything within the organism at once the means and the end. The rapidity with which organic parts change their composition when they are separated from a complex of living organs differs greatly depending on their level of independence and the nature of their materials. The blood of animals, which is modified in many ways within the various classes, suffers transformations sooner than do the juices of plants. Sponges, on the whole, decay faster than the leaves of trees, the flesh of muscles more easily than true skin (*cutis*).

Bones, the elementary structure of which has only in recent times come to our understanding, animal hair, the wood of plants, fruit rinds, and the downy pappus are not inorganic, not without life, yet even while still living they approach the condition that they will display upon their separation from the rest of the organism. The higher the degree of vitality or sensitivity in an animate material, the more conspicuous or rapid the process of change in its composition following the separation. "The sum of the cells is an organism, and the organism lives as long as the parts are active in the service of the whole. Compared to inanimate Nature, the organism seems to be self-determining" (Henle, *Allgemeine Anatomie*, 1841, pp. 216–219). The difficulty in satisfactorily tracing back the vital phenomena of the organism to physical and chemical laws (much like the prediction of meteorological processes in the atmosphere) lies largely in the complexity of the phenomena, in the great number of simultaneously active forces, and in the particulars of their activity.

In *Kosmos* I have remained faithful to the same method of delineation, the same considerations regarding the so-called life forces and the vital affinities (Pulteney in the *Transactions of the Royal Society of Edinburgh*, vol. XVI, p. 305), and regarding the creative force and an organizational vitality. In the first volume, page 67, it says: "The myths of imponderable materials and of intrinsic life forces in any sort of organism evolve and obscure the view of Nature. Among various circumstances and forms of cognition, the weighty burden of our accumulated and now so quickly growing partial knowledge moves laboriously. Contemplative reason attempts boldly and with variable success to shatter the old forms by which, as by mechanical constructions and symbols, we are accustomed to gain mastery over recalcitrant materials." Further on in the same volume, on page 367: "A physical description of the world may well include the reminder that within the inorganic crust of the Earth lie the very same basic materials from which the frames of the animal and plant worlds are constructed. It teaches us that in both of these, the same forces dominate—the forces that bind or separate materials, that render them solid or liquid within organic tissues; subordinate are all complicated circumstances that, unfathomed, are systematically grouped (according to more or less accurately suspected analogies) under the very uncertain designation 'effects of the life forces.'" (Cf. Schleiden's critique of the acceptance of intrinsic life forces in his *Botanik als inductive Wissenschaft*, vol. I, p. 60, and in the just-published, excellent *Untersuchungen über thierische Elektrizität* by Emil du Bois-Reymond, vol. I, pp. XXXIV–L.

8

The Plateau of Cajamarca,
the Old Residential City of the Inca Atahualpa;
First Sight of the Pacific
from the Ridge of the Andes Chain

After spending an entire year upon the spine of the Anti or Andes chain,[1] within 4° of the equator on the high plains of New Granada, Pastos, and Quito at heights of eight to twelve thousand feet, it is a relief to descend gradually through the cinchona forests of Loja into the plain of the Upper Amazon, an unfamiliar world rich in magnificent forms of vegetation. The small city of Loja has given its name to the most efficacious of all fever barks: *quina*, or *cascarilla fina de Loja*. This bark is the exquisite product of the tree that we botanically described as *Cinchona condaminea*, which had earlier been named *Cinchona officinalis*, under the erroneous assumption that all commercial quinine came from a single species of tree. Fever bark was not brought to Europe until the middle of the seventeenth century, either in 1632 to Alcala de Henares, as Sebastian Badus maintains, or in 1640 to Madrid upon the arrival of the Peruvian viceroy's wife, the Countess of Chinchon,[2] who was accompanied by her personal physician, Juan del Vego, and who had been healed of malaria in Lima. The excellent *quina* of Loja grows two to three miles southeast of the city in the mountains of Uritusinga, Villonaco, and Rumisitana, atop mica schist and gneiss at moderate altitudes between 5,400 and 7,200 feet: about the same elevation as the Grimsel Hospital and the Great St. Bernard pass. The actual limits of the *quina* are the small rivers Zamora and Cachiyacu.

The tree is felled during the first blossoming stage, that is, in the fourth or the seventh year, depending on whether it is the product of a healthy sucker or of a seed. We heard with astonishment that, at the time of my trip, only 110 centners of the fever bark of the *Cinchona condaminea*, by official royal reckoning, were brought in annually from around Loja by the *quina* harvesters (*cascarilleros*, or quinine hunters, *cazadores de quina*). None of this wondrous product was put on the market then; the entire supply was sent via the Pacific harbor of Payta, around Cape Horn, and on to Cadiz for the use of the court. To deliver this modest amount of 11,000 Spanish pounds, eight to nine hundred cinchona trees were felled annually. The older and thicker trunks are becoming ever scarcer, but the abundance of growth is so great that the younger trunks now being used, with a diameter of barely 6 inches, often reach heights of as much as 50 or 60 feet. This appealing tree, adorned with leaves 5 inches long and 2 wide, forever strives when growing in a wild thicket to lift itself above its neighbors. The higher foliage, tossed erratically by the wind, gives off a peculiar reddish shimmer that is recognizable from a great distance. The average temperature in the copses of *Cinchona condaminea* oscillates between 12.5° and 15° Réaumur; this is close to the average annual temperature of Florence and of the island of Madeira, but Loja never reaches the extremes of heat and cold that are observed in these places in the temperate zone. The comparisons of climates at greatly different degrees of latitude to the climate of the high tropical plateaus are by their very nature seldom satisfying.

To travel from the knot of mountains around Loja south-southeast to the hot valley of the Amazon, one must cross the *paramos* of Chulucanas, Guamani, and Yamoca: mountainous wildernesses that we have already considered in other works, and that in the southern reaches of the Andes chain are all encompassed by the term *puna* (a word from the Quechua language). Most of them rise to over 9,500 feet; they are stormy, often wrapped for days in dense fog or afflicted by terrible hailstorms in which the water coalesces not only into multiform hailstones, most of them flattened by rotation, but also into thin, laterally moving individual disks (*papa-cara*) that injure the face and hands. During these meteorological processes I occasionally saw the thermometer sink to 7° or 5° (above freezing) while the electrical tension in the atmosphere, measured by a voltaic electrometer, switched in a few minutes from positive to negative. Below 5°, snow falls in large, widely spaced flakes. It disappears after a few hours. The ragged branching of the small-leaved, myrtlelike shrubbery, the size and abundance of the blossoms, and the eternally fresh leaf organs saturated in the moist air give to the treeless vegetation of the *paramos* a unique physiognomic character. No zone of alpine vegetation in the temperate or frigid regions of the Earth can compare to that of the *paramos* in the tropical Andes.

The solemn impression evoked by the wilderness of the cordilleras is increased in a strange and unexpected way, for still preserved within it are the extraordinary remains of the Inca Road, the enormous construction by which, over a stretch of more than 250 geographical miles, all of the provinces of the empire were connected. Placed here and there along the way, at mostly regular intervals, there are dwellings constructed of well-hewn freestone, caravansaries of a sort called *tambos* or *Inca-pilca* (from *pircca*, "the wall"?). Some are surrounded like fortresses; others are set up for baths with conduits for warm water, the largest intended for the family of the ruler himself. I had already carefully measured and sketched such well-preserved buildings (called *aposentos de Mulalo*[3] by Pedro de Cieza in the 16th century) at the foot of the Cotopaxi volcano near Callo. On the Andean pass between Alausi and Loja, which is known as the *Paramo del Assuay* (14,568 feet above the sea, a well-used route across the *Ladera de Cadlud* at almost the same elevation as Mont Blanc), we encountered great difficulties on the high plain *del Pullal* leading our heavily laden mules across the swampy ground, while beside us over the course of more than a German mile our eyes were constantly drawn to the magnificent remains of the 20-foot-wide Inca Road. It had a deep base layer and was paved with well-hewn blackish-brown trap-porphyry. What I had seen of Roman roads in Italy, Southern France, and Spain was no more imposing than this work of the ancient Peruvians, and this latter lay, by my barometric measurements, at an elevation of 12,440 feet. This height thus surpasses the summit of the peak of Tenerife by more than a thousand feet. Equally high up on Assuay lie the ruins of the so-called palace of the Inca Tupac Yupanqui, which are known by the name *Paredones del Inca*. From these ruins the road continues southward in the direction of Cuenca to the small but well-preserved fortress of Cañar,[4] probably from the same period as Tupac Yupanqui or his warlike son Huayna Capac.

We saw still grander ruins of the ancient Peruvian road on the trail between Loja and the Amazon River at the Baths of the Inca on the *Paramo de Chulucanas*, not far from Huancabamba, and around Ingatambo near Pomahuaca. Of these ruins, the last-mentioned stand at such a low point that I found the difference in elevation between the Inca Road at Pomahuaca and the Inca Road of the *Paramo del Assuay* to be more than 9,100 feet. The distance in a straight line by astronomical latitude comprises exactly 46 geographical miles, and the road ascends to 3,500 feet higher than the elevation of the pass of Mount Cenis above Lake Como. Of the two paved road systems, cobbled with flat stones or, occasionally, covered with cemented[5] gravel (macadam), one crossed the wide and arid plain between the coast and the Andes chain, while the other went along the slopes of the cordilleras themselves. Milestones declared the distances at regular intervals. Bridges of three sorts—stone, wood, and rope (*puentes de hamaca* or *de maroma*)—crossed over bridges and chasms; water conduits ran to the *tambos* (hostelries) and fortresses. Both systems of maintained roads were directed toward the central point of Cuzco, the royal seat of the great empire (13°31′ south lat.); the elevation of this capital city, according to Pentland's map of Bolivia, is 10,676 feet (Parisian measure) above sea level. Since the Peruvians drove no sorts of vehicles, the roads were intended only for marching troops, porters, and groups of lightly burdened llamas; thus they are found here and there, with the great steepness of the mountains, to be broken by long rows of steps upon which rest areas were installed. Francisco Pizarro and Diego Almagro, who used to such great advantage the military roads of the Incas on the long marches of their own armies, encountered difficulty, especially for the Spanish cavalry, in those places where the road was interrupted by tiers and stairsteps.[6] The hindrance was all the greater because the Spanish at the beginning of the *conquista* employed only the horse, and not the deliberate mule, who appears to consider his every step when he is in the mountains. Only later did the cavalry begin to use the mule.

Sarmiento, who saw the Inca Road when it was still in complete repair, asks himself in a *relacion* that long lay unused, buried in the Biblioteca del Escorial: "how were a people without the use of iron in high, rocky country able to complete such magnificent works (*caminos tan grandes y tan sovervios*) from Cuzco to Quito and from Cuzco to the coast of Chile?" He adds: "Emperor Carlos with all his power would not accomplish a fraction of what the well-equipped leadership of the Incas was able to demand of the obedient tribes." Hernando Pizarro, the most civilized of the three brothers, who atoned for his misdeeds with a 20-year imprisonment at Medina del Campo and, at the age of 100, died in an air of saintliness (*en olor de santidad*), proclaims: "In the whole of Christendom are no such glorious roads to be seen as the ones we are admiring here." The two important residential cities of the Incas, Cuzco and Quito, are in direct alignment (SSE to NNW) and (putting aside the many twists and turns of the road) 225 geographical miles apart; including the windings of the road, Garcilaso de la Vega and other conquistadors reckon the distance at about 500 *leguas*. According to the completely credible assertion of the licentiate Polo de Ondegardo, Huayna Capac, whose father had conquered Quito, had certain building materials for the royal buildings (Inca quarters) brought from Cuzco, despite the length of the road. I too found this legend widely circulated among the natives in Quito.

Wherever Nature presents to humankind, through the formation of the land, tremendous hindrances to overcome, there will grow, among the tribes who undertake the enterprise, both the courage and the strength to do so. The despotic centralization of the Inca rulership required both security and speed of communication, especially of troop movements—thus the construction of maintained roads and the perfecting of postal facilities. With peoples who stand at the most differing stages of civilization, one perceives with particular preference national activities moving in specific directions; but the conspicuous development of such isolated activities in no way determines the condition of the entire culture. Egyptians, Greeks,[7] Etruscans, and Romans, Chinese, Japanese, and Indians show us these contrasts. How much time was required to complete the Peruvian Road is difficult to determine. The great works of the northern part of the Inca Empire on the highlands of Quito certainly must have been completed in less than 30 or 35 years—within the short epoch that falls between the conquest of the ruler of Quito and the death of the Inca Huayna Capac; meanwhile, a deep darkness reigns over the age of the southern part of the Peruvian Road that actually lies in Peru.

Generally, the mysterious appearance of Manco Capac is set at 400 years before the landing of Francisco Pizarro upon Puná Island (1532), thus around the middle of the 12th century, almost 200 years before the founding of Mexico City (Tenochtitlan); some Spanish writers give a figure of as much as 500 or even 550 years. But the imperial history of Peru knows only thirteen reigning princes of the Inca dynasty, which, as Prescott correctly notes, could not fill so long a period as 400 or 550 years. Quetzalcoatl, Bochica, and Manco Capac are the three mythical figures to whom are tied the beginnings of the cultures of the Aztecs, Muiscas (actually Chibchas), and Peruvians. Quetzalcoatl, bearded, clothed in black, high priest of Tula, later a penitent on a mountain near Tlaxapuchicalco, comes from the coast of Panuco, that is, the eastern coast of Anahuac on the Mexican plateau. Bochica, or more accurately, the bearded, long-robed Messenger of God,[8] Nemterequeteba (a Buddha of the Muiscas), comes to the plateau of Bogota from the grassy steppes east of the Andes. Civilization already reigned along the picturesque shore of Lake Titicaca before Manco Capac. The fortress of Cuzco on the hill of Sacsayhuaman was modeled on the older buildings of Tiwanaku. In the same way, the Aztecs imitated the pyramid construction of the Toltecs, and the Toltecs that of the Olmecs (Hulmeks); climbing gradually, one reaches historic ground in Mexico extending back as far as the 6th century of our reckoning. The stepped pyramids of the Toltecs in Cholula are said by Siguenza to repeat the form of the Olmec stepped pyramids of Teotihuacan. So it is that in passing through any layer of civilization, one penetrates into an older one. And while consciousness did not awaken simultaneously in the peoples of the Old and New Continents, the fantastic realm of myth always immediately precedes historical certainty in any people.

Despite the great admiration that the first conquistadors had for the improved roads and water conduits of the Peruvians, these structures were not only neglected but willfully destroyed. This happened more quickly in the littoral (even though it engendered crop failure through lack of water), where the conquistadors used the ready-made stones for new building, than it did on the ridge of the Andes chain or in the deep, gorgelike valleys that cut into this chain. We were compelled during the long day's journey from the syenite rocks of Saulaca to the fossil-rich valley of San Felipe (situated at the foot of the icy *Paramo de Yamoca*) to wade across the Rio Huancabamba 27 times, thanks to its many windings; meanwhile, we were once again continually able to see, on a steep wall of rock nearby, the remains of the straight, high-walled Road of the Incas and its *tambos*. The small torrent, only 120 to 140 feet wide, had a current so fast that our heavily loaded mules were often in danger of being carried away. They were carrying our manuscripts, our dried plants, everything we had collected over the course of a year. At such times, one waits on the far bank in a state of anxiety until the long string of 18 to 20 pack animals is no longer in danger.

The same Rio Huancabamba, in its lower course where it has many waterfalls, is employed in a curious way for correspondence with the Pacific coast. In order to move more quickly the few letters from Trujillo destined for the Province Jaén de Bracamoros, a swimming postal carrier is employed, *el correo que nada*. In two days, the postman (usually a young Indian) swims from Pomahuaca to Tomependa, first along the Rio Chamaya (the name for this lower end of the Rio Huancabamba), and then along the Amazon. He carefully lays the few letters entrusted to him in a wide cotton towel that he wraps around his head like a turban. At the waterfalls, he leaves the river and walks around the falls through the nearby bushes. So that the long swim does not tire him excessively, he often clings with one arm to a block of lightweight wood (*ceiba; palo de balsa*) of the family of the Bombaceae. Sometimes a friend will accompany the swimmer as a partner. The two need not worry about food, for wherever they are, they receive a hospitable welcome in the scattered huts surrounded abundantly by fruit trees in the attractive *Huertas de Pucara* and *Cavico*.

The river is fortunately free of crocodiles; in the upper run of the Amazon too, they are not encountered until one has gone downriver past the cataracts of Mayasi. The ponderous beast prefers the quieter waters. By my measurement, the Rio Chamaya, from the ford (*paso*) of Pucara to its confluence with the Amazon below the village of Choros, a short distance of 13 geographical miles, has a vertical descent of no less than 1,668 feet.[9] The governor of the Province of Jaen de Bracamoros assured me that in using this peculiar water mail system, the letters are rarely soaked or lost. Indeed, shortly after my return from Mexico, I myself received letters in Paris that had come from Tomependa by this very route. Many of the wild Indian tribes that live near the banks of the Upper Amazon make their journeys in a similar fashion, swimming companionably downstream together. I had the opportunity to see some 30 to 40 heads (of men, women, and children) of the Jivaro tribe in the channel arriving at Tomependa. The *correo que nada* makes his way back by land on the arduous paths of the *Paramo del Paredon*.

When one approaches the hot climate of the Amazon basin, one encounters with pleasure a graceful, in some places very luxuriant vegetation. More beautiful citrus trees than those of the *Huertas de Pucara*—mostly sweet oranges (*Citrus aurantium* [Risso]), and in lesser numbers the sour *C. vulgaris* (Risso)—we had never before seen, not even in the Canary Islands or on the hot coastal region of Cumana and Caracas. Laden with many thousands of golden fruits, they reach a height here of 60 feet. Rather than the rounded top, they had branches that strive upward in a manner almost like the laurel. Not far away, near the ford of Cavico, we were surprised by a quite unexpected sight. We saw a copse of small trees, barely 18 feet tall, covered not with green but apparently with completely pink leaves. It was a new species of the genus *Bougainvillaea* that the elder Jussieu had first categorized according to a Brazilian specimen in the Commerson herbarium. The trees had almost no true leaves; what we had thought at a distance were leaves were densely crowded, light pink bracts (blossom-leaves). The sight, in terms of purity and coloration, was quite different from the appearance so gracefully presented by many of our forest trees in autumn. From the Proteaceae family of Southern Africa, a single species, *Rhopala ferruginea*, climbs down from the cold heights of the *Paramo de Yamoca* to the hot plain of Chamaya. We also often found here the delicately pinnate *Porlieria hygrometrica* (of the Zygophylleae), in which the closing of the small leaves, more than by any other member of the Mimosaceae, indicates that a change in the weather is soon to come, especially approaching rain. It seldom misled us.

In Chamaya we founds rafts (*balsas*) ready to take us to Tomependa, so that we might determine (it being of some importance to the geography of South America, due to an old observation of La Condamine)[10] the longitudinal difference between Quito and the mouth of the Chinchipe. We slept as usual under the open sky on the sandy bank (*Playa de Huayanchi*) at the confluence of the Rio Chamaya and the Amazon. The next day, we navigated down the river as far as the cataracts and narrow torrent of Rentema, where cliffs of coarse-grained sandstone (conglomerate) rise like towers, creating a rock dam in the river. I took a baseline measurement on the flat and sandy bank, and found that at Tomependa the Amazon, which to the east becomes so mighty, was only 1,300 feet wide. At the famous narrows of Manseriche between Santiago and San Borja—a mountain gorge that in some places, due to the overhanging cliffs and the roof of foliage, is but poorly lit, and in which all driftwood, including a vast number of tree trunks, is dashed to pieces and

disappears—the stream is only 150 feet wide. The cliffs that form all such narrow torrents, or *pongos* (in the Quechua language *puncu*, "door; gateway")
have been subjected to many changes over the course of the centuries. A portion of the Pongo de Rentema mentioned above, for example, was pulverized
by the high flood a year before my journey; indeed, among the natives along
the Amazon, tradition preserves a lively memory of the collapse of the then
very high rock mass of the entire *pongo* at the beginning of the 18th century.
The course of the river, due to the collapse and the subsequent damming,
was suddenly arrested, and in the village of Puyaya lying below the Pongo de
Rentema, the natives looked on with horror as the broad riverbed emptied of
water. After a few hours, the river broke through again. It is not believed that
earthquakes were the cause of this extraordinary event. On the whole, the
mighty stream works without ceasing upon improving its bed; one may well
imagine what sort of power it might wield when one sees how it occasionally
swells, despite its tremendous width, over 25 feet in 20 to 30 hours.

We stayed in the hot valley of the Upper Marañón or Amazon River for
17 days. In order to leave this valley and come to the coast, one must crest
the Andes at the place between Moyobamba and Cajamarca (6°57' south lat.,
80°56' west long.) where, according to my observations of magnetic inclination, the magnetic equator cuts across the mountain chain. Continuing
to climb, one reaches the celebrated silver mines of Chota and there begins
to descend (with occasional interruptions) by way of old Cajamarca, where,
316 years ago now, the bloody drama of the Spanish *conquista* was played out,
and by way of Aroma and Gangamarca, down into the Peruvian lowlands.
Here, as almost everywhere in the Andes and in the mountains of Mexico, the
greatest heights are scenically characterized by towering outcroppings of porphyry and trachyte, the ones of porphyry particularly being split into mighty
pillars. Such masses give portions of the spine of the mountains the appearance of cliffs in one place, of domes in the next. Here they have broken through
a limestone formation that extends enormously on both sides of the equator
in the New Continent and that, according to Leopold von Buch's outstanding
analyses, belongs to the chalk strata. Between Huambos and Montan, twelve
thousand feet above the sea, we found pelagic shell fossils[11] (ammonites with a
14-inch diameter, the large *Pectin alatus*, oyster shells, sea urchins, isocardias,
and *Exogyra polygona*). We collected a species of cidaris, which according to
Leopold von Buch is no different from one found by Brongniart in the old
chalk by the *Perte du Rhône*, both at Tomependa in the Amazon basin and
near Moyobamba at an elevation of no less than 9,900 feet. In the Amuich
mountain chain in Caucasian Dagestan, the chalk rises in the same fashion
from the banks of the Sulak, barely 500 feet above sea level, up to Tschunum,

at a full 9,000 feet of elevation, while on the 13,090-foot peak of Mount Shah-dagh are found *Ostrea diluviana* (Goldf.) and the same chalk layer. Abich's excellent observations on the Caucasus accordingly corroborate Leopold von Buch's brilliant geognostic views on the alpine distribution of chalk.

From Montan's lonely manor farm surrounded by llama herds, we climbed farther southward on the eastern slope of the cordillera and came to a high plain in which the silver mountain of Hualgayoc, center of the famous mines of Chota, presented to us at nightfall a marvelous vista. The *Cerro de Hualgayoc*, separated by a deep, gorgelike valley (*quebrada*) from the limestone moun-tain Cormolache, is an isolated cliff of siliceous quartz, shot through with innumerable and often conjoining veins of silver, and deeply, almost vertically shorn off on the northwest face. The highest of the mines lie 1,445 feet above the floor of the gallery, *socabon de espinachi*. The outline of the mountain is interrupted by numerous tower- and pyramidlike points and spires. Its peak also bears the name *Las Puntas*. These deposits most decidedly contrast with the "gentle exterior" that the miner generally ascribes to metal-rich regions. "Our mountain," said a wealthy mine owner whom we encountered, "stands there like an enchanted castle, *como si fuese un castillo encantado.*" Mount Hualgayoc is somewhat reminiscent of a dolomite cone, but even more of the cloven spine of Montserrat in Catalonia, which I also visited and which was so gracefully described by my brother. The silver mountain Hualgayoc is not only perforated even to its very highest reaches by several hundred mine tun-nels bored into it on all sides; the massif of the granular stone itself presents natural fissures through which the vault of the sky, a very deep blue at this mountainous height, may be seen by the onlooker standing at the foot of the mountain. The native people call these openings windows, *las ventanillas de Hualgayoc*; we were shown similar windows in the trachyte walls of the vol-cano Pichincha, similarly named the *ventanillas de Pichincha*. The strange-ness of such a sight is increased still more by the many tunnel portals and workers' cabins that hang like nests on the slopes of the fortresslike mountain, wherever a small clearing allows. The mineworkers carry the ore in baskets down steep and dangerous footpaths to the amalgamation areas.

The value of the silver delivered by the mines in the first 30 years (from 1771 to 1802) probably far exceeds 32 million piasters. Despite the hardness of the stone with its high content of quartz, the Peruvians, even before the coming of the Spanish (as evidenced by old tunnels and shafts), mined for galena rich in silver on the *Cerro de la Lin* and on Chupiquiyacu, and for gold on Curimayo (where there is also natural sulfur in the quartz rock, as in the itacolumite in Brazil). We resided near the mine in the small mountain town of Moyobamba, which sits 11,140 feet above sea level and where, even though it is only 6°43' away from the equator, the water in every residence freezes nightly throughout a large portion of the year. In this waste, devoid of vegetation, there live three to four thousand people, to whom all provisions are delivered from the warm valleys, as they are able to raise only some varieties of cabbage and very fine lettuce. As in every Peruvian mountain town, boredom drives the wealthier (though not better educated) inhabitants to extremely hazardous games of cards and dice. Quickly won riches are even more quickly lost. It is all reminiscent of the soldier of Pizarro's army who, after plundering the temple of Cuzco, lamented having lost "a great piece of the sun" (a sheet of gold plate) in a game of chance. The thermometer in Moyobamba at 8 in the morning showed just 1° Réaumur, by noon, 7°. Amidst the slender ichu grass (perhaps our *Stipaerio stachya*) we found a beautiful Calceolaria (*C. sibthorpioides*), which we would not have expected to see at such an elevation.

Close to the mountain town of Moyobamba, on a plateau called *Llanos* or *Pampa de Nevar*, tremendous amounts of red gold ore and wirelike pure silver in "swirls, spikes, and spreading veins" (*remolinos, clavos, y vetas manteadas*) have been extracted from a piece of ground of more than ¼ square geographical mile, only 3 to 4 fathoms under the turf, as if growing together with the roots of the alpine grass. Another plateau, west of Purgatorio and close to the *Quebrada de Chiquera*, is called *Choropampa*, the "Field of Shells": *churu* is Quechua for "shell," especially small, edible varieties like oysters or mussels. The name refers to fossils in the chalk strata, where they are found in such numbers that very early on they drew the attention of the natives. There, from close to the surface, a treasure of pure gold was extracted, richly interwoven with silver. A deposit of this sort shows the dependence of many of the ores that erupt from the interior of the Earth in fissures and channels upon the nature of the surrounding stone and the relative age of the formations penetrated. The stone of the *Cerro de Hualgayoc* and Fuentestiana is very rich in water, but an absolute aridity reigns in Purgatorio. I found there, to my astonishment, and despite the altitude of the land above the sea, that the temperature in the mine was 15.8° Réaumur, while in the *Mina de Guadalupe* nearby, the water in the mine was only about 9°. Since the outside temperature rose to only 4.5°, the naked and hardworking mineworkers call the subterranean heat of the Purgatorio stifling.

The narrow path from Moyobamba to the old Incan city of Cajamarca is difficult, even for the mules. The name of the city was originally *Cassamarca* or *Kazamarca*, i.e., "City of Frost"; *marca*, when referring to a place, belongs to the northern dialect, Chinchaysuyo or Chinchasuyu, while in the standard Quechua language it means "stories of a house," also "protector" and "guarantor." The path led us for five or six hours through a series of *paramos*, on which one is exposed almost without interruption to the wrath of the storms and that sharp-edged hail that is so peculiar to the slopes of the Andes. The elevation of the path keeps itself for the most part to between nine and ten thousand feet. This induced me to make a magnetic observation of general interest: to determine the point at which the northern inclination of the needle changes over to the southern inclination, that is to say, at what point travelers crossed the magnetic equator.[12]

When one has finally reached the last of those mountain wildernesses, the *Paramo de Yanahuanga*, one looks all the more joyfully down into the fertile valley of Cajamarca. It is a charming view, for the valley, with a small river snaking its way across it, forms a tableland of oval shape of 6 to 7 square geographical miles. This tableland is similar to the Savanna of Bogota and, like it, was probably an old lakebed. The only thing missing here is the myth of the miraculous Bochica or Idacanzas, high priest of Iraca, who opened a way for the waters through the cliffs at Tequendama. Cajamarca lies 600 feet higher than Santa Fé de Bogota, thus almost as high as the city of Quito, but due to the protection of the surrounding mountains has a much milder and more pleasant climate. The ground is extremely fertile, filled with crop fields and gardens with avenues of willows and large-blossomed red, white, and yellow *Datura* varieties, interspersed with mimosas and the lovely *quinuar* trees (our *Polylepis villosa*, a Rosacea along with *Alchemilla* and *Sanguisorba*). The wheat in the *Pampa de Cajamarca* produces on the average a 15- or 20-fold yield, but the hopes for rich harvests are occasionally dashed by nighttime frosts not noticeable in the roofed dwellings and caused by the heat being radiated toward the clear sky into the thin, dry mountain air.

Small domes of porphyry (probably once islands in the ancient lake) rise in the northern part of the plateau and break through widely spread beds of sandstone. On top of one of these porphyry domes, the *Cerro de Santa Polonia*, we enjoyed a captivating view. The old residence of Atahualpa is bordered on this side by fruit orchards and irrigated, pasturelike fields of alfalfa (*Medicago sativa, campos de alfalfa*). In the distance rise steam pillars from the hot baths of Pultamarca, which still bear the name *baños del Inca*. I found the temperature of these sulfur springs to be 55.2° Réaumur. Atahualpa spent part of the year at these baths, where some fragile ruins of his palace escaped the *conquistadores'* thirst for destruction. The large, deep water basin (*el tragadero*) in which tradition says one of the golden sedan chairs was sunk (which is forever being sought in vain) seemed to me, considering the regularity of its round form, to have been artificially carved into the sandstone over the mouth of one of the springs.

Likewise, only fragile remnants of the fortress and palace of Atahualpa remain in the city, which is now adorned with beautiful churches. The destruction was accelerated by the fury with which those driven by the lust for gold knocked over walls and weakened the foundations of all of the residential areas, even before the end of the 16th century, in their attempt to dig up the treasures buried deep beneath the ground. The palace of the Inca stood on a hill of porphyry that on the surface (that is, on the outcropping of the stone strata) had originally been carved and hollowed out in such a manner as to surround the primary residence almost like a wall. A city jail and a community meeting house (*la casa del Cabildo*) have been erected on a portion of the ruins. These are the most sizable of the ruins, yet they stand only 13 to 15 feet tall, across from the Cloister of St. Francis; they consist, as one may observe in the apartment of Caciquen, of well-cut ashlar blocks of 2 to 3 feet in length, laid upon one another with no cement, much the same as the *Inca-Pilca* or fortress of Cañar on the plateau of Quito.

Into the porphyry rock a shaft was sunk which once led to underground chambers and a gallery (tunnel entrance), which, it is believed, led to another porphyry dome already mentioned above, the one at Santa Polonia. These preparations indicate concern about a secure means of escape during times of war. The burying of treasures was, by the way, a very widespread custom of the ancient Peruvians. Beneath many private living quarters in Cajamarca one can still find underground rooms.

We were shown the so-called Footbath of the Inca (*el lavadero de los piés*) and steps that had been carved into the cliff. Such a washing of the sovereign's feet was accompanied by burdensome courtly ceremonies.[13] Some of the neighboring buildings that, according to tradition, were intended for the servants of the Inca are also constructed with ashlar blocks and built with gables, while others are made with well-formed tiles alternating with concrete (*muros y obra de tapia*). The latter include vaulted facings (recessed walls), the great age of which I long doubted, probably erroneously.

In the main building one may still view the room in which the unfortunate Atahualpa was held prisoner[14] for nine months, starting in November 1532; travelers are also shown the wall upon which he made the mark up to which he would fill the room with gold if he were set free. Xerez, in *Conquista del Peru* (which Barcia preserved for posterity), Hernando Pizarro in his letters, and other writers of the time give widely varying descriptions of the height of this mark. The tormented prince said that "the gold bars, sheets, and vessels should be piled as high as he could reach with his hand." Xerez describes the room as 22 feet long and 17 wide. Garcilaso de la Vega, who had left Peru in 1560, his 20th year, estimates that all that was gathered together leading up to the fateful day of the 29th of August, 1533 (the day of the Inca's death), from the treasures of the Sun Temples of Cuzco, Hualyas, Huamachuco, and Pachacamac, had a value of 3,838,000 *Ducados de Oro*.[15]

In the chapel of the city jail, which, as I mention above, is built upon the ruins of the Inca palace, gullible persons are shown with a shudder the stone upon which there are "irremovable bloodstains." It is a very thin slab, 12 feet long, that lies before the altar, and was probably taken from the porphyry or trachyte of the region. An exact analysis involving taking a chip is not allowed. The three or four supposed stains seem to be hornblende or pyroxene-rich constrictions within the groundmass of the rock type. The licentiate Fernando Montesinos, though he visited Peru barely one hundred years after the conquest of Cajamarca, already begins to spread the tale: Atahualpa was beheaded in the prison, and bloodstains are still visible on the rock upon which the execution took place. It is indisputable, and supported by many eyewitnesses, that the betrayed Inca willingly let himself be baptized under the name Juan de Atahualpa by his fanatical persecutor, the Dominican monk Vicente de Valverde, in order to avoid being burned alive. Strangulation (*el garrote*) finally ended his life, in public under the open sky. Another legend maintains that the chapel was erected upon the stone where the strangling took place, and that Atahualpa's body rests under the stone. The putative bloodstains, then, would clearly remain unexplained. The corpse, however, never lay under this stone; after a mass for the dead and a solemn funeral, which the Pizarro brothers attended in mourning attire (!), it was brought first to the churchyard of the *Convento de San Francisco* and later to Quito, Atahualpa's birthplace. This last relocation was carried out according to the express wishes of the dying Inca. His personal enemy the wily Rumiñaui (called The Eye of Stone due to the loss of one of his eyes to a wart: *rumi*, stone, and *ñaui*, eye in Quechua), with craftiness and political ambition, staged a ceremonious funeral in Quito.

There are still descendants of the monarch living in Cajamarca, in the sad architectural remnants of the glory of days gone by. They are the family of the Indian Cacique (or *Curaca* in the Quechua idiom) Astorpilco. They live in conditions of great need, yet frugally and without complaint, fully resigned to a hard fate that came through no fault of their own. Their descent from Atahualpa by the feminine line is denied nowhere in Cajamarca, but their traces of beard may indicate an infusion of Spanish blood. Those two sons of the great (if, for a Son of the Sun, somewhat free-thinking[16]) Huayna Capac, Huascar and Atahualpa, who both reigned before the Spanish invasion, left behind no acknowledged sons. Huascar became Atahualpa's prisoner on the plains of Quipaypan and was shortly thereafter murdered on Atahualpa's order. From the other two brothers of Atahualpa, the insignificant young

Toparca, whom Pizarro (in the autumn of 1533) had crowned as Inca, and the likewise crowned but subsequently rebellious Manco Capac, there also came no male offspring. Atahualpa left a son with the Christian name don Francisco, who died very young, and a daughter, doña Angelina, with whom Francisco Pizarro produced, in the midst of a wild life of war, a son that he deeply loved, the grandson of the executed ruler. Along with the family of Astorpilco, with whom I spent time in Cajamarca, the Carguaraicos and the Titu-Buscamaytas were also distinguished as relatives of the Inca dynasty, but the Buscamayta line has since died out.

The son of the cacique Astorpilco, a friendly young person of 17 years who accompanied me through the ruins of the old palace in his home city, in his situation of great want had filled his imagination with images of the magnificent subterranean treasures of gold beneath the heaps of rubble upon which we stumped about. He related how one of his forefathers had once bound the eyes of his wife and led her through several deviating pathways carved into the rock, down into the subterranean garden of the Inca. There she saw, artfully replicated in purest gold, trees complete with leaves and fruit, birds sitting on the branches, and the much-sought-after golden sedan chair (*una de las andas*) of Atahualapa. The man commanded his wife to touch nothing, for the long-awaited time (of the reinstatement of the Incas) had not yet come. Anyone who acquired any of this wealth before then must die that very night. Such golden dreams and fantasies as those of this boy are founded upon the memories and traditions of prehistory. The luxury of artificial gardens of gold (*jardines ó huertas de oro*) has been often described by eyewitnesses: by Cieza de Leon, Sarmiento, Garcilaso, and other early historians of the *conquista*. They were found beneath the sun temples in Cuzco, in Cajamarca, and in the graceful valley of Yucay, a favorite seat of the royal family. Where the golden *huertas* were not underground, there stood living plants next to the artful replicas. The tall stalks and the fruits of maize on cobs (*mazorcas*) are said to have been especially convincing.

The morbid certainty with which the young Astorpilco proclaimed that below me, somewhat to the right of where I stood, a large-blossomed datura tree, a *guanto*, artfully fashioned from sheets and wires of gold, sheltered the couch of the Inca with its branches, made upon me a melancholy impression. Here again, fanciful visions and delusion are comfort for great privation and earthly suffering. "Do you and your parents not feel," I asked the boy, "since you so firmly believe in the existence of this garden, an occasional desire, in light of your want, to dig for the treasures that lie so near?" The boy's answer was so simple, so much the expression of the quiet resignation that characterizes the aboriginal people of this land, that I put it down in my journal in

Spanish: "Such a desire (*tal antojo*) does not come to us; my father says that it would be a sin (*que fuese pecado*). If we had all of the golden branches with all of their golden fruit, then our white neighbors would hate us and harm us. We possess a little field and good wheat (*buen trigo*)." I believe that few of my readers will fault me for remembering here the words of the young Astorpilco and his golden dreams.

The belief, so very widespread among the natives, that to take possession of buried treasures that might have belonged to the Incas is a punishable offense and would bring down misfortune upon an entire people is connected to another belief, especially prevalent in the 16th and 17th centuries, in the eventual reinstatement of an Inca empire. Every oppressed nationality hopes for liberation and a renewal of the old regime. The flight of Manco Inca, brother of Atahualpa, into the forests of Vilcapampa on the slope of the eastern cordillera and the times when Sayri Tupac and Inca Tupac Amaru were forced to abide in those wild reaches left lasting memories. It was believed that descendants of the deposed dynasty might have settled between the rivers Apurimac and Beni, or perhaps even farther east in Guyana. The myth that traveled from the West to the East of El Dorado and the golden city of Manoa increased such dreams. Raleigh's imagination was so inflamed by it "that he launched an expedition in the hope of conquering the imperial and golden island city, installing a garrison of three to four thousand Englishmen, and levying from the emperor of Guyana, who is descended from Huayna Capac and maintains his court with the same magnificence, an annual tribute of 300,000 pounds sterling as the price of the promised restoration in Cuzco and Cajamarca." For as far as the Peruvian Quechua language has spread, traces of such expectations of a returning Inca dominion[17] have been preserved in the heads of many natives somewhat versed in the history of their fatherland.

We stayed for five days in the city of the Inca Atahualpa, which at that time had but seven or eight thousand inhabitants. Our departure was delayed by the great number of mules required for the transport of our collections and the careful selection of the drivers who were to lead us over the Andes as far as the entrance to the long but narrow Peruvian sand desert (*Desierto de Sechura*). The transversal of the cordillera was from northeast to southwest. One has hardly left the ancient lakebed of the lovely plateau of Cajamarca when one is moved to astonishment, while climbing at an elevation of barely 9,600 feet, by the appearance of two grotesque porphyry domes, Aroma and Cunturcaga—a favorite perch of the mighty vulture that we normally call the condor (*cuntur*), *kacca* being, in the Quechua, "the rock." The domes consist of 35- to 40-foot pillars with 4 to 7 sides, some bent into curves, some split into vertical sections. The porphyry dome of *Cerro Aroma* is especially picturesque. With its division of overlapping, often converging rows of pillars, it looks like a building with two levels, topped like a cathedral by a thick, rounded rock mass not broken into pillars. Such porphyry and trachyte outcroppings are, as mentioned above, quite characteristic of the high ridge of the cordilleras and give to them a physiognomy very different from that of the Swiss Alps, the Pyrenees, and the Siberian Altai.

From Cunturcaga and Aroma, one zigzags a full 6,000 feet down a steep rock face into the crevasselike Magdalena Valley, the floor of which still sits 4,000 feet above sea level. A few pitiful huts, surrounded by the same cotton trees (*Bombax discolor*) that we first saw by the Amazon, are referred to as an Indian village. The meager vegetation of the valley is similar to that of the Jaen de Bracamoros Province, but we regretfully noted the absence of the red *Bougainvillaea*. The valley is among the deepest I know in the Andes. It is a cleft, a true transversal valley running east-west, hemmed in by the *Altos de Aroma* on one side and *Huangamarca* on the other. Here once more begins the quartz formation that for so long puzzled me, the one we had already observed on the *Paramo de Yanahuanga* between Moyobamba and Cajamarca at an elevation of 11,000 feet, and which attains a mighty size of several thousand feet on the western slope of the cordillera. Since Leopold von Buch demonstrated that the chalk is widespread, even in the highest reaches of the Andes on both the near and far sides of the Isthmus of Panama, that quartz formation, its texture transformed perhaps by volcanic action, belongs to the ashlar sandstone between the upper chalk layer and the gault and greensand. From the mild Magdalena Valley, we now had to climb for three and a half hours, in a westerly direction, the 4,800-foot wall opposite the porphyry groups of the Alto de Aroma. The change in climate was even more palpable there on the rock wall, for we were often wrapped in a cold fog.

After 18 months spent uninterruptedly traversing the restrictive interior of a mountainous country, the longing finally to enjoy once more the unconstrained view of the ocean was heightened by frequent illusions. From the peak of the volcano Pichincha, looking out across the dense forests of the *Provincia de las Esmeraldas*, one cannot clearly discern an ocean horizon due to the height and the great distance to the littoral. One looks out, as if encased in a ball of air, into emptiness; one might have a sense of something but can differentiate nothing. Later, when we reached the *Paramo de Guamani*, where many buildings of the Incas lie in ruins, the mule drivers assured us that beyond the plain, past the depressions of Piura and Lambayeque, we would be able to see the ocean, but a thick fog lay on the plain and on the distant littoral. We saw only rock masses of many shapes, alternately protruding like islands above the billowing sea of fog and then disappearing again, a sight similar to that which we enjoyed on the peak of Tenerife. We encountered nearly the same beguilement of our expectations on the Andean pass of Huangamarca, the traversal of which I will describe here. Each time we had climbed another hour, toiling with suspenseful hope against the slope of the mighty mountain, our guides, not entirely familiar with the track, would promise that our hopes would soon be fulfilled. The layer of mist that enfolded us would seem occasionally to open, but our view would soon be cruelly restricted by heights that yet lay before us.

One's longing to see certain objects is not predicated only on their grandeur, their beauty or importance; it is interwoven in each person with many incidental impressions from their youth, with an early predilection for individual pursuits, with an inclination toward distant places and a life in motion. The improbability of seeing a wish fulfilled is what gives it its peculiar charm. The traveler appreciates in advance the joy of the moment when first he beholds the Southern Cross and the Magellanic Clouds that circle the South Pole, the snow of Chimborazo and the smoke columns of the volcanoes of Quito, a stand of arborescent ferns, or the tranquil Pacific Ocean. The days upon which the fulfillment of such wishes occurs mark the epochs of life with indelible impressions, exciting feelings of such vitality as needs no rational justification. Within the longing for the sight of the Pacific from the spine of the Andes is intermingled the interest with which the small boy listened to the narrative of the bold expedition of Vasco Nuñez de Balboa,[18] the fortunate man who was the first European (followed by Francisco Pizarro) to see, from the Sierra de Querequa on the Isthmus of Panama, the eastern Pacific Ocean. The reed-covered shores of the Caspian Sea, which I first spied from the delta of the Volga, are certainly not what one would call picturesque, and yet that first sight of them was to me all the more joyous for my having been attracted from my earliest youth to the shape of this Asiatic inland sea on maps. That which is awakened within us, whether by childlike impressions or through the element of mere chance in the circumstances of life,[19] will later turn in a more serious direction and will often become a motivation for scientific work, for far-reaching ventures.

When, after many undulations of the ground on the craggy mountainside, we finally reached the highest point of the *Alto de Huangamarca*, the long-veiled vault of the sky suddenly cleared. A sharp southwest wind chased away the mist. The deep blue of the thin mountain air appeared between the highest feathery clouds. The entire western slope of the cordillera near Chorillos and Cascas, covered in monstrous blocks of quartz 12 to 15 feet long, the plains of Chala and Molinos all the way to the seashore at Trujillo lay, as though wondrously close, before our eyes. For the first time, we were seeing the Pacific

Ocean; we saw it clearly: next to the littoral, a great body of light, reflecting, ascending to a horizon now more than merely sensed. The joy that my companions, Bonpland and Carlos Montufar, avidly shared caused us to forget to open the barometer on the *Alto de Huangamarca*. By the measurement that we took nearby, somewhat lower than the summit at an isolated farm (the *Hato de Huangamarca*), the point from which we first saw the ocean must lie at only 8,800 or 9,000 feet.

For one who owed part of his education and the nature of many of his wishes to his association with a companion of Captain Cook, there was something solemn about the sight of the Pacific. Georg Forster was familiar with the general outline of my travel plans early on, when I had the privilege of making my first visit to England (now more than a half-century ago) under his guidance. Forster's enchanting descriptions of Tahiti had awakened, especially in Northern Europe, a general and, I might say, yearning interest in the isles of the South Sea. At that time these islands had the good fortune to be seldom visited by Europeans. I too was able to nurture the hope of soon touching some of them, for the purpose of my voyage to Lima was of a dual nature: to observe the transit of Mercury across the disc of the sun, and to fulfill the promise I had made to Captain Baudin upon my departure from Paris, that I would join him in his circumnavigation, as soon as the French republic could deliver the funding it had earlier promised for the purpose.

North American newspapers had spread the report in the Antilles that both corvettes, *le Géographe* and *le Naturaliste*, were to sail around Cape Horn and land in Callao de Lima. In Havana, where I found myself after the completion of the Orinoco journey, upon hearing this news I abandoned my original plan to cross Mexico on my way to the Philippines. I quickly rented a ship that carried me from the Island of Cuba to Cartagena de Indias. But the Baudin expedition took a route quite different from the one that had been announced and expected: it did not sail around Cape Horn, as had been the earlier plan when Bonpland and I had appointed ourselves to meet it; instead, it sailed around the Cape of Good Hope. One goal of my Peruvian journey and the last crossing of the Andes chain thus did not materialize, but I did have the rare good fortune, during an unpropitious time of year, to experience a clear day in the misty country of lowland Peru. I observed the transit of Mercury across the disc of the sun in Callao, an observation that has proved to be of some importance to the accurate determination of the longitude of Lima[20] and the southwestern part of the New Continent. Thus does there often lie, within the entanglement of grave circumstances in life, the germ of a satisfying compensation.

Annotations and Additions

1. Upon the spine of the Anti or Andes chain

The Inca Garcilaso, who had a rich understanding of the language of his fatherland and lingered gladly over etymologies, consistently refers to the Andes chain as *las Montañas de los Antis*. He says with certainty that the great mountain range east of Cuzco got its name from the tribe of the Antis and the Anti Province, which lies in the eastern portion of the royal seat of the Inca. The quaternary division of the Peruvian Empire according to the four regions of the world as determined by looking out from Cuzco did not borrow its terminology from the quite cumbersome words derived from the sun which mean East, West, North, and South in Quechua language (*intip llucsinanpata, intip yaucunanpata, intip chaututa chayananpata,* and *intip chaupunchau chayananpata*) but from the names of the provinces and tribes (*Provincias llamadas Anti, Cunti, Chincha, y Colla*) that are positioned to the east, west, north, and south of the center of the empire (the city of Cuzco). The four parts of the Inca theocracy are accordingly called *Antisuyu, Cuntisuyu, Chinchasuyu,* and *Collasuyu*. The word *suyu* means "strip" or "part." Despite the great distance, Quito belonged to Chinchasuyu; when the Incas, through their religious wars, spread their beliefs, their language, and their constringent form of government, these *suyu* took on larger and unequal dimensions. Thus the concept of regions of the world became attached to the names of nearby provinces. "The naming of those parts is the same thing," says Garcilaso, "as saying 'to the east' or 'to the west.'" The snowy chain of the Antis was therefore viewed as an eastern chain. "The Province of Anti gives the *Montañas de los Antis* their name. The name *Antisuyu* signifies the part to the east, and thus the entire great range of *sierra nevada* that runs through the eastern part of Peru was called *Anti*, thus showing that it lay to the east" (*Comentarios Reales*, part I, pp. 47 and 122). Recent authors have asserted that the name of the Andes chain is derived from *anta*, which is "copper" in the Quechua language. This metal was indeed of great importance to a people who used not iron for their cutting tools but a mixture of copper and tin; the name "Copper Mountains," however, would not be extended to such a large chain, and *anta*, as Professor Buschmann has quite correctly remarked, retains the *a* in compound forms. Garcilaso states clearly: *Anta*, "cobre," *y Antamarca*, "Provincia de Cobre." In general, the word form and compounding in the old language of the Inca Empire (Quechua) is so simple that one cannot speak of a transition from *a* to *i*; thus *anta* (copper) and *Anti* or *Ante* (the land or an inhabitant of the Andes, or the mountain chain itself—*la tierra de los Andes, el Indio hombre de los Andes, la Sierra de los Andes,* as local dictionaries explain it)—are and remain completely different words. The interpretation of the proper name through some other concept is obscured in the darkness of the ages. Compound forms using *Anti*, along with the aforementioned *Antisuyu*, are *Anteruna*, a native-born Andes inhabitant, *Anteunccuy* or *Antionccoy*, Andes-sickness (*mal de los Andes pestifero*).

2. The Countess of Chinchon

She was the wife of the Viceroy don Geronimo Fernandez de Cabrera, Bobadilla y Mendoza, Conde de Chinchon, who administered Peru from 1629 to 1639. The

Vicereine's cure took place in 1638. A tradition that has spread in Spain, but which I often heard disputed in Loja, names a corregidor of the *Cabildo de Loja*, Juan Lopez de Cañizares, as the person through whom the cinchona bark was first brought to Lima and generally recommended as a remedy. In Loja I heard it said that the beneficial powers of the tree had been known in the mountains for a long time, if not widely. Directly after my return to Europe I expressed doubt that the discovery had been made by the natives of the region around Loja, because even today the Indians of the nearby valleys, where many remittent fevers are prevalent, will scorn the cinchona bark. (Cf. my article on the cinchona forests in the *Magazin der Gesellschaft naturforschender Freunde zu Berlin*, vol. I, 1807, p. 59.) The myth of how the natives learned of the curative power of the cinchona from the lions, who "cure themselves of malaria when they gnaw on the bark of the cinchona tree" (*Histoire de l'Acad. des Sciences Année 1738*, Paris, 1740, p. 233), seems to be of purely European origin and an apocryphal tale. Regarding "fever among lions" there is nothing known in the New Continent, for the so-called great American lion (*Felis concolor*) and the small mountain lion (*Puma*), whose footprints I have seen in the snow, are never tamed there as a subject for observation; also, the various species of cat on both continents are not in the habit of stripping the bark from tree trunks. The name "Countess Powder" (*Pulvis Comitissae*), which was inspired by the distribution of the medicament by the Countess of Chinchon, was later changed to Cardinal's or Jesuit's Powder because the procurator general of the Jesuit order, Cardinal de Lugo, distributed the medicine while traveling through France, and recommended it all the more urgently it to Cardinal Mazarin when the brothers of the order began to do a lucrative business in South American cinchona bark, which they had learned from missionaries how to procure. It is hardly necessary to remark that among Protestant doctors religious intolerance and hatred for Jesuits were mixed into the long debate over the efficacy or harmfulness of the fever bark.

3. Aposentos de Mulalo

Regarding these *aposentos* (apartments, shelters; in the Quechua language *tampu*, hence the Spanish *tambo*), cf. Cieça, *Chronica del Peru*, cap. 41 (ed. de 1554, p. 108), and my *Vues des Cordillères*, pl. XXIV.

4. Fortress of Cañar

Not far from Turchi, at an elevation of 9,984 feet. I included an illustration of it in the *Vues des Cordillères*, pl. XVII (cf. Cieza, chap. 44, part I, p. 120). Not a great distance from the *Fortaleza del Cañar*, in the celebrated Valley of the Sun, *Inti-Guaycu* (Quechua: *huayco*), may be found the cliff upon whose face the natives believe to see an image of the sun, and a curious bank of stone that they call *Inga-Chungana* (*Inca-chuncana*), the "Game of the Inca." I made drawings of both; see *Vues des Cordillères*, pl. XVIII and XIX.

5. Roads covered with cemented gravel

Cf. Velasco, *Historia de Quito*, 1844, T. I, pp. 126–128, and Prescott, *Hist. of the Conquest of Peru*, vol. I, p. 157.

6. Where the road was interrupted by tiers and stairsteps

Cf. Pedro Sancho in *Ramusio*, vol. III, fol. 404, and excerpts from handwritten letters of Hernando Pizarro that the great historian was able to consult in Prescott, vol. I, p. 444. "The road of the mountains is something to see, for in truth, seldom on a terrain so rough in all Christendom have such beautiful roads been seen, and most of it paved."

7. Greeks and Romans show us these contrasts

"If the Hellenes," writes Strabo (lib. V, p. 235 Casaub.), "in the building of their cities expected success and good fortune, especially through setting beauty and stability as their goals, the Romans then were primarily concerned with what these others ignored: the paving of streets, the installation of aqueducts and drainage systems which could flush away the filth of the city into the Tiber. They paved all of their roads so that freight wagons might be comfortably able to retrieve the cargo from the trade ships."

8. The messenger of God, Nemterequeteba

The civilization in Mexico (the Aztec land of Anahuac) and that in the Peruvian theocracy, in the Incas' heliadic Empire of the Sun, have so captured the attention of Europe that a third luminous point of dawning enlightenment, that of the mountain peoples of New Granada, has long been almost completely overlooked. I have already discussed this in detail in the *Vues des Cordillères et Monumens des peuples indigènes de l'Amerique* (octavo edition), vol. II, pp. 220–267. The form of government among the Muiscas of New Granada is reminiscent of the constitution of Japan, of the relationship of the worldly leader (*kubo* or *shogun* in Jedo) to the holy personage of the dairi in Miyako. When Gonzalo Jimenez de Quesada pushed on to the Savanna of Bogota (*Bacata*, i.e., the outermost of the cultivated fields, probably because of the nearness of the wall of mountains), he found there three authorities, the mutual subordination of which remains somewhat unclear. The spiritual leader was the elected high priest of Iraca or Sagamoso (*Sugamuxi*, the place of the disappearance of Nemteraqueteba); the secular princes were the *zake* (zaque of Hunsa or Tunja), and the *zipa* of Funza. This latter prince seems originally to have been subordinate to the zaque in the feudal system.

The Muiscas had an ordered reckoning of time, with intercalation, to improve upon the lunar year; they used small discs of poured gold of uniform diameter as coins (for which we still search in vain amongst the highly cultivated Egyptians' remains); they had sun temples with stonelike columns, the remnants of which were quite recently discovered in the valley of Leyva. (Joaquin Acosta, *Compendio historico del Descubrimiento de la Nueva Granada*, 1848, pp. 188, 196, 206, and 208; *Bulletin de la Société de Géographie de Paris*, 1847, p. 114.) Actually, the Muisca tribe should always be referred to by the name Chibchas, for in the Chibcha language *muisca* merely means "people" or "persons." The origin and the elements of immigrated culture were ascribed to two mythical figures, Bochica and Nemterequeteba, who are often confused with one another. The first is even more mythical than the second, for Bochica alone is considered divine and held almost at the same level as the sun. His beautiful companion Chia, or Huythaca, used her skills in magic to cause the flooding of the Bogota Valley, and for this was banned from the Earth by Bochica, now only to circle around it as the moon. Bochica smote the cliff at Tequendama and gave the waters an outlet near the Field of Giants (*Campo de Gigantes*), where, at an elevation of 8,250 feet above sea level, the bones of elephantlike mastodons lie buried, about which Captain Cochrane (*Journal of a Residence in Colombia*, 1825, vol. II, p. 390) and Mr. John Ranking (*Historical Researches on the Conquest of Peru*, 1827, p. 397) report that such animals are still living in the Andes, shedding their teeth there! Nemterequeteba, also called Chinzapogua (*enviado de Dios*), is a human being, a man with a beard who came from the East, from Pasca, and disappeared near Sogamoso. The founding of the sanctuary of Iraca is ascribed sometimes to him, other times to Bochica, and since Bochica is sometimes said to have borne the name Nemqueteba, the confusion, especially on such unhistorical grounds, is easily explained.

By means of the Chibcha language, my friend of many years Colonel Acosta seeks to demonstrate in his comprehensive work (*Compendio de la Hist. de la Nueva Granada*, p. 185) "that, since the potatoes (*Solanum tuberosum*) in Usmè have the native, not Peruvian, name *yomi*, and were found by Quesada in 1537 to be cultivated in the Velez Province at a time when introduction from Chile, Peru, and Quito would have been improbable, the plant may well be considered to be native to New Granada." I recall, however, that the invasion by the Peruvians and the complete occupation of Quito took place before 1525. The southern provinces of Quito had even then already come under the dominion of Tupac Inca Yupanqui at the close of the 15th century (Prescott, *Conquest of Peru*, vol. I, p. 332). In the (unfortunately!) still so obscure history of the first introduction of potatoes to Europe there remains the general tendency to ascribe the credit for the introduction to the seafaring hero Sir John Hawkins, who is said to have obtained them from Santa Fé in 1563 or 1565. It seems more certain that Sir Walter Raleigh planted the first potatoes on his estate of Youghal in Ireland, from whence they came to Lancashire.

Regarding the pisang (*Musa*), which is cultivated in all warmer regions of New Granada since the arrival of the Spanish, Colonel Acosta believes (p. 205) that before the conquest it was to be found only in Choco.

One should also see Joaquin Acosta (p. 189) for information on the subject of the name Cundinamarca, which, in a display of false erudition, was attached to the young republic of New Granada in 1811, a name "full of golden dreams (*sueños dorados*)"— actually *Cundirumarca* (and also not *Cunturmarca*; Garcilaso, lib. VIII, cap. 2). Luis Daza, attached to the small invasion force of the conquistador Sebastian de Belalcazar coming from the south, had heard talk of a distant land of Cundirumarca that was rich in gold and was peopled by the tribe of the Chicas, whose prince had requested support troops from Atahualpa in Cajamarca. The Chibchas or Muiscas of New Granada were mistaken for these Chicas, and thus was the name of the unknown, more southerly land carried over to this place!

9. The vertical descent of the Rio Chamaya
Cf. my *Recueil d'Observ. astron.*, vol. I, p. 304, *Nivellement barométrique*, nos. 236–242. I have sketched the swimming postman as he is tying the towel that holds the letters around his head in the *Vues des Cordillères*, pl. XXXI.

10. Being of some importance to the geography of South America, due to an old observation of La Condamine
It was my intention to chronometrically connect the city of Quito with Tomependa, the point of departure for La Condamine's journey, and with his other locational determinations along the Amazon River. In June of 1743, thus 59 years before me, La Condamine had been in Tomependa, which I determined, over three nights of stellar observation, lay at 5°31′28″ south latitude and 80°56′37″ longitude. The longitude of Quito had been erroneous, as Oltmanns has shown by means of my observations and a painstaking revised calculation of all earlier ones (Humboldt, *Recueil d'Observ. astron.*,

vol. II, pp. 309–359) up to my return to France, by a full 50.5 minutes of arc. The satellites of Jupiter, moon distances, and lunar eclipses concur satisfactorily, and all elements of the calculations have been presented to the public. La Condamine had applied the too-easterly longitude of Quito to Cuenca and the Amazon. "I made," says La Condamine, "my first navigation attempt on a (balsa wood) raft down the Chinchipe River to Tomependa. I had to be content with determining the latitude and longitude by using the roads. There, I made my political testament, writing an extract of my most important observations." (*Journal du Voyage fait à l'Équateur*, 1751, p. 186.)

11. Twelve thousand feet above the sea, we found pelagic shell fossils

Cf. my *Essai géognostique sur le Gisement des Roches*, 1823, p. 236, and for the first zoological determination of the fossils contained in the old chalk formation of the Andes, Leop. de Buch, *Pétrifications recueillies en Amérique par Alex. De Humboldt et Charles Degenhardt*, 1839, (in-fol.) pp. 2–3, 5, 7, 9, 11, 18–22. Pentland found shell fossils from the Silurian formation in Bolivia on the Nevado de Antakaua at an elevation of 16,400 Parisian feet (Mary Somerville, *Physical Geography*, 1849, vol. I, p. 185).

12. Where the magnetic equator cuts across the Andes chain

Cf. my *Relation hist. du Voyage aux Régions équinoxiales*, part III, p. 622, and *Kosmos*, vol. I, pp. 191 and 432; a misprint in that edition, however, has the longitude once at 48°40′, then 80°40′, instead of 80°54′.

13. Accompanied by burdensome courtly ceremonies

According to ancient court ceremonial, Atahualpa never spat on the ground but into the hand of one of the most distinguished ladies of his retinue—"all," says Garcilaso, "for the sake of his Majesty." *El Inca nunca escupia en el suelo, sino en la mano de una Señora mui principal, por Magestad* (Garcilaso, *Comment. Reales*, part II, p. 46).

14. Atahualpa was held prisoner

The imprisoned Inca, shortly before his execution, was taken outside at his command, that he might be shown a comet. The comet, "greenish-black and the thickness of a man" (Garcilaso says, part II, p. 44: *una cometa verdinegra, pocomenos gruesa que el cuerpo de un hombre*), which Atahualpa observed before his death, thus in July or August of 1533, and which he considered to be the same malignant comet that had appeared at the death of his father, Huayna Capac, is certainly the one observed by Appian (Pingré, *Cométographie*, part I, p. 496, and Galle, *Verzeichnis aller bisher berechneten Cometenbahnen in Olbers Leichtester Methode, die Bahn eines Cometen zu berechnen*, 1847, p. 206) and which on the 21st of July, standing high in the North in the region of Perseus, seemed to portray the sword that Perseus holds in his right hand (Mädler, *Astronomie*, 1846, p. 307; Schnurrer, *Chronik der Seuchen in Verbindung mit gleichzeitigen Erscheinungen*, 1825, part II, p. 82). Robinson considers the year of the Inca Huayna Capac's death to be uncertain, but according to the investigations by Balboa and Velasco, it falls in 1525, and the information from Hevelius (*Cometographia*, p. 844) and Pingré part I, p. 485) would be corroborated by Garcilaso's account (part I, p. 321) and by tradition that had been preserved among the *amautas* ("who are the philosophers for all of the Republic").

I would like to insert here one additional observation: Oviedo alone, and almost certainly inaccurately, asserts in the unedited supplement to his *Historia de las Indias* that the actual name of the Inca was not Atahualpa but Atabaliva (Prescott, *Conquest of Peru*, vol. I, p. 498).

15. Ducados de Oro

The sum given in the text is that of Garcilaso de la Vega in the *Comentarios reales de los Incas*, part II, 1722, pp. 27 and 51. The amounts given by Father Blas Valera and by Gomara, *Historia de las Indias*, 1553, p. 67, however, deviate considerably. Cf. my *Essai politique sur la Nouvelle-Espagne* (éd. 2), vol. III, p. 424. Indeed, it is also just as difficult to ascertain the value of the *ducado, castellano*, or *peso de oro* (*Essai pol.*, vol. III, pp. 371 and 377; Joaquin Acosta, *Descubrimiento de la Nueva Granada*, 1848, p. 14). The insightful historian Prescott was able to use a manuscript with the highly promising title *Acta de Repartícion del Rescate de Atahualpa*. While he estimates the entirety of the Peruvian spoils divided by the Pizarro brothers and Almagro to have the exaggerated value of 3.5 million pounds sterling, the figure does include the gold of the ransom as well as the loot from the various temples of the sun and the enchanted gardens, the *huertas de oro* (Prescott, *Conquest of Peru*, vol. I, pp. 464–477).

16. The great (if, for a Son of the Sun, somewhat free-thinking) Huayna Capac

The nightly absence of the sun evoked in the Inca all manner of philosophical doubts regarding the world dominance of this celestial body. Father Blas Valera recorded what the Inca said about the sun: "Many maintain that the sun is alive and is the author of all creation (*el hacedor de todas las cosas*); but he who wishes to accomplish something must keep at the task before him. Now, much occurs when the sun is absent; thus is he not the maker of everything. One may also doubt that he is a living thing, for in his circling he never tires (*no se cansa*). Were he a living thing, then he would grow tired as we do, and if he were really a free being, he would certainly come into those parts of the sky where we never see him. The sun is thus like an animal tied to a rope, forever to make the same circuit (*como una res atada que siempre hace un mismo cerco*); or like an arrow, which only goes where it is sent, not wherever it will." (Garcilaso, *Comment. Reales*, part I, lib. VIII, cap. 8, p. 276.) The natural observation of the circling of a celestial body as though it were on a rope is truly remarkable. Since Huayna Capac, by the way, had already died in Quito in 1525, seven years before the arrival of the Spanish, and divided his empire between Huascar and Atahualpa (the first of these names means "rope" or "cord"; the second, like *hualpa* alone, means "chicken" or "rooster"), then it is certain that Huayna Capac used, instead of *res atada* ("tied animal"), the common expression "animal on a rope"; but in Spanish as well, *res* by no means implies only cattle, but any domesticated livestock. What Father Valera may have injected from his own sermons into the heresies of the Inca in order to push the natives away from the official dynastic sun worship, the court religion, is not to be investigated here. That the lower classes were to be strictly sheltered from such doubts lay, incidentally, in the strictly conservative political acuity of Inca Roca, conqueror of the Charcas Province. He founded schools for the upper classes only and forbade, on pain of severe punishment, that common people be educated in any way, "lest they grow overly bold and unsettle the State!" (*No es lecito que enseñen á los hijos de los plebeios las Ciencias, porque la gente baja no se eleve y ensobervezca y menoscabe la Republica*; Garcilaso, part I, p. 276.) This was the theocracy of the Incas, much like the politics in the southern states of free North America, the slave states.

17. A returning Inca dominion

I have addressed this subject in detail in another work (*Relation hist.*, vol. III, pp. 703–705 and 713). Raleigh was convinced that in Peru, an old prophecy prevails: "that from Inglaterra those Ingas shoulde be againe in time to come restored and deliuered from the seruitude of the said Conquerors. I am resolued that if there were but a smal army a foote in Guiana marching towards Manoa the chiefe Citie of Inga, he would yield her Majesty by composition so many hundred thousand pounds yearely, as should both defend all enemies abroad and defray all expences at home, and that he woulde besides pay a garrison of 3,000 or 4,000 soldiers very royally to defend him against other nations. The Inca wil be brought to tribute with great gladnes" (Raleigh, *The Discovery of the large, rich and beautiful Empire of Guiana, performed in 1595*, taken from the edition by Sir Robert Schomburgk, 1848, pp. 119 and 137)—a true project of restoration, which promised sweet satisfaction for both sides; all that was lacking for its success was (alas!) a dynasty to restore, and from whom to receive tribute.

18. Of the bold expedition of Vasco Nuñez de Balboa

I have already pointed out elsewhere (*Examen critique de l'histoire de la Géographie du Nouveau Continent, et de progrès de l'Astronomie nautique aux 15ème et 16ème siècles*, vol. I, p. 349) that Columbus, long before his death and a full ten years before Balboa's expedition, was aware of the existence of the Pacific and its close proximity to the eastern coast of Veragua. He came to this awareness not through theoretical speculations on the configuration of East Asia but from the specific local accounts that he collected from the natives on his fourth voyage (11 May, 1502 through 7 November 1504). This fourth voyage took the Admiral from the coast of Honduras to the *Puerto de Mosquitos*, and on to the western end of the Isthmus of Panama. Regarding what the natives related, Columbus comments in the *Carta rarissima* of 7 July 1503 that "not far from the Rio de Belen, the other ocean (the Pacific) turns (*boxa*) toward the mouths of the Ganges, so that the lands of the *Aurea* (i.e., that of the *Chersonesus aurea* of Ptolemy) sit relative to the eastern coasts of Veragua as Tortosa (at the mouth of the Ebre) does to Fuenterrabia (on the Bidasoa) in Biscay, or as Venice does to Pisa." While soon thereafter, on 25 September, Balboa first saw the Pacific from the heights of the *Sierra de Quarequa* (Petr. Martyr, Epist. DXL, p. 296), it was not until several days later that Alonso Martin de don Benito, who had found a passage from the Quarequa range to the Gulf of San Miguel, rode out onto the Pacific in a canoe. (Joaquin Acosta, *Compendio hist. del Descubrimiento de la Nueva Granada*, p. 49.)

Since the possession of a considerable portion of the western coast of the New Continent on the part of the United States of North America and the call of the wealth of gold in New California (now called Upper California) has in very recent times intensified more than ever the desire for a connection between the Atlantic states with the West by way of the Isthmus of Panama, I feel it is my duty to here again point out that the shortest route that the natives showed to Alonso Martin de don Benito for reaching the Pacific lies in the eastern part of the isthmus and leads to the *Golfo de San Miguel*. We know that Columbus (*Vida del Almirante por Don Fernando Colon*, chap. 90) was searching for an "*estrecho de tierra firme*," and in the official documents that we possess from 1505, 1507, and especially 1514, the sought-after opening (*abertura*) and the pass (*paso*) are mentioned which would lead directly to the "Indian land of spices."

Occupied as I have been for more than forty years with the means of communication between these two oceans, in my published works and in the various memoirs that I am honored to have had entrusted to me by the free states of Spanish America, I have always insisted: the entire length of the isthmus should be hypsometrically explored, especially there, where it connects to the mainland of South America at the Darién Gap and the uninhabitable former *Provincia de Biruquete*, and where, between the Atrato River and the Bay of Cupica (on the Pacific littoral), the mountain range of the isthmus virtually disappears. (Cf. my *Atlas géographique et physique de la Nouv. Espagne*, pl. IV, and the *Atlas de la Relation historique*, pls. XXII and XXIII; *Voyage aux Régions équinoxiales du Nouveau Continent*, vol. III, pp. 117–154, and *Essai politique sur le royaume de la Nouvelle-Espagne*, vol. I, 2nd ed., 1825, pp. 202–248.)

At my request, General Bolivar had a careful leveling survey of the isthmus between Panama City and the mouth of the Rio Chagres completed by Lloyd and Falmarc in 1828 and 1829 (*Philosophical Transactions of the Royal Society of London for the Year 1830*, pp. 59–68). Other measurements have since been done by knowledgeable and experienced French engineers, as are done for projects involving canals and railroads with locks and tunnels—but always in the meridional direction between Portobello and Panama City or to the west, toward Chagres and Cruces. The most important points on both the eastern and southeastern portions of the isthmus on both coasts have been ignored! As long as this portion has not been geographically charted according to exact (but easily and quickly obtained) determinations of latitude and chronometrical longitude, and had its topographical features hypsometrically mapped by means of barometric measurements of elevation, I will continue to view the currently (1849) oft-repeated assertion "the Isthmus is not suited to the construction of an oceanic canal (a canal with fewer locks than the Caledonian Canal), nor to an unimpeded crossing, independent of the season, with such ships as come from Chile and California, from New York and Liverpool," as unfounded and thoroughly ill-advised.

On the Antillean littoral of the isthmus, the *Ensenada de Mandinga* presses so deeply southward (according to research that the *Dirección* of the *Deposito hidrografico* in Madrid has incorporated into their maps since 1809) that it seems to be only about 4 to 5 geographical miles (15 making one equatorial degree) away from the Pacific littoral east of Panama City. The isthmus is cut into almost equally far on the Pacific shore by the Golfo de San Miguel, into which the Rio Tuira flows, along with its tributary, the Chuchunque (Chucunaque). The latter river, in its upper course, also comes to within four geographical miles of the Antillean coast west of Cape Tiburon. For more than 20 years, companies willing to commit considerable monetary resources have approached me with questions regarding the Isthmus of Panama, but the simple advice that I give has never been followed. Every scientifically educated engineer knows that in the tropics, even without corresponding observations, good barometric measurements with attention to hourly variations can yield a degree of accuracy of 70 to 90 feet. It would thus be easy to set up two fixed corresponding barometer stations on the two coasts for a few months and to compare the portable instruments used in the previously completed leveling survey with one another and with the fixed stations several times. One would preferably seek to do this at the place where the dividing mountains, as they approach the continental mass of South America, dwindle down to hills. Considering the importance that the object has for world trade, one cannot remain captive, as has been the case up to now, to a narrow point of view. Only a great amount of work, encompassing the entire eastern isthmus—toward any type of possible installation, whether a canal project or a railroad, for they are equally useful—can reach a positive or negative decision on this problem which has been the object of so much discussion. In the end they will do those things that, had my advice been followed, they should have done at the beginning.

19. That which is awakened within us through mere chance in the circumstances of life

See the sources of motivation for the study of Nature in *Kosmos*, vol. II, p. 5.

20. Of importance to the accurate determination of the longitude of Lima

At the time of my expedition, the longitude of Lima according to the observations of Malaspina in the maps published by the *Deposito hidrografico de Madrid* was accepted to be $5^h16'53''$. The transit of Mercury across the disc of the sun on November 9, 1802, which I observed in Callao, in Lima Harbor (in the northern turret, the *Torreon del Fuerte de San Felipe*), gave, by means of the touching of both edges, a figure of $5^h18'16.5''$ for Callao—by the outer contact alone, $5^h18'18''$ ($79°34'30''$). This result of the Mercury transit was corroborated by Lartigue, Duperrey, and Captain FitzRoy on the expedition of the *Adventure* and the *Beagle*. Lartigue found for Callao $5^h17'58''$, Duperrey $5^h18'16''$, and FitzRoy $5^h18'15''$. With the longitudinal difference between Callao and the Cloister of San Juan de Dios in Lima determined through four trips with a chronometer, the observation of the transit of Mercury gives $5^h17'51''$ ($79°27'45''$) for Lima. (Cf. my *Recueil d'Observations astron.*, vol. II, pp. 397, 419 and 428 with *Relat. hist.*, vol. III, p. 592.)

Potsdam, June 1849

Index

Yakut al-Hamari, and "Mountains of the Moon," 93

Yaruro people (native Am.), 144; "strangers" to Christian doctrine, 145

yellow fever, 218

yew (*Taxus*), 207, 221; found in Himalayas, 220; longevity of, 193–94; wide range of, 222

yucca, 167, 191, 228–29

Zambo people (Spanish-speaking native Am.), 145

Zend-Avesta, 196

Zimmermann, Carl, 65, 95–96

Zuccarini, J. G., 217, 220